Imperial Science

In the second half of the nineteenth century, British firms and engineers built, laid, and ran a vast global network of submarine telegraph cables. For the first time, cities around the world were put into almost instantaneous contact, with profound effects on commerce, international affairs, and the dissemination of news. Science, too, was strongly affected, as cable telegraphy exposed electrical researchers to important new phenomena while also providing a new and vastly larger market for their expertise. By examining the deep ties that linked the cable industry to work in electrical physics in the nineteenth century – culminating in James Clerk Maxwell's formulation of his theory of the electromagnetic field – Bruce J. Hunt sheds new light both on the history of the Victorian British Empire and on the relationship between science and technology.

Bruce J. Hunt is an Associate Professor of History at the University of Texas at Austin.

SCIENCE IN HISTORY

Series Editors
Simon J. Schaffer, University of Cambridge
James A. Secord, University of Cambridge

Science in History is a major series of ambitious books on the history of the sciences from the mid-eighteenth century through the mid-twentieth century, highlighting work that interprets the sciences from perspectives drawn from across the discipline of history. The focus on the major epoch of global economic, industrial and social transformations is intended to encourage the use of sophisticated historical models to make sense of the ways in which the sciences have developed and changed. The series encourages the exploration of a wide range of scientific traditions and the interrelations between them. It particularly welcomes work that takes seriously the material practices of the sciences and is broad in geographical scope.

Imperial Science

Cable Telegraphy and Electrical Physics
in the Victorian British Empire

Bruce J. Hunt

University of Texas at Austin

CAMBRIDGE
UNIVERSITY PRESS

CAMBRIDGE
UNIVERSITY PRESS

University Printing House, Cambridge CB2 8BS, United Kingdom

One Liberty Plaza, 20th Floor, New York, NY 10006, USA

477 Williamstown Road, Port Melbourne, VIC 3207, Australia

314–321, 3rd Floor, Plot 3, Splendor Forum, Jasola District Centre,
New Delhi – 110025, India

79 Anson Road, #06–04/06, Singapore 079906

Cambridge University Press is part of the University of Cambridge.

It furthers the University's mission by disseminating knowledge in the pursuit of
education, learning, and research at the highest international levels of excellence.

www.cambridge.org
Information on this title: www.cambridge.org/9781108830669
DOI: 10.1017/9781108902700

First published 2021

A catalogue record for this publication is available from the British Library.

Library of Congress Cataloging-in-Publication Data
Names: Hunt, Bruce J., author.
Title: Imperial science : cable telegraphy and electrical physics in the Victorian
British empire / professor Bruce J. Hunt, University of Texas, Austin.
Description: Cambridge, United Kingdom ; New York, NY, USA : Cambridge
University Press, 2021. | Series: Science in history | Includes bibliographical
references and index.
Identifiers: LCCN 2020047668 (print) | LCCN 2020047669 (ebook) |
ISBN 9781108830669 (hardback) | ISBN 9781108902700 (ebook)
Subjects: LCSH: Telegraph – Great Britain – History – 19th century. |
Electromagnetism – Research – Great Britain – History – 19th century.
Classification: LCC TK5157 .H86 2021 (print) | LCC TK5157 (ebook) |
DDC 621.3830941/09034–dc23
LC record available at https://lccn.loc.gov/2020047668
LC ebook record available at https://lccn.loc.gov/2020047669

ISBN 978-1-108-83066-9 Hardback

Contents

List of Figures		*page* vi
Acknowledgments		viii
List of Abbreviations		x
	Prologue: "An Imperial Science"	1
1	"An Ill-Understood Effect of Induction": Telegraphy and Field Theory in Victorian Britain	3
2	Wildman Whitehouse, William Thomson, and the First Atlantic Cable	37
3	Redeeming Failure: The Joint Committee Investigation	97
4	Units and Standards: The Ohm Is Where the Art Is	144
5	The Ohm, the Speed of Light, and Maxwell's Theory of the Electromagnetic Field	181
6	To Rule the Waves: Britain's Cable Empire and the Making of "Maxwell's Equations"	216
	Epilogue: Full Circle	272
	Bibliography	276
	Index	299

Figures

1.1 *Goliath* laying the short-lived first Channel cable in 1850 *page* 11
1.2 The 1851 English Channel cable, as retrieved in 1859 12
1.3 Lines of force around wires carrying electric currents 19
1.4 William Thomson's curves showing how an electric current
    rises in a cable 35
2.1 Cyrus W. Field, John Watkins Brett, Charles Tilston Bright,
    and Wildman Whitehouse 39
2.2 The route of the 1858 Atlantic cable 42
2.3 Wildman Whitehouse's telegraphic recorder and marked
    paper tapes 49
2.4 William Thomson in 1852 51
2.5 Crewmen coiling the first Atlantic cable on the US warship
    Niagara 67
2.6 Wildman Whitehouse's magneto-electrometer 71
2.7 William Thomson's mirror galvanometer 79
2.8 William Thomson's marine galvanometer 82
2.9 A souvenir piece of the 1858 Atlantic cable,
    sold by Tiffany's 85
2.10 Theodor Linde's depiction of the Newfoundland
    cable station in 1858 88
2.11 Captain Frederic Brine's map of Valentia harbor, 1859 93
3.1 The routes of the Red Sea and India cables, 1859–60 115
3.2 Title page of the *Joint Committee Report*, 1861 126
3.3 William Thomson reading a letter from Fleeming Jenkin
    in 1859 130
3.4 Specimens of cable types 135
4.1 Latimer Clark and Cromwell Fleetwood Varley 147
4.2 Fleeming Jenkin's self-portrait, 1859 154
4.3 The British Association spinning coil resistance apparatus 171
4.4 A British Association resistance coil 172
5.1 Maxwell's vortex model of the ether 184
5.2 Maxwell's apparatus for measuring the ratio of units (v) 210

5.3 Portrait of Maxwell with the British Association spinning coil
    resistance apparatus   214
6.1 The *Great Eastern* grappling for the broken end of the Atlantic
    cable   225
6.2 "Awaiting the Reply" in 1866   229
6.3 *Vanity Fair* caricature of "cable king" John Pender   240
6.4 Plaque showing the vector form of "Maxwell's equations"   257
6.5 Oliver Heaviside at his home in Devonshire in 1893   258
7.1 The global cable network, circa 1901   273

Acknowledgments

I have incurred many debts in writing this book, and I would like to thank the many people and institutions who have helped along the way. Like any historian, I relied heavily on libraries and archives and on the skilled staff who work in them. For access to the materials they hold and for assistance in my research, I would like to thank the staff and Syndics of Cambridge University Library, as well as the Science and Technology Facilities Council; the Archives of the Institution of Engineering and Technology; the PK Porthcurno Museum of Global Communications and the PK Trust; the Merseyside Maritime Museum of the National Museums Liverpool; the New York Public Library, particularly its Rare Book Collection (Astor, Lenox, and Tilden Foundations); and the libraries and special collections of Trinity College Dublin, University College London, and the University of Glasgow. I would also like to thank my own institution, the University of Texas at Austin, for maintaining its excellent library collections, including online access to the historical newspaper collections that I have found so invaluable, and for supporting my research over many years. I owe special thanks to Allan Green for the extended loan of his photocopy of the Atlantic Telegraph Company Minute Book (1856–58) and to Bill Burns for the remarkable collection of materials he has gathered on atlantic-cable.com and for the generous help he has offered me on many occasions, including the use of maps and images that were available nowhere else. Bill's website led me to the remarkable photograph of sailors on the *Niagara* that appears in Chapter 2, and I would also like to thank Page and Bryan Ginns for granting me permission to reproduce it. I would also like to thank Kyle Hedrick for giving me the map showing the route of the 1858 Atlantic cable; Beth Hedrick for the souvenir piece of that cable as well as the *Vanity Fair* print of John Pender; and Linda Henderson for the copy of Louis Figuier's *Merveilles de la Science*, from which I drew several illustrations. Lucy Rhymer, my editor at Cambridge University Press, has always been very patient and supportive; I would like to thank her, as well as Emily Sharp, Natasha Whelan, Anjana Karikal Cholan, and Judieth

Sheeja, for their help in bringing this book to fruition. Some chapters of this book incorporate material that appeared in earlier forms in several journal articles: Bruce J. Hunt, "Michael Faraday, Cable Telegraphy, and the Rise of British Field Theory," *History of Technology* (1991) *13*: 1–19 (Chapter 1); Hunt, "Scientists, Engineers, and Wildman Whitehouse: Measurement and Credibility in Early Cable Telegraphy," *British Journal for the History of Science* (1996) *29*: 155–69 (Chapter 2); Hunt, "The Ohm Is Where the Art Is: British Telegraph Engineers and the Development of Electrical Standards," *Osiris* (1994) *9*: 48–63 (Chapter 4); and Hunt, "Maxwell, Measurement, and the Modes of Electromagnetic Theory," *Historical Studies in the Natural Sciences* (2015) *45*: 303–39 (Chapter 5). I would like to thank the editors and publishers of these journals for permission to include this material here.

Over the years, many friends and many audiences have heard me talk about the history of cable telegraphy and electrical physics, and I would like to thank them for their patience and support. On the British side, I would especially like to thank Simon Schaffer, Richard Noakes, Crosbie Smith, and Jim and Anne Secord, and on the American side Tom Hankins, Bob Kargon, Russell McCormmach, Bob Rosenberg, Robert Smith, Bruce Hevly, and Megan Raby. My deepest thanks go of course to Beth, Peter, and Emma.

Abbreviations

ATC Minute Book	Minute Book of the Atlantic Telegraph Company, October 26, 1856, to April 9, 1858, BICC Archive, Merseyside Maritime Museum, Liverpool
BA Report	*Report of the Annual Meeting of the British Association for the Advancement of Science*
C&W	Cable and Wireless Archives, PK Porthcurno Museum of Global Communications
CUL	Cambridge University Library
CUL-RGO	Royal Greenwich Observatory Archives, Cambridge University Library
Elec.	*The Electrician*
Further Corr. 269	*Further Correspondence Respecting the Establishment of Telegraphic Communications in the Mediterranean and with India*, British Parliamentary Papers, 1860, LXII.269
Further Corr. 461	*Further Correspondence Respecting the Establishment of Telegraphic Communications in the Mediterranean and with India*, British Parliamentary Papers, 1860, LXII.461
Heaviside, *EP*	Oliver Heaviside, *Electrical Papers*, 2 vols. (London: Macmillan, 1892)
Heaviside, *EMT*	Oliver Heaviside, *Electromagnetic Theory*, 3 vols. (London: Electrician Co., 1893–1912)
IET	Institution of Engineering and Technology Archives, London
Joint Committee Report	*Report of the Joint Committee to Inquire into the Construction of Submarine Telegraph Cables*, British Parliamentary Papers, 1860, LXII.591
Maxwell, *SLP*	James Clerk Maxwell, *Scientific Letters and Papers of James Clerk Maxwell*, ed. P. M.

	Harman, 3 vols. (Cambridge: Cambridge University Press, 1990–2002)
Maxwell, *SP*	James Clerk Maxwell, *Scientific Papers of James Clerk Maxwell*, ed. W. D. Niven, 2 vols. (Cambridge: Cambridge University Press, 1890)
Maxwell, *Treatise*	James Clerk Maxwell, *Treatise on Electricity and Magnetism*, 2 vols. (Oxford: Clarendon Press, 1873)
ODNB	*Oxford Dictionary of National Biography*
Phil. Mag.	*Philosophical Magazine*
Phil. Trans.	*Philosophical Transactions of the Royal Society of London*
Proc. ICE	*Proceedings of the Institution of Civil Engineers*
Proc. RS	*Proceedings of the Royal Society of London*
Red Sea Contract	*Electric Telegraph Companies. Copies of All Correspondence between the Electric Telegraph Companies Under Contract with the Government Respecting the Failure to Lay Down or Keep in Working Order the Electric Wires*, British Parliamentary Papers, 1860, LXII.211
Smith, *Reports*	F. E. Smith, ed., *Reports of the Committee on Electrical Standards Appointed by the British Association for the Advancement of Science* (Cambridge: Cambridge University Press, 1913)
Thompson, *Kelvin*	Silvanus P. Thompson, *The Life of William Thomson, Baron Kelvin of Largs*, 2 vols. (London: Macmillan, 1910)
Thomson, *MPP*	William Thomson, *Mathematical and Physical Papers*, 6 vols. (Cambridge: Cambridge University Press, 1882–1911)
WC-NYPL	Wheeler Collection of Electricity and Magnetism, Rare Book Collection, New York Public Library

Prologue: "An Imperial Science"

In January 1889, in the wake of Heinrich Hertz's dramatic discovery of electromagnetic waves, the British physicist Oliver Lodge declared that with this experimental confirmation of James Clerk Maxwell's electromagnetic theory of light, "the whole domain of Optics is annexed to Electricity, which has thus become an imperial science."[1] Lodge had hit on a very up-to-date way to express the preeminence electrical science had achieved by the last decades of the nineteenth century. But in 1889 electricity was an imperial science in a less metaphorical sense as well: it lay at the scientific heart of submarine telegraphy, one of the characteristic technologies of the Victorian British Empire. Often described as the "nervous system of the empire," the web of undersea wires that British firms stretched around the globe in the second half of the nineteenth century carried information streaming toward London from the far reaches of the world and commands flowing back out, extending and reinforcing Britain's military, naval, commercial, and cultural power.[2] Cable telegraphy "annihilated space and time," in the phrase of the day, and cut the time it took to send a message from, say, London to Australia from weeks or months to hours or, for some messages, mere minutes. For the first time, the world was wired up, setting in motion changes in commerce, government, the dissemination of news, and the workings of everyday life that continue to play themselves out today. The advent of global telecommunications stands as one of the watersheds of modern history.

It was a watershed that transformed science as well. One of the perennial questions in the history of both science and technology concerns the

[1] Oliver J. Lodge, "Modern Views of Electricity," *Nature* (January 31, 1889) *39*: 319–22, on 322. This was the last installment of Lodge's long series of articles on the subject, collected and published later that year as *Modern Views of Electricity* (London: Macmillan, 1889); revised editions appeared in 1893 and 1907. The "imperial science" remark appears on p. 309 of the 1889 edition.

[2] See, e.g., George Peel, "Nerves of Empire," in *The Empire and the Century* (London: John Murray, 1905), 249–87.

relationship between the two. Should technology be understood simply as "applied science," with scientists, driven purely by their own curiosity, making discoveries that engineers and entrepreneurs then turn to practical use? Or is the pursuit of scientific knowledge instead sometimes shaped in fundamental ways by the technological context in which it is produced? Which way does the arrow of influence run? It is well known, for example, that electromagnetic field theory initially drew adherents only in Britain, and only in the mid-nineteenth century. But why then, and why there? The answer, as we will see, lies not just in the details of the lives and ideas of a few great men but also in the principal context in which electricity was studied in the middle decades of the nineteenth century: the telegraph industry, particularly submarine telegraphy. British firms dominated the world cable industry from its beginnings in the 1850s until well into the twentieth century, and the demands and opportunities presented by submarine telegraphy steered British electrical research into areas – particularly the study of how pulses of current travel along insulated wires – that fostered the spread of a field approach. Michael Faraday had been developing his field ideas since the 1830s, but they attracted little support until the 1850s, when physicists and engineers began to take them up in connection with submarine telegraphy. The Maxwellian field theory that grew to fruition in the next few decades was also shaped in important ways by the British cable industry, as was the whole system of electrical units and standards that is still used around the world today. By examining the links in the nineteenth century between cable telegraphy and the rise of electrical science, particularly field theory, we will be able not only to shed light on several important episodes in the history of physics but also to broaden and clarify our understanding of how science and technology interact.

1 "An Ill-Understood Effect of Induction"
Telegraphy and Field Theory in Victorian Britain

The men who made and laid the first submarine telegraph cables in the early 1850s gave little thought to electrical theory or precision measurement. Over the preceding decade they had watched as networks of landlines – most of them little more than bare wires strung from poles – had spread across the United States, Britain, and Continental Europe and had seen how the new telegraph systems were making money for their owners while also transforming the operation of markets and the dissemination of news. The prospect of extending such networks across the English Channel, and perhaps eventually across the Atlantic, was tantalizing. The promoters of the first submarine cables thought if they could just contrive a suitable way to insulate their wires, they would find it just as easy, and at least as profitable, to send signals under the sea as it had been to send them along overhead telegraph lines.

It turned out not to be so simple. Signaling through a submarine cable, particularly a long one, proved to be a far more complex and challenging task than sending dots and dashes along an overhead wire. Enormous resources were brought to bear on the problem in the 1850s and 1860s, and the effort to make cable telegraphy a success drove some of the most important advances in electrical science in the second half of the nineteenth century. This was especially true in Britain. As the leading industrial, commercial, maritime, and imperial power of the day, Britain had both the greatest effective demand for transoceanic telegraphy and the greatest resources with which to try to make it work. British firms dominated the cable industry from its beginnings, and British scientists and engineers were the first to encounter the new phenomena presented by submarine telegraphy. They responded by devising new electrical instruments, new measuring techniques, and even the basic units of ohms, amps, and volts that are still used today. More profoundly, their exposure to the peculiarities of submarine telegraphy, particularly the distortion or "retardation" that signals suffered in passing along undersea cables, led British scientists and engineers to rethink the basic nature of electric and magnetic phenomena. Instead of focusing on the supposed flow of electrical fluids within the conducting wires, they shifted their attention to the subtle interplay of energies in the surrounding space – that is, to what came to be called the "electromagnetic field." It has long been recognized that British physicists were the first to take up and extend this new field approach to electromagnetism, at a time when most of their counterparts on the Continent continued to treat electric and magnetic phenomena in terms of particles acting directly on each other from a distance. What explains this sharp difference in national scientific styles, which persisted from the 1850s until at least the 1890s? As we shall see, much of the answer can be found in differing technological contexts. To put the point simply and a bit too baldly, the British did field theory because they had submarine cables, and the French and Germans did not because they had none.

Technology has often been conceived, or even defined, as "applied science." On this view, scientists, driven mainly by their own curiosity, make discoveries in their studies and laboratories that engineers and inventors then take up and turn to practical use. Many cases could be cited to illustrate this pattern; indeed, the telegraph itself grew out of laboratory discoveries Alessandro Volta and Hans Christian Oersted made in the early 1800s, and in the two centuries since then such science-based technologies have become increasingly common. But sometimes the arrow of influence runs the other way, with scientific advances growing out of encounters with new technologies and the new phenomena they produce. It is widely recognized that the science of thermodynamics emerged in the mid-nineteenth century in large part from efforts to understand the workings of steam engines.[1] In a similar way, important developments in electrical science in the second half of the nineteenth century, particularly the adoption and extension of field theory, were driven in part by the demands and opportunities presented by cable telegraphy. James Clerk Maxwell, the Scottish theoretician who was the first to formulate the mathematical laws of the electromagnetic field, recognized this connection very clearly. In his seminal *Treatise on Electricity and Magnetism* (1873), he remarked:

The important applications of electromagnetism to telegraphy have ... reacted on pure science by giving a commercial value to accurate electrical measurements, and by affording to electricians the use of apparatus on a scale which greatly transcends that of any ordinary laboratory. The consequences of this demand for electrical knowledge, and of these experimental opportunities for acquiring it, have been already very great, both in stimulating the energies of advanced

---

[1] D. S. L. Cardwell, *From Watt to Clausius: The Rise of Thermodynamics in the Early Industrial Age* (Ithaca: Cornell University Press, 1971); Bruce J. Hunt, *Pursuing Power and Light: Technology and Physics from James Watt to Albert Einstein* (Baltimore: Johns Hopkins University Press, 2010).

electricians, and in diffusing among practical men a degree of accurate knowledge which is likely to conduce to the general scientific progress of the whole engineering profession.[2]

In what follows, we will trace how cable telegraphy shaped the development of electrical science in the nineteenth century, particularly in Britain, and seek to shed light on broader patterns in the interaction between science and technology.

Telegraph Business

The telegraph did not spring fully formed from the mind or workshop of any one person. The basic idea of using electricity to convey messages dates back to the eighteenth century, but practical ways to do it were not found until after Oersted discovered in 1820 that electrical currents can produce magnetic effects. By the early 1830s several inventors had devised rudimentary electromagnetic telegraphs, and after witnessing a demonstration of one in Munich in 1836, W. F. Cooke, a young English medical student, set out to produce a commercially viable system. Back in London, he teamed up with Charles Wheatstone, professor of natural philosophy at King's College London, and in 1837 they secured a patent on a telegraph that used pulses of electric current to deflect magnetized needles and so indicate letters.[3] By 1839, Cooke and Wheatstone had completed a short line along a railway just west of London, insulating the wires with varnished cotton and burying them beside the tracks in iron pipes. Such lines were expensive to build and slow to catch on, but by 1846, when the Electric Telegraph Company was formed to take over Cooke and Wheatstone's patents, underground lines had largely given way to cheaper overhead wires strung from poles, and the telegraph network began spreading rapidly across England.[4]

In the United States, the enterprising artist Samuel F. B. Morse heard of European work on electric telegraphy and in the mid-1830s set about devising a system of his own. His "canvas stretcher" telegraph was remarkably clumsy and in its initial form barely worked across a room;

[2] Maxwell, *Treatise, 1*: vii–viii. In nineteenth-century usage, "electrician" did not mean an electrical tradesman but rather an expert in electrical theory or experiment.

[3] Brian Bowers, *Sir Charles Wheatstone* (London: Her Majesty's Stationery Office, 1975), 103–23. On the role the culture of electrical performance played in the origins of telegraphy, see Iwan R. Morus, "Telegraphy and the Technology of Display: The Electricians and Samuel Morse," *History of Technology* (1991) *13*: 20–40.

[4] Jeffrey Kieve, *The Electric Telegraph: A Social and Economic History* (Newton Abbot: David & Charles, 1973), 46–51. A thorough account of the "Electric" and other early telegraph companies in Britain, with many original documents and images, can be found on Steven Roberts's website "Distant Writing": distantwriting.co.uk.

what later became known as the "Morse" system, using a simple key to tap out dots and dashes, was devised later by Morse's young associate Alfred Vail.[5] Whatever skill Morse lacked as a practical inventor, however, he made up for as an energetic promoter. After a long lobbying campaign, in 1843 he secured a Congressional grant to build a demonstration line from Baltimore to Washington. Morse initially planned to use copper wires insulated with varnished cotton and buried in lead tubes, but such underground lines proved slow and expensive to construct, and in spring 1844 he switched to bare wires strung from poles. The work went ahead quickly after that and the line reached Washington in late May. The Democratic convention was then meeting in Baltimore, and Morse created a sensation in Washington by reporting the nomination of James K. Polk for president hours ahead of messengers arriving by rail.[6] The federal government declined to make telegraphy part of the postal system, however, and its further development in the United States, as in Britain, was left to private companies. Lines soon spread up and down the East Coast and by 1848 reached as far west and south as Chicago and New Orleans. Service in the early years was often erratic, but the "lightning wires" were nonetheless widely hailed as one of the wonders of the age.

Not everyone was impressed. In a passage of *Walden* that he drafted in 1849, Henry David Thoreau dismissed the telegraph, along with most other new inventions, as "but improved means to an unimproved end." "We are in great haste to construct a magnetic telegraph from Maine to Texas," he said, "but Maine and Texas, it may be, have nothing important to communicate."[7] Whatever the ultimate significance of what they had to say, however, people were eager to say to it, and the telegraph companies soon found their wires filled with traffic. Newspapers clamored to obtain even brief accounts of distant events and began to feature columns of "News by Electric Telegraph." Around 1846 a group of New York newspapers began to cooperate to share the costs of gathering and distributing telegraphic news reports, and the New York City Associated Press and similar "wire services" soon came to play leading

---

[5] Kenneth Silverman, *Lightning Man: The Accursed Life of Samuel F. B. Morse* (New York: Alfred A. Knopf, 2003), 148; on Vail's contributions, see 163–64, 235–36; Robert Luther Thompson, *Wiring a Continent: The History of the Telegraph Industry in the United States, 1832–1866* (Princeton: Princeton University Press, 1947), 10–11; and J. Cumming Vail, *Early History of the Electro-Magnetic Telegraph, from Letters and Journals of Alfred Vail* (New York: Hine Brothers, 1914).

[6] Silverman, *Lightning Man*, 237–38.

[7] Henry David Thoreau, *Walden; or, Life in the Woods* (Boston: Ticknor & Fields, 1854), 57. The text first appears in this form in Version C, drafted in 1849; see the "Fluid Text" of *Walden* on the "Digital Thoreau" website, www.digitalthoreau.org.

roles in mediating the delivery of news in both the United States and Europe.[8]

The effects of the telegraph on commerce were even more far-reaching. The telegraph companies in both the United States and Britain found merchants and traders to be their best customers, and from the first most of the traffic on their wires concerned commodity prices, sales orders, and similar business information. By the early 1850s the spread of telegraph networks had helped to reduce transaction costs, integrate wider regional markets, and facilitate the growth of commodity exchanges and futures markets.[9] Merchants were now able to place orders directly from a distance, reducing the need for middlemen and warehouses holding large inventories. As early as 1847 *The Republican* of St. Louis declared that "The Magnetic Telegraph has become one of the *essential means of commercial transactions.* ... Steam is a means of commerce – the Magnetic Telegraph is now another, and a man may as well attempt to carry on successful trade by means of the old flatboat and keel, against a steamboat, as to transact business by the use of mails against the telegraph."[10] Operating in increasingly close concert with the railroads, the growing telegraph network soon transformed large sectors of the American and British economies, working both to smooth and expand the operation of markets and to promote the growth of large and often monopolistic enterprises.[11]

In 1850, there remained one great obstacle to the spread of the electric wires: water. An ordinary river could be spanned by a bridge; a wider one by wires strung from high masts erected on either shore. But the Hudson at New York City was too wide even for that, and attempts to cross it with a wire insulated with tarred cotton and laid on the riverbed proved abortive, as the insulation quickly failed, even when ships' anchors did not snag and snap the line. For years, telegraph messages that had come hundreds of miles at lightning speed had to be taken down in Jersey City and carried to Manhattan by boat.[12] The English Channel presented an

[8] Richard A. Schwarzlose, *The Nation's Newsbrokers. Volume 1: The Formative Years: From Pretelegraph to 1865* (Evanston, IL: Northwestern University Press, 1989), 94–95, 108; Donald Read, *The Power of News: The History of Reuters, 1849–1989* (Oxford: Oxford University Press, 1992).

[9] Richard B. Du Boff, "The Telegraph and the Structure of Markets in the United States, 1845–1890," *Research in Economic History* (1983) 8: 253–77; William Cronon, *Nature's Metropolis: Chicago and the Great West* (New York: W. W. Norton, 1991), 120–23.

[10] *Republican* (St. Louis), September 18, 1847, quoted in Richard B. Du Boff, "Business Demand and the Development of the Telegraph in the United States, 1844–1860," *Business History Review* (1980) 54: 459–79, on 471–72.

[11] Richard B. Du Boff, "The Telegraph in Nineteenth-Century America: Technology and Monopoly," *Comparative Studies in Society and History* (1984) 26: 571–86.

[12] Thompson, *Wiring a Continent*, 45–46. The Mississippi posed a similar obstacle at St. Louis, and when telegraph lines first reached the east side of the river in

even more daunting barrier. The prospect of linking the enormous markets at London and Paris was enticing, not least to those trading shares in the City and on the Bourse, and for years before the technical means to accomplish it were available, ambitious entrepreneurs dreamt of somehow laying a telegraph line across the Channel.

The first practical steps toward spanning the Channel were taken by an unlikely pair of English brothers. John Watkins Brett was a painter and an art dealer who had cultivated a wide acquaintance among the moneyed elite of both Britain and France while also dabbling in some slightly shady financial dealings. He shared a house in London with his younger brother Jacob, who in 1845 enthusiastically took up an American-designed printing telegraph and began to lay grand plans for a system of "ocean telegraphs" to connect England not only with "Continental Kingdoms" but also with America, India, and the entire British Empire.[13] The elder Brett was soon also swept up by the idea of undersea telegraphs and proceeded to put matters on a slightly more practical footing, drawing on his art world connections to secure an exclusive concession from the French government to land a telegraph line on the coast near Calais.[14] (The British government declined to grant such monopoly rights but gave the Bretts permission to land a line near Dover, if they could manage it.) The Bretts' first concession lapsed before they could build anything, but in 1849 Louis Napoleon (later Napoleon III), the newly elected president of France and an enthusiast for all things electrical, granted them a ten-year monopoly on telegraphic communication between England and France, provided they succeeded in transmitting messages across the Channel by September 1, 1850. This would be a challenge, however, as the Bretts as yet had little idea how actually to make or lay a submarine cable.

John Watkins Brett had already sold much of his art collection to raise money for the new project. He now set about rounding up more financial backers and hiring engineers and a contractor. In perhaps their most important step, he and his engineers turned to the Gutta Percha

December 1847, messages had to brought across to St. Louis by boat. Wires strung from high masts were used for a time but often broke. An insulated cable was first laid across the river in 1850, and an improved one, using a surplus length from the first Atlantic cable, was laid in 1859; see J. Thomas Scharf, *History of Saint Louis City and County, From the Earliest Periods to the Present Day*, 2 vols. (Philadelphia: Louis H. Everts and Co., 1883), 2: 1422–30.

[13] Roger Bridgman, "John Watkins Brett," *ODNB*; Steven Roberts, "The Moving Fire: A Biography of Johns Watkins Brett, Father of Submarine Telegraphy," atlantic-cable.com//CablePioneers/Brett. The provisional registration of the "General Oceanic Telegraphic Company," filed by Jacob Brett in 1845, is reproduced in John W. Brett, *On the Origin and Progress of the Oceanic Electric Telegraph* (London: W. S. Johnson, 1858), 44.

[14] Correspondence concerning the French concessions, as well as the Bretts' contacts with the British government, is reproduced in Brett, *Origin and Progress*, 15–25.

Company to provide the twenty-five nautical miles of insulated wire they would need to span the Channel. Gutta-percha is a natural plastic derived from the latex of the *Palaquium gutta* trees of Malaya. Although chemically similar to rubber, it is not springy; when heated above 60°C (140°F), it softens and can be easily molded. Malayans had been using gutta-percha for centuries to make hats, whips, and knife-handles, but it did not become an item of European commerce until 1843, when William Montgomerie, an East India Company surgeon, sent samples to London and published an account of its uses.[15] Tons of gutta-percha were soon being shipped out of Singapore, and British firms began offering bottle stoppers, chess pieces, golf balls, and other products made of the new material. In 1848 Michael Faraday drew attention to its excellent electrical-insulating properties and urged "working philosophers, both juvenile and adult," to use gutta-percha in their electrical experiments.[16] British telegraphers soon adopted it as well, finding gutta-percha to be far superior to the tarred or varnished cotton with which they had earlier tried to insulate their underground lines. By 1849 they were stringing wires covered with gutta-percha through damp tunnels and burying them beneath the streets in cities where overhead lines were not allowed. The Bretts looked upon gutta-percha as a godsend. They knew they would not be able to realize their dream of spanning the Channel, and eventually the oceans, by telegraph until they could find a suitable form of insulation, and now it seemed that British industry – and the Malayan forests – had provided just what they needed, and at just the right time.

The 1840s were the heyday of "economic botany," as Europeans sought to gather and study plant specimens from around the world for possible commercial exploitation.[17] The availability of gutta-percha was a product of that effort and of British imperial expansion into Southeast Asia. The cable network that would spread around the globe in the second half of the nineteenth century relied heavily on this exotic latex from Malaya, and British firms' control of the gutta-percha market in Singapore gave them a virtual monopoly over what would become a strategically important material. This became a self-reinforcing loop, as Britain's imperial and commercial power gave it favored access to Malayan gutta-percha supplies, and so facilitated the construction of

[15] Bruce J. Hunt, "Insulation for an Empire: Gutta-Percha and the Development of Electrical Measurement in Victorian Britain," in Frank A. J. L. James, ed., *Semaphores to Short Waves* (London: Royal Society of Arts, 1998), 85–104.

[16] Michael Faraday, "On the Use of Gutta Percha in Electrical Insulation," *Phil. Mag.* (1848) *32*: 165–67.

[17] Lucile Brockway, *Science and Colonial Expansion: The Role of the British Royal Botanic Garden* (New York: Academic Press, 1979); Daniel R. Headrick, "Botany, Chemistry, and Tropical Development," *Journal of World History* (1996) *7*: 1–20.

a cable network that, in turn, strengthened the Empire and British commerce – including British control of the gutta-percha trade. Ecologically, however, the cycle was not so sustainable; gutta-percha was typically harvested by felling rather than just tapping the trees, and by 1900 the *Palaquium gutta* trees had been virtually wiped out over much of their range.[18]

All of that still lay far in the future when the Bretts placed their initial order with the Gutta Percha Company. The company was still working out how best to insulate such great lengths of wire, and the coils it eventually produced were, by later standards, quite crude, with an unevenly applied layer of gutta-percha and misshapen joints where the wire had been spliced.[19] The completed wire was what would later be called a "core" rather than a "true cable," since it lacked an outer armoring of wire rope; it had little tensile strength and could be easily snapped by an errant anchor. The electrical tests made on it were rudimentary at best; the Bretts' chief engineer, Charlton Wollaston, reportedly sometimes used his tongue to check for the flow of current.[20]

After a delay while the Bretts lined up last-minute financing, the coils of insulated wire were shipped out of London on the steam tug *Goliath* and finally unspooled from Dover to Cap Gris Nez on August 23, 1850, just ahead of the September 1 deadline (Figure 1.1). After sending a few simple signals to establish that the line was working, the operators hooked up Jacob Brett's printing telegraph and tried to exchange some congratulatory messages, but found they were garbled in transmission. By the next day, the line had gone dead, either abraded on the rocky shore at Cap Gris Nez or, by some accounts, snagged and cut by a Boulogne fisherman.[21] The few signals that had gotten through were enough, however, to secure a renewal of the Bretts' ten-year concession, as well as to prompt *The Spectator* to imagine what the "transmarine telegraph" might eventually accomplish:

[18] John Tully, "A Victorian Ecological Disaster: Imperialism, the Telegraph, and Gutta-Percha," *Journal of World History* (2009) *20*: 559–79; Daniel R. Headrick, "Gutta-Percha: A Case of Resource Depletion and International Rivalry," *IEEE Technology and Society Magazine* (December 1987) 6: 12–16. On the gutta-percha trade, see especially Helen Godfrey, *Submarine Telegraphy and the Hunt for Gutta Percha: Challenge and Opportunity in a Global Trade* (Leiden: Brill, 2018).

[19] Willoughby Smith, *The Rise and Extension of Submarine Telegraphy* (London: J. S. Virtue, 1891), 1–3. Smith was with the Gutta Percha Company from its earliest days and witnessed every stage in the development of cable manufacturing methods.

[20] Charles Bright, *Submarine Telegraphs: Their History, Construction, and Working* (London: Lockwood, 1898), 6n.

[21] Bright, *Submarine Telegraphs*, 9n.

Figure 1.1 The steam tug *Goliath*, accompanied by the *Widgeon*, laying the short-lived first cable across the English Channel in August 1850. (From Louis Figuier, *Merveilles de la Science*, Vol. 2: 189, 1868.)

The electric telegraph is laid down across the Channel between England and France; the salt sea is traversed by instantaneous communication. We stand on the threshold of an improvement that may hasten the progress of our race more rapidly than any other. . . . [In the future] a merchant may have in London a wire to his countinghouse in Calcutta, and address his clerk down at the antipodes as he would in the countinghouse below stairs . . . ; a man in London might sign a bill in Calcutta, transmit it for endorsement in St. Petersburg, and receive cash for it on authority from Cairo, in the space of an hour or so.[22]

The first telegraph line across the Channel may have failed ignominiously, but the potential rewards for eventual success remained enticing.

The Bretts regrouped and tried again the next year, backed this time by the railway engineer and inventor T. R. Crampton, who contributed not only funds but also an improved design: four copper wires separately insulated with gutta-percha and wrapped with tarred yarn, all encased in a sturdy armor of iron wire rope. This was a true cable, and after being successfully laid across the Channel on September 25, 1851, it worked

[22] "Transmarine Telegraph," *The Spectator* (August 31, 1850) *23*: 831.

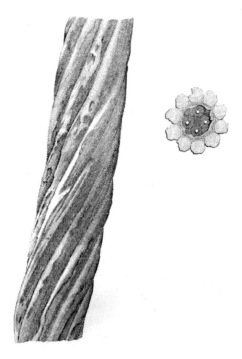

Figure 1.2 The cable laid across the English Channel in 1851 set the basic pattern for almost all that were to follow, with copper conductors covered by an insulating layer of gutta-percha and protected by an outer armoring of iron wires. When a length of it was retrieved in 1859, its electrical condition was found to be excellent.
(From *Joint Committee Report*, appendix 9, plate 8, 1861.)

admirably for many years[23] (Figure 1.2). The Submarine Telegraph Company, as the Bretts' enterprise was now called, earned substantial profits, as its monopoly on telegraphic communication between England and France allowed it virtually to name its price for urgent messages. Not surprisingly, the company later reported that for some hours each weekday its traffic consisted almost entirely of stock transactions passing between the London and Paris exchanges.[24]

The success of the Channel cable set off a flurry of other cable-laying efforts, not all of which succeeded. The first three attempts to lay a cable

[23] "The Submarine Telegraph," *Times* (October 4, 1851), 5.
[24] Steven Roberts, "The Submarine Telegraph Company between Great Britain and the Continent of Europe," quoting 1857 courtroom testimony concerning company operations in 1854, distantwriting.co.uk/competitorsallies.html.

from Britain to Ireland failed before the young engineer Charles Tilston Bright of the Magnetic Telegraph Company finally managed the feat in May 1853.[25] That same month, the Submarine Telegraph Company completed a cable from Dover to Ostend in Belgium, reinforcing its domination of electrical communications between Britain and the Continent.[26] The Electric Telegraph Company, the oldest and largest of the British telegraph companies, was not content to be locked out of this lucrative market, however, and over the summer of 1853, working through a subsidiary, the International Telegraph Company, it laid three cables from the Suffolk coast to the Netherlands.[27] Plans for more and longer cables were soon in the works. John Watkins Brett made preparations to lay cables across the Mediterranean, intending eventually to extend them all the way to India, while he and others began to speak confidently of laying a cable across the Atlantic.

The availability of gutta-percha insulation led several British companies to revive the idea of laying long landlines underground, reasoning that buried wires would prove less vulnerable to the depredations of vandals and the vagaries of the weather than ones strung overhead. The Magnetic Telegraph Company, based in northwestern England, was the first to construct long underground lines, linking Liverpool and Manchester in August 1852 and extending its buried wires south to London the following summer. At about the same time, Brett launched the European Telegraph Company, partly to expand the catchment area for his Channel cable; it laid underground wires from Dover to London in August 1852 and later extended them to Birmingham and Manchester. Not to be outdone by upstart rivals, the Electric Telegraph Company laid its own all-weather "express" underground lines from London to Liverpool and Manchester in the autumn of 1853.[28]

After a halting start, undersea and underground telegraphy was, by the end of 1853, an established success in Britain. Several firms, particularly the Submarine Telegraph Company, were turning solid profits, and the prospects for future growth appeared bright. As longer and longer lines were laid down, however, troubling phenomena turned up that were to have far-reaching consequences not just for telegraphy but for electrical science as a whole.

---

[25] Edward Brailsford Bright and Charles Bright, *The Life Story of the Late Sir Charles Tilston Bright*, 2 vols. (London: Archibald Constable, 1899), *1*: 78–88.

[26] Brett, *Origin and Progress*, vii, 22–23.   [27] Bright, *Submarine Telegraphs*, 16.

[28] Kieve, *Electric Telegraph*, 83; for a more detailed account, see Steven Roberts's "Distant Writing" website, distantwriting.co.uk/competitorsallies.html.

Retardation

Pulses of current sent along overhead wires seemed to travel almost instantaneously – "at the speed of lightning," as it was often put. More importantly, the pulses passed along the wires crisply and with minimal distortion. However closely they followed one another, the pulses remained distinct and readable at the far end, so that the only real constraint on how much traffic an overhead line could handle was the speed with which an operator could tap out dots and dashes on a Morse key or work the handles of a Cooke and Wheatstone two-needle telegraph. On the submarine cables and long underground lines that were laid in the early 1850s, however, telegraphers found that their signals were slightly delayed in transmission and substantially stretched out, so that if they tried to send them in very rapid succession, the pulses ran together into an indecipherable blur. As the telegraph engineer Cromwell Fleetwood Varley later remarked, unless a way could be found to reduce such "retardation," long underground lines and submarine cables "would only work at speeds too slow for commercial purposes."[29]

In retrospect, it was evident that the retardation of signals was responsible for some of the problems that had been encountered on the earliest cables, including the garbling of messages sent through the short-lived Channel cable of 1850.[30] It was not until 1852, however, that Latimer Clark, then a young engineer with the Electric Telegraph Company, first clearly identified and closely studied the phenomenon. The "Electric" was then considering proposals to lay underground lines from London to Manchester and cables from England to the Netherlands, and it sent Clark to the Gutta Percha Company works to perform tests on the great lengths of insulated wire being prepared there. A brief notice of his experiments appeared in a short-lived trade journal, the *Chemical Record and Journal of Pharmacy*, in March 1852, and Clark gave a much fuller account in the official *Report of the Joint Committee to Inquire into the Construction of Submarine Telegraph Cables* in 1861.[31] In the latter, he emphasized what a serious problem retardation posed:

---

[29] C. F. Varley, "On Improvements in Submarine and Subterranean Telegraph Communications," *BA Report* (1854), part 2, 17. Varley described his "reversing key," which sped up the discharge of the cable and so allowed somewhat faster signaling rates.

[30] Smith, *Rise and Extension*, 10. The Brett printing telegraph used on the 1850 cable was especially susceptible to disruption by retardation effects; simpler needle telegraphs could tolerate much more retardation before the signals became unreadable.

[31] "Experiments on the Transmission of Voltaic Electricity through Copper Wires Covered with Gutta Percha," *Chemical Record and Journal of Pharmacy* (March 27, 1852) 2: 234–35; Latimer Clark, "Report," *Joint Committee Report*, 293–335, on 303–6. The account of Clark's experiments in the *Chemical Record* did not mention his observation

There is no phenomenon in electricity that has a more important bearing on the electric telegraph than that of induction, and none which interferes more with the commercial success of telegraphic enterprise. If it were not for this evil presenting itself in the form known as retardation of the current, any telegraph cable, however long, could be worked at almost any speed; and although much may be done to reduce its effects, there is at present no known method of avoiding them altogether.[32]

At the Gutta Percha Company works, Clark sent pulses of current through both 100 miles of insulated wire kept immersed in the Regent's Canal and 175 miles stored dry in the factory. He "instantly noticed that the current took a very perceptible time in travelling the 100 miles" along the immersed wire, but virtually none to traverse the even greater length of dry wire. The effect was especially striking when he used a Bain telegraph receiver, in which the current made marks directly on a chemically treated paper tape that was drawn along beneath a stylus. Pulses of current sent through the dry wire made sharply defined dots and dashes on the paper tape, while those that had passed through the immersed wire produced long drawn-out lines that "were not only slow in appearing, but . . . tailed off gradually to a point." If the signals were sent in rapid succession, they simply ran together into one continuous smear.[33]

Clark quickly surmised that this retardation of the pulses was caused in some way by electrostatic induction. The insulated wire immersed in water was acting, he said, as an enormous cylindrical Leyden jar or condenser; when the battery was connected to the central conducting wire, the charge that flowed into it induced an equal and opposite charge on the outer surface of the gutta-percha insulation, drawing that charge from the slightly conductive surrounding water. This process evidently took some time, resulting in the observed retardation and blurring of the signals. Since air does not conduct electricity, the dry wire experienced very little of this inductive effect or of the resulting retardation. It was not yet clear, however, exactly what determined the observed degree of retardation, nor how the effect might be reduced or perhaps eliminated. Clark continued to ponder these questions, but he published nothing about them at the time.[34]

of signal retardation but focused on his demonstration of the working of an electrochemical telegraph receiver through the immersed wire. Clark was later a noted bibliophile. After his death his large collection of rare books and ephemera on the history of electricity and telegraphy was acquired by Schuyler Skaats Wheeler; as the Wheeler Collection, it is now held by the New York Public Library and provided important sources for this book.

[32] Clark, "Report," *Joint Committee Report*, 303.

[33] Clark, "Report," *Joint Committee Report*, 304.

[34] A transcription of Clark's notes of his early experiments is in IET ScMSS 22/22–24. Clark's reluctance to publish his findings at the time may have been related to his pursuit

Perhaps surprisingly, retardation problems were first brought to public attention in part by an astronomer. George Biddell Airy, the Astronomer Royal and director of the Greenwich observatory, had begun working with telegraphers in the late 1840s to distribute precise time signals as aids to navigation. By the early 1850s the observatory was connected via telegraph lines to "time ball" towers that had been erected at Deal and other seaports where they could be clearly seen by ships at anchor. At precisely 1 pm each day, the observatory sent a pulse of current out over the wires to each tower. This released the ball, and by watching for the moment it began to fall, ships' captains could be sure their chronometers were set exactly to "Greenwich time," as they needed to be when used to find longitude at sea.[35] (The ball that is dropped in Times Square at midnight every New Year's Eve is a vestige of New York's version of the system.)

Besides distributing time signals, Airy also wanted to use the telegraph for the direct determination of longitudes. By connecting Greenwich telegraphically with other observatories, astronomers in widely separated locations could synchronize their observations of star transits and so find their differences in longitude very precisely. Early in 1853 Airy used overhead lines to connect Greenwich with observatories at Cambridge and Edinburgh, and he began laying plans to use the Channel cable to connect to the Paris observatory. Lord de Mauley, chairman of the Submarine Telegraph Company, pledged his enthusiastic cooperation, declaring that once they had settled their landline connections in England, the company would be happy to offer Airy "unimpeded communication with Paris" and thereafter with "Brussels, Berlin, Vienna, and ... St. Petersburgh, en route for the Moon, and distant Planets." Complications soon arose at the Paris observatory, however, and the telegraphic determination of the difference in longitude between Greenwich and Paris would remain a notorious problem for years.[36]

of a patent, filed December 20, 1853, for "an improvement in insulating wire used for electric telegraphs, with a view to obviate the effects of return or inductive currents"; see patent 2956, *Repertory of Patent Inventions* (1854) *23*: 474. An extended description of Clark's experiments, clearly based on information he supplied, appeared in F. R. Window, "On Submarine Electric Telegraphs," *Proc. ICE* (January 1857) *16*: 188–202, on 198–201.

[35] On time balls, see Derek Howse, *Greenwich Time and the Discovery of the Longitude* (Oxford: Oxford University Press, 1980), 99–102.

[36] Lord de Mauley to G. B. Airy, January 6, 1852, CUL-RGO 6/610.236. On arrangements to connect Greenwich telegraphically with Cambridge and Edinburgh for longitude determinations, see the extensive file in CUL-RGO 6/633. On the difficulties in determining the difference in longitude between Paris and Greenwich, see Michael Kershaw, "'A Thorn in the Side of European Geodesy': Measuring Paris-Greenwich Longitude by Electric Telegraph," *British Journal for the History of Science* (2014) *47*: 637–60.

His plans for Paris stymied, Airy made preparations in 1853 to use the new cable to Ostend to connect to the Brussels observatory, and also to use the underground lines the Electric Telegraph Company was then laying to Manchester and Liverpool to connect Greenwich to observatories in those cities. Clark took a strong interest in Airy's work, lending him apparatus and helping arrange his use of the underground wires. On October 13, 1853, however, he wrote to tell Airy some bad news, and to issue an important invitation:

We have lately observed a great and variable retardation of the Electrical Current when sent through long lengths of underground wire, which would much interfere with your use of Subterranean or Submarine wires for purpose of determining longitudes, and I think it right therefore to call your attention to it. Professor Faraday will attend at Lothbury on Saturday evening next at 5 o'clock to repeat some experiments in which this retardation will be well exhibited, and we should be happy for you to be present at the same time if you should consider it desirable.[37]

The English soil in which the underground wires were buried was evidently damp enough to produce the same retardation effects on them that Clark had earlier seen on insulated wires immersed in the Regent's Canal.

It is not clear when Clark first invited Michael Faraday to see his retardation experiments, but his reasons for doing so were straightforward: besides wishing to bring an interesting new phenomenon to Faraday's attention, Clark wanted advice from Britain's leading electrical experimenter on how to combat what threatened to become a serious practical problem. Clark had first observed retardation more than a year earlier at the Gutta Percha Company works; now he sought to use the interest stirred up by Airy's longitude studies, and by the completion of the new underground lines to Manchester and Liverpool, to bring his discovery to the attention of the scientific community. Faraday would be his first point of contact.

The son of a London blacksmith, Faraday had had to work hard to make his way in the scientific world, and he always remained somewhat outside its mainstream, even when he was later showered with honors.[38] Although he had almost no formal education, he was intent on improving

---

[37] Latimer Clark to G. B. Airy, 13 Oct. 1853, CUL-RGO 6/468.173.

[38] On Faraday, see Geoffrey Cantor, *Michael Faraday: Sandemanian and Scientist* (Basingstoke: Macmillan, 1991). Faraday and his family were members of what he later described as "a very small and despised sect of Christians known, if known at all, as Sandemanians"; see Michael Faraday to Augusta Ada Lovelace, October 24, 1844, in F. A. J. L. James, ed., *Correspondence of Michael Faraday*, 6 vols. (London: IEE/IET, 1991–2011), *3*: 266. Faraday's simple but demanding faith had a strong effect on his life and thought and also posed a further obstacle to his participation in many areas of English social life.

himself and contributing to the advancement of science, and as a young man he managed to land a job as an assistant in the chemical laboratory at the Royal Institution, a privately funded research center in London that doubled as a venue for popular scientific lectures. Faraday gradually worked his way up to become director of the Institution's laboratory while also establishing a reputation as a prolific if somewhat idiosyncratic experimenter, especially on electricity and magnetism. He made his most important discovery in 1831, when he found that he could produce an electric current by moving a coil of wire near a magnet. Such electromagnetic induction remains the basis of almost all electric power production to this day.

Faraday had no grasp of mathematics, and the mathematically sophisticated theories of electricity and magnetism that were developed in the 1820s by A.-M. Ampère in France and later by Wilhelm Weber in Germany were a closed book to him.[39] These theories assumed the existence of electrical particles that exerted attractive and repulsive forces on one another directly from distance, on the model of Newton's theory of gravitation. Unable to follow Ampère's and Weber's equations, Faraday instead developed his own intensely visual and tactile approach to electric and magnetic phenomena, picturing curved "lines of force" stretching across space in patterns like those that iron filings form around a bar magnet or a wire carrying an electric current (Figure 1.3). Unlike Weber and most other electrical researchers of the day, Faraday did not look on the space around magnets, charges, and currents as empty and inert but instead as filled with energy and activity. He regarded what he came to call the electromagnetic *field* as the primary physical reality; magnets, charges, and currents, he said, are not the sources of electromagnetic effects, but simply points and regions where lines of force converge.[40]

Although other scientists held Faraday in high regard as an experimenter and a man of exemplary moral character, most regarded his ideas about lines of force and the electromagnetic field as little more than a mental crutch for one who was unable to handle a proper mathematical theory. He had no immediate disciples at the Royal Institution, and his younger associate and eventual successor there, John Tyndall, though a great admirer of Faraday's abilities as an experimenter, spoke rather

---

[39] On Ampère and Weber, see Olivier Darrigol, *Electrodynamics from Ampère to Einstein* (Oxford: Oxford University Press, 2000), 6–15, 54–66, and Christa Jungnickel and Russell McCormmach, *Intellectual Mastery of Nature: Theoretical Physics from Ohm to Einstein*, 2 vols. (Chicago: University of Chicago Press, 1986), *1*: 137–46 and *2*: 74–77.

[40] See, for example, Michael Faraday, "A Speculation Touching Electric Conduction and the Nature of Matter," *Phil. Mag.* (February 1844) *24*: 136–44.

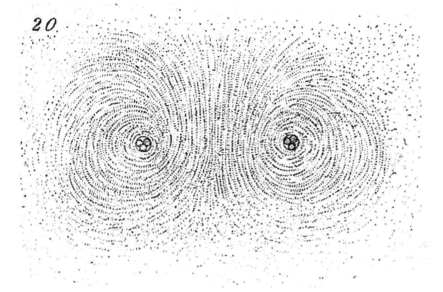

Figure 1.3 The lines of magnetic force around two bundles of wires carrying electric currents, as revealed by iron filings sprinkled around them.
(From Michael Faraday, "Experimental Researches in Electricity, 29th Series," *Philosophical Transactions*, Vol. 142, 1852; courtesy University of Texas Libraries.)

patronizingly of the "mistiness" of his theoretical ideas.[41] Airy was personally friendly with Faraday, but he too dismissed the idea of lines of force as a childish fancy; theories based on direct action at a distance gave a complete and accurate account of electric and magnetic effects, he said, and no one who understood the mathematics of the situation would "hesitate an instant in the choice between this simple and precise action, on the one hand, and anything so vague and varying as lines of force, on the other hand."[42] James Clerk Maxwell, though himself sympathetic to Faraday's approach, noted in 1855 that "an impression

[41] Roland Jackson, "John Tyndall and the Early History of Diamagnetism," *Annals of Science* (2015) 72: 435–89, on 439, for Tyndall's remark in a letter to T. A. Hirst, November 5, 1855, on the "mistiness" of Faraday's field ideas.
[42] G. B. Airy to John Barlow, February 7, 1855, in Henry Bence Jones, *Life and Letters of Faraday*, 2 vols. (London: Longmans, Green, and Co., 1870), 2: 352. An anonymous reviewer rendered an even harsher judgment in 1857, declaring that Faraday's lack of mathematics rendered his theoretical ideas worthless; see *Athenæum* (March 28, 1857) 397–99.

still prevails that there is something vague and unmathematical about the idea of lines of force," as compared to the clear and rigorous methods of "professed mathematicians."[43]

Clark and other telegraph engineers were less concerned with how mathematicians might view Faraday's theoretical ideas than with drawing on his wealth of experience with electrical phenomena for help in addressing the looming problem of signal retardation. For his part, Faraday saw in these new telegraphic phenomena a body of evidence that he could use to illustrate and substantiate his theoretical views and perhaps win more serious consideration for them from previously dismissive mathematical physicists, as well as from telegraph engineers. Cable telegraphy offered in effect a new *market* for Faraday's ideas in which he might hope to put them to use and so make their value more evident.

Faraday took an intense interest in the phenomena Clark showed him, first at the Gutta Percha Company works on Wharf Road in Islington on October 4, 1853, and then at the offices of the Electric Telegraph Company at Lothbury in central London later that day and on October 15, when Airy and others joined them. Though later often ascribed by others to Faraday himself, the experiments were in fact all conceived and performed by Clark, together with Samuel Statham of the Gutta Percha Company; Faraday was, as he often acknowledged, merely an observer.[44]

The phenomena Faraday witnessed at the Gutta Percha Company works were essentially those Clark had first observed there more than a year earlier: the charging and discharging of 110 miles of insulated wire immersed in a canal; the appreciable retardation of pulses sent along this "water wire," in contrast to their virtually instantaneous passage along 100 miles of wires stored in air; and the stretching out of pulses sent along the water wire, so that what had started as sharply defined dots and dashes emerged as an unreadable blur. In a particularly striking experiment, Clark and Statham showed Faraday how, by applying a battery to one end of the long water wire, with the far end put to earth, they could start

---

[43] James Clerk Maxwell, "Abstract of Paper 'On Faraday's Lines of Force,'" in Maxwell, *SLP 1*: 353–66, on 353–54; Maxwell, "On Faraday's Lines of Force," *Transactions of the Cambridge Philosophical Society* (1864) *10*: 27–83, repr. in Maxwell, *SP 1*: 155–229, on 157. Maxwell presented this paper in two parts on December 10, 1855, and February 11, 1856, but it was not printed until February 1857 (see Maxwell to P. G. Tait, February 15, 1857, in Maxwell, *SLP 1*: 494–95, on 495) and the full volume of the *Transactions* did not appear until 1864.

[44] Michael Faraday, "On Electric Induction – Associated Cases of Current and Static Effects," *Proceedings of the Royal Institution* (1854) *1*: 345–55, and in *Phil. Mag.* (March 1854) *7*: 197–208, repr. in Faraday, *Experimental Researches in Electricity*, 3 vols. (London: Taylor/Quaritch, 1839–55), *3*: 508–20. Faraday thanked Clark (misspelling his name as "Clarke") on 508.

a current flowing into the wire and then, by putting the near end quickly to earth through a galvanometer, draw nearly all of it back out; that is, they could in effect stop the current in its tracks and pull it back to its starting point.[45] As Faraday noted, these experiments gave a clear demonstration of the identity of static and dynamic electricity: the galvanic current from the battery flowed into the water wire until it had established a substantial static charge on the surface of the wire, and the later gradual discharge of the wire produced all of the same effects, both magnetic and electrochemical, as an ordinary battery current.

Clark next took Faraday to the Electric Telegraph Company offices to show him similar phenomena on the newly laid underground lines to Liverpool, and on October 15 gave more elaborate demonstrations there, with Airy and several telegraph men also in attendance.[46] This time the eight underground lines to Manchester were all looped together, giving a total of nearly 1600 miles of insulated wire, with galvanometers connected on the return loops in London to show the passage of the current along the way. The retardation of the signals was obvious: when Clark touched a key and started a current flowing from the London end, the galvanometers on the looped circuit deflected one by one rather than all at once. The stretching and blurring of the pulses of current was also clearly visible, as what started out as an abrupt deflection of the first needle became a more and more gradual swing on the later ones. This could be seen even more strikingly when the galvanometers were replaced with Bain electrochemical receivers, whose paper tapes gave a permanent and visible record of the retardation of the signals. Faraday remarked in his notes that "the wave of power" clearly moved more "sluggishly" along the underground wires than it did along a parallel set of overhead wires, on which the pulses of current returned sharply and almost instantaneously.[47]

Airy was also impressed with the phenomena he was shown and pondered how they might affect his efforts to use the telegraph to measure longitudes. He and his assistants were then completing their preparations

[45] See Clark's "Memoranda of Experiments" in IET ScMSS 22/24–41, and Faraday's notes in Thomas Martin, ed., *Faraday's Diary*, 7 vols. (London: G. Bell and Sons, 1932–36), 7: 393–408.

[46] According to Latimer Clark's notes in IET ScMSS 22/26, the October 15, 1853, session was attended by himself, his brother Edwin, Faraday, Airy, "and others." These included Latimer Clark's assistant (later his brother-in-law), W. H. Preece, who took some of the notes. When Preece later headed the British Post Office telegraph department, he often spoke of having sat "at the feet of Faraday" during these experiments; see, for example, W. H. Preece, "Inaugural Address," *Journal of the Institution of Electrical Engineers* (January 1893) 22: 36–68, on 61.

[47] Faraday, in *Faraday's Diary*, 7: 398, 408.

to connect to the observatory at Brussels, and in tests performed in November and December 1853, they found it took about a tenth of a second for a pulse of current to pass through the 200 miles of underground and submarine wires connecting Greenwich to Brussels – far more than the very slight retardation they had found on the nearly 400-mile circuit between Greenwich and Edinburgh, almost all of which ran on overhead wires. Airy made a point of drawing attention to this telegraphic retardation in an article in the *Athenæum* on January 14, 1854, ascribing it to "an ill-understood effect of induction."[48] This was the first published mention of a phenomenon that would soon become central to both cable telegraphy and British electrical physics.

After a return visit to the Gutta Percha Company works in early January to be shown a few further experiments, Faraday delivered a Friday evening lecture at the Royal Institution on January 20, 1854, in which he described what he had seen and sought to explain its significance. Faraday's lectures always attracted substantial attention, and this one drew more than most, including a report in the *Morning Chronicle*.[49] The *Philosophical Magazine* published what it called a "verbal copy of an abstract" of this "important paper" in March, and it also appeared in the *Proceedings of the Royal Institution* and in the third volume of Faraday's *Experimental Researches in Electricity*, published in 1855.[50] It was later often cited as the foundation of the theory of telegraphic retardation.

Faraday began his lecture by staking a claim on behalf of "philosophers" like himself to credit as discoverers, while also acknowledging that practical technology often served to expand the bounds of scientific knowledge: "when the discoveries of philosophers and their results are put into practice," he said, "new facts and new results are daily elicited," as recent work in telegraphy had amply demonstrated.[51] After thanking Clark and Statham, he then turned to the heart of his argument: their experiments, he said, provided "remarkable illustrations of some fundamental principles of electricity, and strong confirmation of the truthfulness of the view which I put

---

[48] [G. B. Airy], "Telegraphic Longitude of Brussels," *Athenæum* (January 14, 1854), 54–55; signed "A. B. G." For confirmation that this was written by Airy, see CUL-RGO 6/663.22–35. Note that if the delay was reasonably constant, most of its effect could be factored out; it thus had less effect on the precision of longitude determinations than Clark and Airy had initially feared.

[49] "Royal Institution," *Morning Chronicle* (January 23, 1854), 3.

[50] Faraday, "On Electric Induction."

[51] Quoted from notes of Faraday's lecture taken by W. H. Preece, IET ScMSS 22/13–15. These notes differ significantly from Faraday's published paper "On Electric Induction," which should not be regarded as simply a transcription of his lecture.

forth sixteen years ago [in 1838]" on the relations between "induction, conduction, and insulation" – a view that had hitherto attracted little support from other scientists.[52]

Faraday proceeded to describe the retardation phenomena he had been shown and to explain them in terms of his own theories of electrical action. Theories based on direct action at a distance between particles of electricity could account for retardation only by introducing additional hypotheses, and they gave no obvious reason why the electrical fluid should move so differently along an insulated wire running under water than it did along one strung overhead on poles. Since Faraday regarded the space *around* the wire as the real seat of electromagnetic phenomena, however, he had no trouble explaining why pulses of current sent along an underwater cable should be retarded and distorted. There was an intimate connection, he said, between induction and conduction, the conduction of a current along a wire always being preceded by the induction of a state of strain in the insulating space or medium (the dielectric) around it, with the consequent storage of a certain amount of charge at the surface where the conductor and the dielectric met. When the inductive capacity was large, as in a long cable, it took an appreciable time to induce this strain and store the resulting charge, and the rise of the current in the conductor was thus somewhat delayed and stretched out, as was the subsequent fall of the current when the circuit was broken and the charge was released. Conversely, when the inductive capacity was small, as in ordinary overhead wires, the strain was induced almost instantaneously, very little charge was stored, and no retardation was observed.

All of these phenomena depended crucially on the nature and arrangement of the surrounding dielectric – a thin layer of gutta-percha sandwiched between the central conducting wire and the surrounding iron armoring and seawater in the case of a cable, twenty or thirty feet of insulating air in the case of an overhead wire. Cable telegraphy showed in a striking and palpable way that it took time for electrical actions to *propagate* through space – something that was central to Faraday's conception of electrical phenomena, but that he had previously been able to assert only theoretically, based on indirect hints from some of his earlier experiments. Moreover, experience with cables and underground wires showed that the speed of this propagation depended as much on the surrounding dielectric as it did on the conductor itself. In particular, it

---

[52] Faraday, "On Electric Induction," *Phil. Mag.* (March 1854) *7*: 197–98. On other scientists' criticism of the views on induction and conduction that Faraday had expressed in 1838, see Darrigol, *Electrodynamics*, 93–96.

brought out the importance of what Faraday called the "specific inductive capacity" of different dielectrics, by which electrical phenomena, such as the charge stored by a condenser, are found to depend not just on the arrangement and state of the conductors but on the composition of the surrounding medium as well.[53] Glass, for example, has a specific inductive capacity (or "relative permittivity," as it is now called) about four times as great as that of air, while that of gutta-percha is about three times as great. The idea that the seemingly inert dielectric played such an active role in electrical phenomena had been widely dismissed when Faraday first proposed it in 1837, but the advent of cable telegraphy a little more than a decade later made specific inductive capacity the focus of intense practical and scientific interest. As Clark wrote in the official *Report of the Joint Committee to Inquire into the Construction of Submarine Telegraph Cables* in 1861, "At the date of Faraday's interesting researches [on specific inductive capacity], it could little be foreseen that such an obscure phenomenon should be destined to become one day, as it has now, a consideration of high national importance, and one which has a direct and most important bearing on the commercial value of all submarine telegraphs."[54]

Prussian Underground

A brief digression will illustrate how cable telegraphy helped focus attention on the role of the dielectric and will reinforce the point that signal retardation was a specifically British concern in the mid-1850s and 1860s. As he would later acknowledge, Clark had in fact not been the first to observe the inductive charging of underground wires.[55] In 1848, Werner Siemens, then a young lieutenant in the Prussian army (and later the founder what remains one of the largest electrical firms in the world), was ordered to lay a telegraph line as quickly as possible from Berlin to Frankfurt, where the new German national assembly was then meeting. The Prussian authorities wanted to monitor the activities of the Frankfurt assembly and directed Siemens to lay the line underground to protect it from damage by the "turbulent" population. Several hundred miles of wire were hastily covered with gutta-percha and buried two or three feet

---

[53] Michael Faraday, "Experimental Researches in Electricity, Eleventh Series," *Phil. Trans.* (1838) *128*: 1–40, esp. 37–40.

[54] Clark, *Joint Committee Report*, 313. Clark said Faraday had first presented his findings on specific inductive capacity to the Royal Society in 1842; in fact he had done so in 1837–38.

[55] Clark, *Joint Committee Report*, 303; Clark insisted, however, that he rather than Siemens was first to observe the retardation of signals.

deep along roads and railway lines.[56] Siemens soon noticed peculiar charging effects on his buried lines and concluded, as Clark and Faraday would a few years later, that they were acting as enormous condensers or Leyden jars – what he called "jar wires." He mentioned this observation when he presented an account of his telegraphic work to the French Academy of Sciences in 1850, but it attracted little notice at the time.[57]

Siemens's experience with the jar wires led him to adopt a somewhat Faraday-like view of electrical action in which he traced inductive effects to "molecular polarization" of the surrounding medium rather than to direct action at a distance between particles of electricity. Siemens was too busy with his business affairs to write up a full account of his new views at the time, but in 1857 he published a long paper, "Ueber die electrostatische Induction und die Verzögerung der Stroms in Flaschendrähten," in the *Annalen der Physik*, the leading German physics journal.[58] Here he laid out in detail the view, partly drawn from Faraday and partly developed by Siemens himself, that electrical actions were produced by the polarization of the surrounding dielectric. Indeed, Siemens concluded his paper by stating that "it is very likely that the seat of the electricity is removed from the conductors to the non-conductors surrounding them, and may be defined as the electrical polarization of the molecules of the latter."[59] This shift of focus from the conductor to the dielectric was central to the field approach that was then gaining ground in Britain, but it remained unusual in Germany, and Weber and other proponents of action-at-a-distance theories criticized Siemens's views.[60]

By the time Siemens's paper appeared, the underground wires that had led him to his new views were long gone. The insulation on the Berlin–Frankfurt line and the rest of the Prussian underground telegraph network had failed soon after the lines were laid, as the use of poorly prepared materials, improper maintenance, and the depredations of foraging rats

---

[56] Werner Siemens, *Inventor and Entrepreneur: Recollections of Werner von Siemens* (London: Lund Humphries, 1966), 71–79; Wolfgang Löser, "Der Bau unterirdischer Telegraphenlinien in Preussen von 1844–1867," *NTM* (1969) 6: 52–67.

[57] An abstract appeared as Werner Siemens, "Sur la telegraphie electrique," *Comptes Rendus* (1850) *30*: 434–37; for a translation of the full paper, see "Memoir on the Electric Telegraph" in Siemens, *Scientific and Technical Papers of Werner von Siemens*, 2 vols. (London: Murray, 1892–95), *1*: 29–64. Faraday included a translation of most of Siemens's abstract and acknowledged Siemens's priority in observing the "jar-wire" effect in Faraday, "On Subterraneous Electro-telegraph Wires," *Phil. Mag.* (June 1854) 7: 396–98; see also Clark, *Joint Committee Report*, 303.

[58] Werner Siemens, "Ueber die electrostatische Induction und die Verzögerung der Stroms in Flaschendrähten," *Annalen der Physik* (September 1857) *102*: 66–122, trans. as "On Electrostatic Induction and Retardation of the Current in Cores" in Werner Siemens, *Scientific and Technical Papers*, 1: 87–135.

[59] Siemens, "On Electrostatic Induction," 135.

[60] Siemens, *Inventor and Entrepreneur*, 163–65.

combined to destroy the gutta-percha. The civilian telegraph authorities dismissed Siemens and pulled up the underground lines; by the early 1850s they had all been replaced with overhead wires.[61] With no more underground lines, and with no submarine cables, German scientists and engineers now had neither the need nor the opportunity to deal with retardation phenomena, and over the next few decades they paid little attention to the more revealing aspects of telegraphic propagation. German physicists focused instead on the precise measurement of phenomena they could produce in the laboratory, for which their orthodox action-at-a-distance theories served admirably. Significantly, Siemens transferred his cable business to his brother Wilhelm, by then known as William, who was based in London.[62]

Lord Cable

Among those whom Faraday's 1854 lecture drew into a concern with cable problems, none would exert a more far-reaching influence than William Thomson. Then a young professor of natural philosophy at the University of Glasgow, Thomson (later Lord Kelvin) would become one of the towering figures of nineteenth-century science.[63] He made seminal contributions to thermodynamics, electromagnetism, and many other branches of physics, as well as to practical technologies. He was especially closely involved with cable telegraphy; he was knighted in 1867 for his work on the Atlantic telegraph, and cable consulting fees and royalties on his patented telegraph instruments provided the foundation of his substantial personal fortune. When he was raised to the peerage in 1892 and had to pick a new name for himself, friends half-jokingly suggested "Lord Cable" would be a fitting choice. In the end, however, he settled on "Kelvin," after a small river that runs near the University of Glasgow.[64]

Born in Belfast, Thomson moved to Scotland as a child when his father was named to the chair of mathematics at the University of Glasgow. Young William was a mathematical prodigy, attending classes at the

---

[61] Löser, "Unterirdischer Telegraphenlinien," 59–61; Siemens, *Inventor and Entrepreneur*, 89–90. The British underground lines laid in the early 1850s suffered a similar fate, as their gutta-percha insulation slowly oxidized and broke down. By the late 1850s most of them had been replaced with overhead wires; Kieve, *Electric Telegraph*, 83.

[62] On Siemens Brothers, the London-based cable firm, see J. D. Scott, *Siemens Brothers, 1858–1958: An Essay in the History of Industry* (London: Weidenfeld and Nicolson, 1958).

[63] Crosbie Smith and M. Norton Wise, *Energy and Empire: A Biographical Study of Lord Kelvin* (Cambridge: Cambridge University Press, 1989); Thompson, *Kelvin*.

[64] On the suggestion that Thomson dub himself "Lord Cable," see Agnes Gardner King, *Kelvin the Man: A Biographical Sketch by His Niece* (London: Hodder and Stoughton, 1925), 106; see also Bruce J. Hunt, "Lord Cable," *Europhysics News* (2004) *35*: 186–88.

university when he was eight and publishing important mathematical papers by the time he was sixteen. He went on to win high mathematical honors at Cambridge and studied experimental physics in Paris before being elected as the professor of natural philosophy at Glasgow in 1846, when he was just 22. He would remain there for the rest of his long career.

Thomson had some early contact with cable telegraphy through his friend Lewis D. B. Gordon, professor of engineering at Glasgow and a business partner of R. S. Newall, the most energetic and adventurous of the early cable manufacturers.[65] By December 1853 Thomson was interested enough in cable propagation to make inquiries about the retardation Airy had observed during his Brussels longitude tests.[66] But it was his reading of the published version of Faraday's Royal Institution lecture in March 1854 that really sparked Thomson's interest in cable telegraphy and its relationship to fundamental questions of electrical theory. In particular, it was evidently what prompted Thomson to dust off two papers on the mathematical laws of electricity he had written in 1842 and 1845 and have them reprinted, with some additional notes, in the *Philosophical Magazine*, where they appeared in June and July 1854.[67] The first of these, published anonymously in the *Cambridge Mathematical Journal* when Thomson was a young undergraduate, concerned a mathematical analogy between the flow of heat and the distribution of electric potential. In it, he used mathematical methods developed by Joseph Fourier to show that, despite their evident physical differences, problems in heat flow and electrical attraction could be expressed interchangeably in terms either of direct action at

---

[65] On Gordon, see Thomas Constable, *Memoir of Lewis D. B. Gordon* (Edinburgh: privately printed, 1877); on Newall, see Bill Burns, "Wire Rope and the Submarine Cable Industry," atlantic-cable.com//Article/WireRope/wirerope.htm. As the list in the *Joint Committee Report*, 512–19, shows, Newall manufactured most of the cables made and laid in the 1850s but largely withdrew from the business after the collapse of the Red Sea cables in 1860.

[66] Thomson made these inquiries through Edward Sabine; see Sabine to G. B. Airy, December 9, 1853 (extract) and Airy to Sabine (pressbook copy), December 19, 1853, Airy Papers, CUL-RGO 6/468.194 and 196. A copy of Airy's reply to Sabine is in the Kelvin Collection, Cambridge, CUL 7342–S8a; Smith and Wise, *Energy and Empire*, 456 n. 30 mistakenly give its date as December 19, 1854.

[67] [William Thomson], "On the Uniform Motion of Heat in Homogeneous Solid Bodies, and Its Connection with the Mathematical Theory of Electricity," *Cambridge Mathematical Journal* (February 1842) *3*: 71–84, repr. in *Phil. Mag.* (supp. 1854) *7*: 502–15, and in William Thomson, *Reprint of Papers on Electrostatics and Magnetism* (London: Macmillan, 1872), 1–14; William Thomson, "On the Mathematical Theory of Electricity in Equilibrium," *Cambridge and Dublin Mathematical Journal* (November 1845) *1*: 75–95, repr. (with additions) in *Phil. Mag.* (July 1854) *8*: 42–62, and in Thomson, *Electrostatics and Magnetism*, 15–37. There is a preprint of Faraday's January 1854 Royal Institution lecture in the Kelvin Collection, Cambridge, CUL 7342–PA412.

a distance or of action propagated through an intervening medium. In his second paper, written while he was in Paris and published in what had by then become the *Cambridge and Dublin Mathematical Journal*, Thomson used these methods to analyze some of Faraday's electrical experiments. Thomson's main aim seems to have been to defend the orthodox theory of action at a distance from the apparent threat posed by experiments by Faraday and others that seemed to contradict it. After performing a series of mathematical transformations, Thomson showed that the results Faraday had explained by lines of force and the action of the field could just as readily be accounted for using "the commonly received ideas of attraction and repulsion exercised at a distance, independently of any intervening medium."[68] Conversely, of course, one could read Thomson's 1845 paper as a demonstration that Faraday's approach was an acceptable alternative way of looking at electric and magnetic phenomena, and even if that was not Thomson's intention at the time, his paper helped to make Faraday's approach more mathematically respectable.[69] In a note he added when his papers were reprinted in 1854, Thomson made a point of saying they constituted "a full theory of the characteristics of lines of force, which have been so admirably investigated experimentally by Faraday."[70]

Whatever role reading Faraday's Royal Institution lecture may have played in prompting Thomson to republish his 1842 and 1845 papers, it certainly lay behind a third paper, intended as a supplement to the first two, that he wrote in June 1854. This concerned the calculation of the electrostatic capacity of a Leyden jar or "a telegraph wire insulated in the axis of a cylindrical . . . sheath" – that is, Thomson said, "the copper wires in gutta-percha tubes under water, with which Faraday has recently performed such remarkable experiments."[71] (Like many others, Thomson persisted in crediting these experiments to Faraday, though they were really Clark's and Statham's.) Using the methods of his 1840s papers, Thomson showed that a submarine cable of quite ordinary length would have a truly enormous electrostatic capacity, far greater than any laboratory apparatus could

---

[68] Thomson, "Electricity in Equilibrium," *Phil. Mag.* (July 1854) 8: 59; Thomson also cited experiments by William Snow Harris.

[69] In "Electricity in Equilibrium," *Phil. Mag.* (July 1854) 8: 51, Thomson said "it may I think be shown that either method of viewing the subject, when carried sufficiently far, may be made the foundation of a mathematical theory which would lead to the elementary principles of the other as consequences."

[70] Thomson, "Uniform Motion of Heat," *Phil. Mag.* (supp. 1854) 7: 502n.

[71] William Thomson, "On the Electro-statical Capacity of a Leyden Phial and of a Telegraph Wire Insulated in the Axis of a Cylindrical Conducting Sheath," *Phil. Mag.* (supp. 1855) 9: 531–35, repr. in Thomson, *Electrostatics and Magnetism*, 38–41; on 532n, Thomson cited Faraday's January 1854 Royal Institution lecture and its publication in the *Philosophical Magazine*.

match. Although he wrote it as an appendix to the reprint of his 1845 paper, this third paper was not published until about a year later. The reason for the delay is not clear, but it may have been related to Thomson's efforts to secure a patent – his first of many – on "Improvements in electrical conductors for telegraphic communication"; concern on that score certainly led him to hold back the release of related work later in 1854.[72]

Just as he was taking up Faraday's lecture early in 1854, Thomson was also beginning an important correspondence with James Clerk Maxwell. Maxwell had completed the mathematics Tripos examination at Cambridge toward the end of January, and having entered what he called "the unholy state of bachelorhood," he wrote a few weeks later to Thomson, whom he knew through various family connections, for advice on what he might take up next.[73] The theory of electricity and magnetism had been excluded from the Tripos curriculum on the grounds that its mathematical foundations were not yet sufficiently secure, but while Maxwell had previously given relatively little attention to the subject, he now evidently judged it to be ripe for further exploration. In fact, during breaks in the Tripos, a grueling eight-day examination, he had reportedly relaxed with friends by doing experiments in his rooms with magnets and gutta-percha.[74] Maxwell may have seen or heard reports of the lecture Faraday was delivering at the Royal Institution at just this time; in any case, he was aware of the recent burst of activity in submarine telegraphy and the intriguing phenomena it had turned up.

Maxwell told Thomson that with the Tripos behind them, he and some of his friends had "a strong tendency to return to Physical Subjects and several of us here wish to attack Electricity." In light of that, he asked for advice on how someone unfamiliar with the subject should proceed; in

[72] William Thomson, with W. J. M. Rankine and John Thomson, 1854 patent 2547, listed in Thompson, *Kelvin*, 2: 1275. On December 1, 1854, Thomson asked Stokes to wait until the patent had been secured before mentioning publicly what Thomson told him "regarding the remedy for the anticipated difficulty in telegraphic communication to America"; see David B. Wilson, ed., *Correspondence between Sir George Gabriel Stokes and Sir William Thomson, Baron Kelvin of Largs*, 2 vols. (Cambridge: Cambridge University Press, 1990), *1*: 182; Smith and Wise, *Energy and Empire*, 453.

[73] Maxwell to Thomson, February 20, 1854, in Maxwell, *SLP 1*: 237–38. This is the first known letter from Maxwell to Thomson. In 1849 Maxwell's cousin Jemima Wedderburn, with whom he had grown up in Edinburgh, had married Hugh Blackburn, professor of mathematics at the University of Glasgow and often described as Thomson's closest friend; see Rob Fairley, "Jemima Blackburn," and A. J. Crilly, "Hugh Blackburn," *ODNB*. On the Cambridge Mathematical Tripos, see Andrew Warwick, *Masters of Theory: Cambridge and the Rise of Mathematical Physics* (Chicago: University of Chicago Press, 2003).

[74] Maxwell to "Miss Cay," January 13, 1854, in Lewis Campbell and William Garnett, *The Life of James Clerk Maxwell* (London: Macmillan, 1882), 195. Jane Cay was Maxwell's maternal aunt.

particular, "If he wished to read Ampère Faraday &c how should they be arranged, and at what stage and in what order might he read your articles in the Cambridge Journal?"[75] Thomson replied with a long letter, now lost, that clearly had a strong effect on Maxwell's thinking; months later Maxwell returned to it, thanking Thomson and saying his letter had helped him clear up what was otherwise "a mass of confusion" about electrical matters.[76] As their correspondence continued, Maxwell drew on Thomson's ideas about alternative ways to formulate electric and magnetic relations, and in May 1855 he wrote that he was working out how to translate between Weber's action-at-a-distance theory (of which he said "I confess I like it not") and Faraday's field approach.[77] This work culminated that December when Maxwell presented to the Cambridge Philosophical Society the first part of his long paper "On Faraday's Lines of Force," to be followed over the next few years by a string of papers that would transform the foundations of electromagnetic theory.[78] This work grew directly out of Maxwell's exchanges with Thomson in 1854–1855, when Thomson was closely engaged with the implications of Faraday's Royal Institution lecture; its deeper roots thus ran back to Clark's experiments on retardation in submarine cables.

Thomson's thinking about cable problems received a strong stimulus, though in a rather roundabout way, from the September 1854 meeting of the British Association for the Advancement of Science, held that year in Liverpool. The meeting featured papers on cable telegraphy by John Watkins Brett, E. B. Bright, Frederick Bakewell, and Cromwell Fleetwood Varley of the Electric Telegraph Company. Varley, who would later become Thomson's partner in a lucrative patent business, drew particular attention to the problems posed by retardation, which had become a pressing issue for telegraph engineers as well as an intriguing one for electrical scientists.[79]

Among those attending the Liverpool meeting was Sir William Rowan Hamilton, the Royal Astronomer of Ireland and director of the Dunsink observatory near Dublin, accompanied by his nineteen-year-old son Archibald. Airy had written to Hamilton the previous year about plans to use the new Irish cable to connect the Greenwich and Dunsink observatories for longitude determinations, while young "Arch" had been doing experiments at Dunsink on telegraphing short distances by

[75] Maxwell to Thomson, February 20, 1854, in Maxwell, *SLP 1*: 237. Maxwell's inquiry about Thomson's "articles in the Cambridge journal" may have contributed to Thomson's decision to have his 1842 and 1845 papers reprinted.

[76] Maxwell to Thomson, November 13, 1854, in Maxwell, *SLP 1*: 254.

[77] Maxwell to Thomson, May 15, 1855, in Maxwell, *SLP 1*: 305.

[78] Maxwell, "On Faraday's Lines of Force," repr. in Maxwell, *SP 1*: 155–229.

[79] Abstracts of these papers can be found in the 1854 *British Association Report*; see especially Varley, "Improvements," 17–18.

conduction directly through the ground without connecting wires.[80] Perhaps inspired by a report at the British Association meeting of experiments on telegraphing across rivers by conduction through the water, Arch wondered whether the same thing might be done all the way across the Irish Sea.[81] As the Liverpool meeting was breaking up, Arch and his father asked Thomson what he thought of the idea, but Thomson had, as he later said, "to run away to get to a steamer by which I was bound to leave for Glasgow," so he handed the Hamiltons off to his friend (and fellow Irishman) George Gabriel Stokes, the distinguished Cambridge mathematical physicist. Stokes had no special expertise in electricity, but he looked into the matter with his customary thoroughness and later sent Arch a long mathematical letter, now lost, in which he showed that relying on water conduction would not work; too little current would make it across the sea to be detectable with even the most sensitive receiver.[82]

Arch Hamilton's question piqued Stokes's interest in the whole subject of telegraphy, and on October 16 he wrote to ask Thomson to clarify just how ordinary cable signaling worked. Stokes's understanding of the process was a bit garbled – he initially assumed that the far end of a cable was kept insulated, when in fact it was always connected through the receiving instruments to the earth – but he managed to identify the resistance of the wire and the induction across the gutta-percha as the main factors that contributed to retardation.[83] His questions had a strong effect on Thomson; as Stokes later put it, they led Thomson to "work out analytical results bearing on an important practical question, possibly even to see the principles in a clearer light than before."[84] Those results were contained in two important mathematical letters to Stokes in which Thomson laid out what would remain for many years the standard theory of telegraphic propagation. By translating Fourier's equations for the diffusion of heat into electrical terms and then combining them with Faraday's ideas about inductive capacity, Thomson was able to derive a simple equation that governed the rise of current and potential at the far end of a submarine cable:

[80] G. B. Airy to W. R Hamilton, September 19, 1853, Hamilton Papers, Trinity College Dublin, MS 7767/1238; on Arch Hamilton's electrical experiments, see Thomas L. Hankins, *Sir William Rowan Hamilton* (Baltimore: Johns Hopkins University Press, 1980), 367.

[81] J. B. Lindsay, "On some Experiments upon a Telegraph for communicating across Rivers and Seas, without the employment of a submerged Cable," *BA Report* (1854), part 2, 157.

[82] William Thomson, "Ether, Electricity, and Ponderable Matter," *Journal of the Institution of Electrical Engineers* (January 1889) *18*: 4–35, on 10, repr. in Thomson, *MPP 3*: 484–515; Archibald Hamilton to "Dixon," January 6, 1859, Hamilton Papers, Trinity College Dublin, MS 7767/1394.

[83] Stokes to Thomson, October 16, 1854, in Wilson, ed., *Correspondence*, *1*: 168–69.

[84] Stokes to Thomson, November 4, 1854, in Wilson, ed., *Correspondence*, *1*: 180.

$\partial^2 v/\partial x^2 = ck \; \partial v/\partial t,$

where v is the potential at a point and c and k are the capacitance and resistance per unit length.[85] This marked a milestone not just in telegraphic theory but in electrical science more broadly. Oliver Heaviside, perhaps the leading British electrical theorist of the late nineteenth century, later called Thomson's theory "the first step toward getting out of the wire into the dielectric" in the treatment of telegraphic propagation – a step Heaviside would himself follow up on in the 1880s when he finally superseded Thomson's theory with his own, in which transmission wires are seen to act not as pipelines but as guides for electromagnetic waves that slip along through the field surrounding them.[86]

Thomson's theory implied that the retardation on a cable would be proportional to the product of its resistance and its capacitance. Since, for a cable of given thickness, both quantities were proportional to its length, the total retardation on a cable would increase with the square of its length – what came to be called the "law of squares." This explained why signals sent along short cables and underground lines suffered relatively little retardation, and why the problem grew so much worse on longer cables. It also suggested a direct way to reduce retardation: use fatter cables. If the thickness of the conducting wire and its insulating layer of gutta-percha were both increased in proportion to the length of the cable, the retardation on a long cable would be no greater than that on a short one. The cost of the materials would be much greater, however, increasing with the cube of the length, and the mechanical obstacles to manufacturing and laying a cable hundreds of miles long that would be thick enough to ensure acceptably little retardation might well prove insuperable. Thomson's theory had clarified the causes of retardation, but it had not yet pointed the way to any really practical solution.

Seeking to put solid numbers to the problem, Thomson wrote to Airy in January 1855, explaining his as yet unpublished theory and asking about the dimensions of the 200 miles of landlines and submarine cable used in the longitude tests between Greenwich and Brussels. Given the retardation of about a tenth of second found in those tests, the law of squares implied that a cable of the same thickness and long enough to stretch

---

[85] Thomson to Stokes, October 28, 1854, in Wilson, ed., *Correspondence*, *1*: 171–75. Thomson also added a term to take into account leakage of current through the gutta-percha, but this had relatively little effect on the result and was often ignored.

[86] Oliver Heaviside, *EP 2*: 79 and 86; on Heaviside's treatment of telegraph signals as electromagnetic waves, see Bruce J. Hunt, *The Maxwellians* (Ithaca: Cornell University Press, 1991), 129–32.

nearly 2000 miles across the Atlantic would suffer a retardation of about ten seconds – enough, Thomson said, to make it "almost useless."[87] Of course any calculation of the retardation between Greenwich and Brussels was complicated by the fact that the route used a combination of underground and overhead wires together with the cable from Dover to Ostend, but Thomson hoped to be able to make at least rough estimates that would indicate whether it might be possible to design a cable that could carry signals for really long distances without needing to be so thick as to be mechanically or financially unwieldy. He also hoped to be able to use accurate retardation measurements to estimate the "ratio of units" that related the strength of the electrostatic force between two charges to the strength of the electromagnetic force between the corresponding currents. This ratio was of fundamental importance in electrical theory, but its value was then wholly unknown.

Airy forwarded Thomson's query to Charlton Wollaston, the engineer of the Submarine Telegraph Company – the same man who had tested the original Channel cable with his tongue – telling him that information about the dimensions of the wires and cable would aid Thomson in "theoretical investigations on the transmission of galvanic currents which appear likely to lead to results of great practical importance."[88] Wollaston wrote back with full particulars, saying he was happy to help and expressing the hope that Thomson would be able to shed light on the whole question of telegraphic transmission, of which he said "little is at present known either theoretically or practically."[89] After Thomson saw the numbers, he sent Airy a long letter in which he laid out his latest thoughts on retardation and the ratio of units, whose value he estimated to lie between 104,000,000 and 419,000,000 feet per second – the first even approximate measurement of this important quantity.[90] Laborious experiments performed over many years by physicists in Britain, Germany, and the United States eventually

---

[87] William Thomson to G. B. Airy, January 19, 1855, Airy Papers, CUL-RGO 6/470.24–25; Thomson wrote again on January 20, 1855, CUL-RGO 6/470.26–28, enclosing curves showing how the current would rise at the far end of the cable.

[88] G. B. Airy to C. S. Wollaston, January 22, 1855 (pressbook copy), Airy Papers, CUL-RGO 6/470.29–31. Airy urged Thomson to come to London to see Latimer Clark demonstrate retardation phenomena, but Thomson was unable to get away from Glasgow; see Airy to Thomson, January 22 and January 26, 1855 (pressbook copies), CUL-RGO 6/470.32–35, and Thomson to Airy, February 2, 1855, CUL-RGO 6/470.39–44.

[89] C. S. Wollaston to G. B. Airy, January 25, 1855, Airy Papers, CUL-RGO 6/470.36–37.

[90] Thomson to Airy, February 2, 1855, Airy Papers, CUL-RGO 6/470.39–44. Thomson's estimate, equivalent to between 32,000,000 and 128,000,000 meters per second, was low by a factor of about four, probably owing mainly to the guesswork involved in relating the observed retardation to the quantities that appeared in his theory; see Airy to Thomson, February 8, 1855 (pressbook copy), Airy Papers, CUL-RGO 6/470.45.

showed that the ratio of units came to just under 300,000,000 meters per second, equal to the speed of light – as the electromagnetic theory of light that Maxwell formulated in the early 1860s said it must.[91] By 1854–1855, Thomson's exploration of cable retardation was clearly coming to touch on some of the deepest questions in electrical physics.

Shortly after hearing from Airy and first calculating the ratio of units, Thomson wrote to tell Stokes he was "anxious to get something brought out" on all of this and seeking his friend's help in doing so.[92] Stokes had recently been appointed one of the secretaries of the Royal Society of London, effectively making him the editor of its journals.[93] Thomson was too busy with other duties, including tending to his ailing wife, to be able to write a fresh paper on the subject, so he asked if Stokes might arrange simply to publish in the *Proceedings of the Royal Society* extracts from the letters Thomson had sent him the previous October. Stokes obliged, and with some additions and revisions, including a few paragraphs from one of Stokes's replies to Thomson, the letters appeared in the *Proceedings* in May 1855 under the title "On the Theory of the Electric Telegraph." The paper reached a wider readership when it was reprinted early the next year in the *Philosophical Magazine*.[94]

The most important addition Thomson made to his earlier letters was a series of carefully drawn "arrival curves" showing in graphic form how the current would rise at the far end of a long cable[95] (Figure 1.4). These showed how the rate at which the current rose depended on the resistance and capacitance of the cable and on the length of time the battery was applied at the sending end. Thomson's theory reproduced in mathematical form what Clark and other engineers had seen on their cables and underground wires: a pulse of current that started out crisp and sharp at the transmitting end was gradually stretched out and worn down as it moved along the line, its strength approaching its maximum value at the far end only after a substantial delay. Thomson's theory also meant that there was no such thing as a definite "velocity of electricity"; rather than zipping along a wire like a bullet, a pulse of current diffused along it like heat creeping along an iron rod, with the measured speed of the current depending on the length and properties of the wire and the surrounding dielectric, as well as on the sensitivity of the instruments used to detect it.

[91] See Chapter 5, and Bruce J. Hunt, "Maxwell, Measurement, and the Modes of Electromagnetic Theory," *Historical Studies in the Natural Sciences* (2015) *45*: 303–39.

[92] Thomson to Stokes, February 12, 1855, in Wilson, ed., *Correspondence*, *1*: 186–88.

[93] David B. Wilson, *Kelvin and Stokes: A Comparative Study in Victorian Physics* (Bristol: Adam Hilger, 1987).

[94] William Thomson, "On the Theory of the Electric Telegraph," *Proc. RS* (May 1855) *7*: 382–99; repr. in *Phil. Mag.* (February 1856) *11*: 146–60 and in Thomson, *MPP 2*: 61–76.

[95] Thomson, "Electric Telegraph," 393.

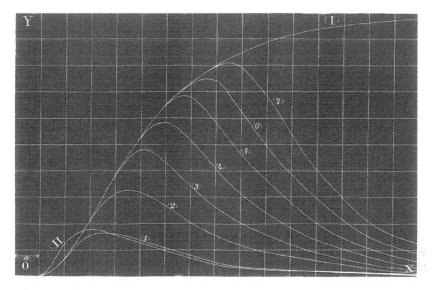

Figure 1.4 In 1854 William Thomson used his theory of electrical diffusion to calculate curves showing how current and potential would rise at the far end of a telegraph cable. The numbered curves correspond to successively longer periods of contact.
(From *Proceedings of the Royal Society of London*, Vol. 7, 1855; courtesy University of Texas Libraries.)

In his original letters to Stokes, Thomson had looked toward using retardation data from a series of cables Brett was then preparing to lay in the Mediterranean; intended to connect France and Italy to Algeria, they would be the longest cables yet laid.[96] The project faced many setbacks, however, and the leg from Sardinia to Algeria failed altogether as the paying-out machinery proved unable to handle laying cable in such deep waters.[97] By the time he revised his letters for publication by the Royal Society, Thomson looked instead to a much longer cable the British government had just laid across the Black Sea from Varna to Balaklava in an effort to speed up communications during the Crimean War.[98] Newall had made and laid this 300-mile cable very hurriedly; to save

[96] Thomson to Stokes, October 30, 1854, in Wilson, ed., *Correspondence*, *1*: 177.
[97] Brett, *Origin and Progress*, vii–x.
[98] On the Black Sea cable, see Bright, *Submarine Telegraphs*, 21, and Walter Peterson, "The Queen's Messenger: An Underwater Cable to Balaklava," *War Correspondent: The Journal of the Crimean War Research Society* (2008) *26*: 31–39; the latter is also available at atlan tic-cable.com/Cables/1855Crimea. The Black Sea cable failed in December 1855, just as the war was ending, and no effort was made to repair it.

time and money, only the shore ends were given the usual outer armoring of iron wires, the rest of the line being left as simply a "core" of copper wire covered with a layer of gutta-percha. Thomson said he had "little doubt but that the Varna and Balaklava wire will be the best yet made for the purpose" of determining the electrical constants related to retardation, and particularly for calculating the ratio of units. Having those values in hand, he said, "will enable me to give all the data for estimating telegraph retardation, without any data from telegraphic operations" – that is, to achieve the scientific ideal of being able accurately to foretell, based solely on general principles, how any cable would perform even before it had been laid.[99]

When in 1852 Latimer Clark first noticed the peculiar way pulses of current behaved when he sent them along an insulated wire submerged in water, he had little inkling that his discovery would have consequences that would reach far beyond the Gutta Percha Company works or even the growing network of telegraph lines. The retardation of signals that Clark and other telegraph engineers observed on their submarine cables and underground wires challenged the prevailing understanding of some important electrical phenomena and served to draw attention to Faraday's previously marginalized ideas about induction, conduction, and the role of the electromagnetic field. Faraday's exposition of retardation effects in turn led Thomson to take up the theoretical investigation of telegraphic propagation and to bring his formidable mathematical powers and physical insight to bear on a problem that was clearly of both practical and scientific importance. It may also have helped draw Maxwell into the study of electromagnetism; in any case, the study of related effects certainly supplied data that were to prove crucial to the development of his electromagnetic theory of light and theory of the electromagnetic field. By the mid-1850s, the seemingly small and out-of-the-way phenomenon of telegraphic retardation had not only become, as Clark would say of specific inductive capacity, "a consideration of high national importance," with a bearing on the successful operation of all submarine cables but was also doing much to set the course electrical physics would follow in Britain over the next decade or more.

[99] Thomson, "Electric Telegraph," 399, 396.

2 Wildman Whitehouse, William Thomson, and the First Atlantic Cable

The first serious effort to span the Atlantic with a telegraph cable was launched in 1856 by a group of entrepreneurs led by Cyrus Field in the United States and John Watkins Brett in England. Theirs was an audacious, not to say foolhardy, enterprise. The nearly 2000-mile-long cable they proposed to lay between Ireland and Newfoundland would be more than six times longer than any yet laid and would lie in waters deeper than any previously attempted. That such an ambitious undertaking was nonetheless launched and quickly drew financial backers is testimony to the technological enthusiasm of the mid-Victorian era, as well as to the prospects for large profits if it should succeed.

The first Atlantic cable proved, however, to be a spectacular failure. It snapped during the first two attempts to lay it in August 1857 and during three more the following June, dashing the hopes of all but its most ardent backers. The success of a final attempt, completed on August 5, 1858, caught almost everyone by surprise and set off rapturous celebrations on both sides of the Atlantic. Troubles soon arose, however, and after just a few weeks of fitful service the cable sputtered to its demise.[1]

Even before the 1858 cable had fallen completely silent, much of the blame for its failure came to be laid at the feet of Edward Orange Wildman Whitehouse, the "electrician-projector" of the Atlantic Telegraph Company and the man in charge of its electrical department. Less than two weeks after the cable was laid and its first messages transmitted, he was summarily dismissed by the board of the company he had helped to found and denounced by its officials as a fool, a fraud, or both. The board dispatched one of its own members, the young University of Glasgow professor of natural philosophy William Thomson (later Lord Kelvin), to

---

[1] On the first Atlantic cable, see the accounts in Charles Bright, *Submarine Telegraphs, Their History, Construction, and Working* (London: Lockwood, 1898), 38–54; Charles Bright, *The Story of the Atlantic Cable* (New York: Appleton, 1903); Thompson, *Kelvin, 1*: 325–96; Bern Dibner, *The Atlantic Cable* (Norwalk, CT: Burndy Library, 1959), 5–45; Crosbie Smith and M. Norton Wise, *Energy and Empire: A Biographical Study of Lord Kelvin* (Cambridge: Cambridge University Press, 1989), 667–75; and Gillian Cookson, *The Cable: The Wire That Changed the World* (Port Stroud: Tempus, 2003).

take Whitehouse's place at the Irish end of the cable and supervise electrical operations there. Thomson, who had advised on electrical arrangements for the project from its early days and sailed on all of its laying expeditions, nursed the cable along for two more weeks but in the end could not keep its insulation from giving way. By early September the cable was effectively dead.

There were many reasons why the first Atlantic cable failed and many reasons why blame for its failure came to be apportioned as it was. Whitehouse certainly deserved to shoulder a substantial share of that blame, but why in the end was almost all of it heaped on him while others who arguably bore comparable responsibility, particularly Field and the young chief engineer Charles Tilston Bright, escaped virtually unscathed? And how did Thomson, who presided over the actual death of the cable, manage to emerge from the debacle with his reputation not just untarnished but substantially enhanced, and to go on to become perhaps the most revered figure in the Victorian cable industry (Figure 2.1)? An important but often overlooked factor in the differing fates of Whitehouse and Thomson, and in the subsequent development of cable telegraphy as a whole, centers on *measurement*. Thomson strongly emphasized the precision measurement of electrical quantities, particularly resistance, and the exercise of strict quality control. He devised important new instruments and measuring techniques and helped lay the foundation for a connected system of electrical units and standards. The measurement practices he pursued and advocated in his work on the first Atlantic cable converged with and reinforced many of those being developed at the same time by leading telegraph engineers. The result by the mid-1860s was the development of a cohesive and effective system for understanding and managing the operation of submarine cables, the main features of which would continue to be followed well into the twentieth century.

Whitehouse was also an avid measurer, but it was not always clear quite what he was actually measuring. He too developed new techniques and instruments, some of considerable delicacy and precision, but his instruments remained his alone, and many of his measurements could not be readily related to those made by more mainstream scientists and engineers. He was thus left isolated and vulnerable, and when things began to go wrong, others on the project were able to throw him overboard with little risk he would take them down with him. Thereafter the technical aspects of cable telegraphy were left in the hands of a group of scientists and engineers whose shared attitudes and practices, particularly concerning electrical measurement, would set much of the direction of British work in electrical science and technology for decades to come.

Figure 2.1 The four "projectors" of the Atlantic Telegraph Company: a: Cyrus W. Field, b: John Watkins Brett, c: Charles Tilston Bright, and d: E. O. Wildman Whitehouse.
(Field, Brett, and Whitehouse from Louis Figuier, *Merveilles de la Science*, Vol. 2: 248, 208, and 253, 1868; Bright from Charles Bright, *Life of Sir Charles Tilston Bright*, frontispiece, 1908.)

Field's Dream

Cyrus Field had barely heard of submarine telegraphy before January 1854, when he first encountered Frederic Gisborne. Gisborne, a British-born telegraph engineer, had been working for years to promote a scheme to speed up transatlantic communication via Newfoundland. He had made some headway by mid-1853 when his financing abruptly collapsed. Facing bankruptcy, he headed to New York to seek new backing and there chanced to meet Matthew Field, a civil engineer, who suggested he speak with his brother Cyrus. Though only 34, Cyrus Field had already made a fortune in the paper business and was now looking for a new outlet for his wealth and restless energies. Gisborne's scheme intrigued him. In the 1850s, it took about ten days for even the most urgent news from England to reach New York by steamship. A ship could beat that by a day or two by putting in at Halifax, Nova Scotia, and sending its messages ahead by overland telegraph – and many businesses and news agencies were willing to pay handsomely for a day's head start on their competitors. Gisborne proposed to shave off another day or more by extending this telegraphic shortcut eastward to St. John's, Newfoundland, using overland lines across the island and a short cable across Cabot Strait from Cape Ray to Cape Breton Island in Nova Scotia. Buoyed by hopes of making Newfoundland a transatlantic communications hub, the colonial legislature in St. John's granted Gisborne valuable concessions, including a thirty-year monopoly on all telegraphs on the island.[2]

Field listened closely to Gisborne's pitch but was not immediately won over. When he later consulted his globe to check the proposed route, however, he was struck by a far grander idea: why stop at Newfoundland? St. John's was about a third of the way from New York to London; why not extend a cable clear across the Atlantic to Ireland, and so link the New World directly to the Old?[3] A cable spanning nearly 2000 miles of open ocean would be an enormous leap over anything yet accomplished or even attempted in submarine telegraphy, but Field knew too little about the technical obstacles to be put off by them. Not long after meeting with Gisborne, he wrote to the pioneering oceanographer Matthew Fontaine Maury of the US Naval Observatory, who told him that recent soundings showed the bed of the North Atlantic would provide an ideal resting place for a submarine cable – Maury even dubbed it the "Telegraphic Plateau." Field also wrote to the telegraph entrepreneur Samuel F. B. Morse, who assured him that an electric current could indeed be made to pass through

[2] On Gisborne's Newfoundland plan, see Donald Tarrant, *Atlantic Sentinel: Newfoundland's Role in Transatlantic Cable Communications* (St. John's, NL: Flanker Press, 1999), 7–17.
[3] Henry M. Field, *The Story of the Atlantic Telegraph* (New York: Scribner's, 1893), 16.

an insulated wire long enough to span an ocean.[4] Neither Maury nor Morse had any real experience with submarine cables, but their assurances were enough for Field, and he pushed ahead at full speed. Backed by Peter Cooper, Moses Taylor, and other New York capitalists, Field quickly organized the ambitiously named New York, Newfoundland, and London Telegraph Company – an American firm aiming to lay a cable to link two parts of the British Empire. By April 1854 the new group had bought up the assets of Gisborne's bankrupt operation and secured more concessions from the government of Newfoundland, including a fifty-year monopoly on landing telegraph cables on the island.[5] This, along with Field himself, would prove to be the New York company's most valuable asset.

Running a telegraph wire across Newfoundland was, Field later said, "a very pretty plan on paper"; one simply drew a line on the map and the job was done.[6] He expected the task to take just a few months and to cost a small fraction of his company's capital. In fact it took his brother Matthew and a crew of workers more than two years to hack a route through the rugged interior of the island, and erecting and maintaining the overhead line from St. John's to Cape Ray, as well as a new overhead line across Cape Breton Island, proved far more costly than Field and his partners had anticipated (Figure 2.2).

While that work was going forward, Field set about securing the cable that was to span Cabot Strait and connect Newfoundland to the North American telegraph network. He could obtain wires and equipment for the landline in the United States and Canada, but a cable was a different matter; for that, he would have to go to Britain, then and for long to come virtually the sole home of submarine telegraph technology.[7] Armed with an introduction from Gisborne, Field arrived in London early in 1855 and called on John Watkins Brett, whose Submarine Telegraph Company had laid the first cable across the English Channel a few years before. Already an enthusiast for oceanic telegraphy, Brett invested $10,000 in the New York company and joined whole-heartedly in its efforts to span the Atlantic.[8] On his advice,

---

[4] Field, *Atlantic Telegraph*, 18–22.   [5] Tarrant, *Atlantic Sentinel*, 20–22.

[6] Field made this remark at a banquet in 1866; see *The Atlantic Telegraph: Report of the Proceedings of a Banquet Given to Mr. Cyrus W. Field by the Chamber of Commerce of New-York, at the Metropolitan Hotel, November 15th, 1866* (New York: privately printed, 1866), 19.

[7] Samuel Bishop of New York reportedly began supplying wires insulated with gutta-percha in 1851, but while his Bishop Gutta Percha Works long remained the chief American cable maker, it produced only short cables for use in rivers and harbors rather than longer ones suitable for sea crossings; see "The Bishop Gutta Percha Works," *The Telegrapher* (January 1, 1870) 6: 145–47.

[8] John W. Brett, *On the Origin and Progress of the Oceanic Electric Telegraph* (London: W. S. Johnson, 1858), 47.

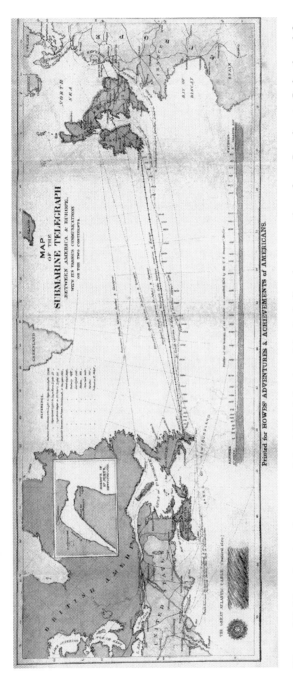

Figure 2.2 Map showing the route of the 1858 Atlantic cable, including its connections from London to Ireland and from Newfoundland to New York.
(From Henry Howe, *Adventures and Achievements of Americans*, frontispiece, 1858.)

Field ordered seventy-five miles of multiconductor cable from Küper and Company (later to become Glass, Elliot and Company) and arranged to have it shipped to Newfoundland in a sailing bark. While in England, Field also first met Charles Tilston Bright, then just 22 but already chief engineer of the Magnetic Telegraph Company. Impressed by the young engineer's drive, Field took steps to enlist him in the Atlantic cable project.[9]

Always attuned to the value of publicity but with little grasp of the practicalities of cable-laying, Field chartered a steamer in New York in August 1855 and invited more than fifty guests, including Morse, Cooper, and their families, on what he expected would be a pleasant summer excursion to Newfoundland.[10] There they were to watch as their vessel towed the bark from which the cable would be uncoiled across Cabot Strait. It all turned into an embarrassing fiasco, however, as the towing arrangement proved unwieldy and a fierce gale blew the vessels off course. After some tense moments when it seemed the bark might be swamped, the cable had to be cut and the attempt to lay it abandoned.[11] The chastened party slunk back to New York, but Field treated the whole episode as simply a lesson learned: he soon ordered a new length of cable from England and had a properly equipped steamship lay it the following summer, this time with little fanfare. By then the landline across Newfoundland was finally complete and Field was ready to tackle the next and far bigger step: spanning the Atlantic.

Field and his New York partners had initially hoped to finance the entire Atlantic telegraph project on their own, but after exhausting so much of their capital extending their lines just to St. John's, they realized the task would be beyond their means. In July 1856 Field sailed again to England, authorized by his partners to take whatever steps he thought would best advance the project. After consulting with Brett, he decided to launch a new company in London, and the two began sounding out potential investors. Field also sought to bolster the credibility of the project by enlisting engineers and scientific men willing to lend it their skills and reputations. He had a particular eye on Bright, "strenuously urging," as Brett later put it, that the young engineer be included as one of the initial "projectors" of the new company.[12] On September 29, 1856, Field, Brett, and Bright signed a document mutually pledging to do all in

[9] Edward Brailsford Bright and Charles Bright, *The Life Story of the Late Sir Charles Tilston Bright, Civil Engineer*, 2 vols. (London: Archibald Constable, 1899), *1*: 109–10.

[10] For a list of the many guests on this excursion, see John Mullaly, *The Laying of the Cable, or The Ocean Telegraph* (New York: Appleton, 1858), 51.

[11] Mullaly, *Laying of the Cable*, 52–75.

[12] John W. Brett, "The Atlantic Telegraph" (letter), *Morning Post* (September 23, 1858), 2.

their power to advance the formation and success of what they proposed to call the Atlantic Telegraph Company.[13]

Field was by all accounts a whirlwind of activity in these months, meeting with cable makers to examine sample designs and discuss costs, pressing the British government to supply ships to assist with the laying of the cable and to guarantee a substantial amount of official business for the cable once it was successfully completed, and constantly touting the merits of the proposed cable to the press and potential investors. Morse came over from America to lend his support, and Field saw to it that a lavish dinner was mounted in his honor, thus winning further publicity for the project.[14]

Field and Brett formally registered the Atlantic Telegraph Company on October 20, 1856, and issued a prospectus a few weeks later. They initially set its capital at £300,000, soon raised to £350,000, to be offered in shares of £1000 each; Field and Brett started by taking twenty-five shares each.[15] Stressing both the grandeur and utility of a cable spanning the Atlantic, the prospectus sought to allay concerns about the practicability of the project by citing Morse and Maury's endorsements, as well as recent experiments on signaling through long underground wires. "It is considered," the prospectus went on, "that little requires to be said in favour of the undertaking," as "the benefits which it will confer upon all classes are too obvious to need mention, and its proved practicability renders its accomplishment a duty." Nor did the prospectus ignore financial practicalities, reporting that cable manufacturers had given assurances that £350,000 would be more than enough to cover all of the costs of making and laying the cable, and stating that "upon a very moderate computation of the probable amount of traffic, and a consideration of the comparatively small working expenses (which are necessarily limited to those of the terminal stations), the net receipts will yield an annual return exceeding 40 per cent. upon the capital" – a truly enticing rate of profit.[16] The pitch was evidently persuasive, as the shares were all subscribed within a few weeks. Liverpool merchants were especially avid to invest;

[13] George Saward, *The Trans-Atlantic Submarine Telegraph: A Brief Narrative of the Principal Incidents in the History of the Atlantic Telegraph Company* (London: privately printed, 1878), 7–8. Saward was for many years the secretary of the Atlantic Telegraph Company.

[14] On the October 1856 London banquet for Morse, see Samuel Irenaeus Prime, *The Life of Samuel F. B. Morse* (New York: Appleton, 1875), 646; on Field's activities, see Field, *Atlantic Telegraph*, 69–71.

[15] Brett, *Origin and Progress*, 48; on the increase in the capitalization of the company, see ATC Minute Book, entry for October 31, 1856, 5. This Minute Book is held by the BICC Archive at the Merseyside Maritime Museum, Liverpool. I thank Allan Green for his generous loan of a copy of it.

[16] "Prospectus of the Atlantic Telegraph Company," November 6, 1856, reprinted in Brett, *Origin and Progress*, 49–51.

many of them were active in the transatlantic trade in cotton and other commodities and were attracted by the prospect of quicker access to American market information. According to the *Times*, when Field showed a sample of the proposed cable to a group of Liverpool business-men, "a broker admiringly exclaimed, 'There's the thing to tell the price of cotton!'"[17] Most of the rest of the shares were sold in London, with a scattering to investors in Manchester, Glasgow, and other British cities. Americans, however, evinced little enthusiasm; although Field had reserved eighty-eight shares, a quarter of the total, to sell in the United States, he could only find buyers there for twenty-seven and was forced to carry the rest himself.[18] He would remain the largest single investor in the Atlantic Telegraph Company.

Although it quickly drew financial backers, the project met with con-siderable skepticism in Britain. Many observers doubted that a cable could be successfully laid in such deep waters, while others thought the capital had been set too low. The eminent engineer Isambard Kingdom Brunel, whose ship the *Great Eastern* would later lay the 1865 and 1866 Atlantic cables, reportedly estimated it would cost about £2,000,000 to do the job properly – very close to what, after many reverses, proved to be the eventual total.[19] Field and Brett judged, however, that £350,000 was the most they could raise in 1856 and so set out to do what they could with that amount.

Beyond its relatively low estimate of costs, perhaps the most striking point in the prospectus was the statement that "it is determined to complete and have the Telegraph in operation during the ensuing summer."[20] That is, the Atlantic cable was to be designed, manufactured, tested, loaded, shipped, laid, and put into service, with all of its associated apparatus and personnel, within just ten months after the company was first launched. It was a stupendously ambitious, not to say wildly unreal-istic, timetable. Brett later said it was Field, with "the go-ahead character of an American," who had insisted on aiming to lay the cable in 1857, and while Brett conceded that promising such quick completion had been necessary, "as a matter of policy," to attract investors when the company

[17] "The Atlantic Telegraph Company," *Times* (November 14, 1856), 12. On how teleg-raphy, and particularly the Atlantic cables, affected the global cotton trade, see Harold D. Woodman, *King Cotton and His Retainers: Financing and Marketing the Cotton Crop of the South, 1800–1925* (Lexington: University of Kentucky Press, 1968), 267, 273, 292–93, and Sven Beckert, *Empire of Cotton: A Global History* (New York: Vintage, 2014), 320, 336.
[18] Saward, *Trans-Atlantic Submarine Telegraph*, 8–9.
[19] Saward, *Trans-Atlantic Submarine Telegraph*, 9.
[20] "Prospectus," in Brett, *Origin and Progress*, 49.

was first floated, he also admitted that the ensuing rush lay behind many of the problems that would bedevil the project.[21]

Even if a cable could be successfully laid beneath the Atlantic, critics doubted it could be made to carry signals at a commercially viable rate. The great obstacle was retardation, as discovered by Latimer Clark in 1852 and brought before the public by Michael Faraday in 1854. Induction effects had already been found to interfere with signaling on some shorter cables, and Thomson had given theoretical grounds for expecting the retardation on a cable of any given thickness to increase with the square of its length. If Thomson was right, the retardation on a cable long enough to span the Atlantic might be so severe as to render it almost useless.

Here was a serious threat to the viability of the whole Atlantic cable project, and in 1856 Field latched onto a new and seemingly unlikely ally to combat it. Wildman Whitehouse proceeded to take up the task with remarkable vigor and soon emerged as one of the most active and controversial figures in the story of the first Atlantic cable.

Wildman Whitehouse

By training and background, Wildman Whitehouse was neither an engineer nor a scientist, but a surgeon. He had built up a thriving practice at Brighton when, in the early 1850s, at the age of about thirty five, he began to experiment with electricity.[22] His efforts to devise a new system of multi-wire telegraphy soon brought him into contact with John Watkins Brett, who later said he saw in Whitehouse "most patient qualities of investigation." Here was just the man, Brett thought, to tackle the threat posed by retardation, and in the spring of 1855 he began providing Whitehouse with hundreds of pounds for equipment and experimental expenses as well as the assistance of James Banks, an experienced technician from Brett's staff.[23] Crucially, Brett also gave Whitehouse access to

---

[21] Testimony of J. W. Brett, December 10, 1859, in *Joint Committee Report*, 58.

[22] The fullest collection of biographical information on Whitehouse, compiled by Bill Burns, Allan Green, and others, can be found on the "Atlantic Cable" website at http://atlantic-cable.com/Books/Whitehouse/eoww.htm. Whitehouse did not confine his electrical studies to telegraphy; in 1852–53 he devised a sensitive galvanometer that his fellow Brighton resident John O. N. Rutter used to investigate electrical phenomena in muscle tissue. See Richard Noakes, *Physics and Psychics: The Occult and the Sciences in Modern Britain* (Cambridge: Cambridge University Press, 2019), 32, and John O. N. Rutter, *Human Electricity: The Means of Its Development, Illustrated by Experiments* (London: John W. Parker and Son, 1854), 117 and frontispiece.

[23] Brett later reported that he supplied Whitehouse with "several hundred pounds" for his experiments; see Brett, "Atlantic Telegraph" (letter), *Morning Post* (September 23, 1858), 2.

two multiconductor cables then being readied for shipment from
Küper's works in East Greenwich: a 150-mile-long cable destined for
the Mediterranean containing six separately insulated conductors, and
the 75-mile-ong cable with three conductors that Field would soon try
and fail to lay from Newfoundland to Nova Scotia.[24] Whitehouse proceeded to spend several weeks performing virtually every test he could
think of on this total of 1125 miles of insulated wire, hoping, as he later
said, to demonstrate the practicability of oceanic submarine telegraphy
or, failing that, at least to "make us better acquainted with the electrical
difficulties to be encountered, and so place us in a position to meet the
enemy with the true indomitable English spirit, – determined to
conquer."[25]

Whitehouse presented a very full account of his experiments to the
Mathematical and Physical Section of the British Association at its
September 1855 meeting in Glasgow. It was his first scientific paper
and he took great pains with it, even having it printed as a pamphlet;
he also sent a shorter version to the *Illustrated London News*.[26] He
opened by seeking to justify himself before his audience – to explain
why he, a surgeon with neither scientific credentials nor practical
engineering experience, should be taken seriously when speaking on
cable telegraphy. "The study of the varied phenomena of Electricity,"
he declared, "is no longer the exclusive privilege of the philosopher";
the recent spread of telegraphy had opened the subject far more
widely, so that even one like himself, "unknown in the world of
science," might now venture to record his electrical experiments
and observations, justifying his intervention in the field by citing its
great practical importance.[27]

---

[24] See Whitehouse's testimony, December 15, 1859, in *Joint Committee Report*, 69;
J. W. Brett, "Atlantic Telegraph" (letter), *Engineer* (October 8, 1858), 6: 267; and
Wildman Whitehouse, *The Atlantic Telegraph: The Rise, Progress, and Development of Its
Electrical Department* (London: Bradley and Evans, 1858), 6, also published in *Engineer*
(September 24, 1858) 6: 230–32.

[25] Wildman Whitehouse, *Report on a Series of Experimental Observations on two lengths of
Submarine Electric Cable, containing, in the aggregate, 1,125 miles of wire, being the substance
of a paper read before the British Association for the Advancement of Science, at Glasgow,
September 14, 1855* (Brighton: privately printed, 1855), 6.

[26] "Mediterranean Telegraph," *Illustrated London News* (October 6, 1855) 27: 423.
Although it was published after the Glasgow meeting, this article was written well before
it. An account of Whitehouse's British Association paper appeared in *Athenæum*
(September 22, 1855), 1091–92, and a brief abstract appeared under the title
"Experimental Observations on an Electric Cable" in the 1855 *BA Report*, Section A,
23–24; extensive excerpts were later published as "The Atlantic Telegraph," *Engineer*
(January 30, 1857), 3: 82–83.

[27] Whitehouse, *Report* (1855), 3.

Whitehouse was unknown, however, not just in the world of science but in the world of practice as well. With no experience on lines or cables in actual use, he had no real standing among the emerging community of telegraph engineers. The facts he could present for consideration were the products of special experiments, not of practical field experience. He evidently concluded that in such circumstances his best hope of achieving some standing in the electrical world and of contributing to the advancement of submarine telegraphy lay in presenting himself as neither a "philosopher" nor a practical engineer but as a Baconian experimenter whose findings might aid practical progress. In his 1855 paper Whitehouse explicitly disavowed the role of the disinterested natural philosopher, asking that, in light of the importance of cable telegraphy, he be "pardoned" for having "investigated the phenomena exhibited by electrical currents in subterranean and submarine wires, as a speciality, and with a direct leaning towards their practical application, rather than in their more general and more extended theoretical aspects."[28]

The basic experimental technique Whitehouse described in his Glasgow paper was straightforward: he sent trains of electrical pulses into an insulated wire and, at its far end, recorded the arriving currents electrochemically on a moving paper tape. This yielded what he called "the handwriting, the autograph, of the current itself" and enabled him to "observe its habits and behaviour," particularly the delay and stretching of signals due to retardation, as he varied the length of wire used and the strength and nature of the pulses sent[29] (Figure 2.3). He found, among much else, that strong alternating pulses from a magneto-electric generator showed much less retardation than did the usual battery currents – indeed, with such magneto pulses, signaling speeds "ample for commercial success" could, he said, be achieved through the full 1125 miles of wire.[30] Notably, Whitehouse made these tests using simple reversals rather than the irregularly spaced pulses that would be needed to spell out actual words. He also reported that "doubling or trebling the mass of conducting metals" carrying the current (which he did by sending it simultaneously along two or three of the separately insulated wires within the cable, *not* by using a single larger conductor) did not reduce the retardation, so that "no adequate advantage would be gained by any considerable increase in the

[28] Whitehouse, *Report* (1855), 5.
[29] Whitehouse, *Report* (1855), 10–11. Whitehouse gave a fuller account of his methods (with illustrations) in "Experiments on the Retardation of Electric Signals, Observed in Submarine Conductors," *Engineer* (January 23, 1857) 3: 62–63.
[30] Whitehouse, *Report* (1855), 21.

Fig. V.—Drum Apparatus.—(Half size.)

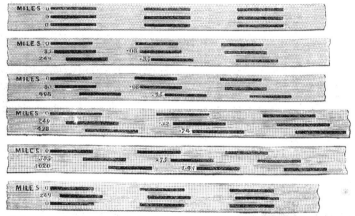

Fig. VI.—The Records.—(Full size.)

Figure 2.3 Wildman Whitehouse's telegraphic recording apparatus and marked paper tapes, showing the degree of retardation experienced by signals passing through different lengths of cable.
(From *Engineer*, Vol. 3: 63, January 23, 1857; courtesy University of Texas Libraries.)

size of the wire" – a claim that would later become the focus of sharp controversy.[31] In short, Whitehouse claimed to have shown that retardation did not pose nearly as serious an obstacle to submarine telegraphy as many had feared, and he was happy to be able to conclude "that India, Australia, and America are readily accessible by telegraph without the

[31] Whitehouse, "Experimental Observations," 24; Whitehouse, *Report* (1855), 20.

use of wires larger than those now commonly employed in submarine cables."[32]

This was just what the promoters of an Atlantic cable wanted to hear, of course, and they were no doubt delighted when the Duke of Argyll, the president of the Glasgow meeting, said in his closing remarks that Whitehouse's paper "deserved especial notice" for its demonstration that "there remained no practical difficulty" to overcoming the ill effects of retardation, even on a cable long enough to span the Atlantic.[33] Brett and Field would later tout Whitehouse's results to potential backers, and Field thought enough of Whitehouse's Glasgow paper to send a copy of it to Morse in March 1856.[34]

Not all of the response to Whitehouse's paper was positive, however. Thomson voiced doubts on hearing it delivered at the Glasgow meeting, and in fact presented a paper of his own at the same session that undercut many of Whitehouse's main claims[35] (Figure 2.4). In "On Peristaltic Induction of Electric Currents in Submarine Telegraph Wires," Thomson examined theoretically how signals would move along parallel insulated wires like those in the multiconductor cables Whitehouse had used in his experiments.[36] Drawing on the mathematical analysis of induction in telegraph wires he had published a few months earlier in the *Proceedings of the Royal Society*, he concluded that electrostatic action between neighboring wires might skew the results of tests like those Whitehouse had performed on multiconductor cables, and warned that "expectations as to the working of a submarine telegraph between Britain and America, founded on such experiments, may prove fallacious."[37] It was a serious shot across Whitehouse's bow.

Reverting to a suggestion he had first made in his Royal Society paper, Thomson said that "to avoid the chance of prodigious losses," those proposing to lay a cable across the Atlantic should look not to flawed experiments like Whitehouse's but instead to the actual performance of existing cables, particularly the single-conductor line recently laid beneath the Black

---

[32] Whitehouse, *Report* (1855), 22.

[33] The Duke of Argyll was quoted in "General Concluding Remarks," *Glasgow Herald* (September 21, 1855), 5; John Tyndall was quoted in "Meetings of the British Association," *Glasgow Sentinel* (September 15, 1855), 4.

[34] The copy of Whitehouse's *Report* (1855) held by the Olin Library at Cornell University is inscribed "March 4, 1856 Sam. F. B. Morse from Cyrus W. Field, Esq."

[35] Thomson's remarks about Whitehouse's paper are quoted in "The British Association," *The Globe* (September 17, 1855).

[36] William Thomson, "On Peristaltic Induction of Electric Currents in Submarine Telegraph Wires," *BA Report* (1855), Section A, 21–22, repr. in Thomson, *MPP 2*: 77–78.

[37] Thomson, "Peristaltic Induction," 22.

Figure 2.4 William Thomson in 1852, at age 28.
(From Silvanus P. Thompson, *Kelvin*, Vol. 1: 232, 1910.)

Sea from Varna to Balaklava to speed communications during the Crimean War. Signaling on this 300-mile-long cable was known to be slow, no more than five words per minute, because of effects its operators ascribed to induction.[38] Citing his mathematical demonstration that, for a given thickness of wire and insulation, the retardation on a cable is proportional to the square of its length, Thomson said that a cable across the Atlantic – six times the distance from Varna to Balaklava – could be expected to suffer thirty-six times the retardation. "If the distinctness of utterance and rapidity of action practicable with the Varna and Balaklava wire are only such as not to be inconvenient," he said, transmission rates on an Atlantic cable of similar gauge would thus be far too slow to be commercially useful. The only solution appeared to be to make the cable six times as thick, thus requiring thirty-six times as much copper and gutta-percha per mile, while still achieving no more than the slow speed of the Black Sea line.[39]

---

[38] Thomson, "Peristaltic Induction," 22. Thomson had first suggested looking to the Black Sea cable in "On the Theory of the Electric Telegraph," *Proc. RS* (May 1855) 7: 382–99, on 99, repr. in Thomson, *MPP 2*: 75. On signaling rates, see Alfred Varley to Douglas Galton, May 4, 1861, in *Joint Committee Report*, 506. The Black Sea line lacked external armoring except at its shore ends and so was not a true "cable."

[39] Thomson, "Peristaltic Induction," 22.

Few engineers or investors could follow the intricacies of Thomson's differential equations, but they could multiply by six – or by thirty-six. If Thomson's analysis was correct, the dream of spanning the Atlantic would be dashed; the required cable would either be too expensive and unwieldy to make and lay, or too slow to pay its way. Unless some answer to Thomson could be found, Field and Brett's project appeared doomed.

Whitehouse was convinced – justifiably, as it turned out – that Thomson had erred on some major points, but he did not rush to offer a public response. He spent much of the year after the Glasgow meeting devising new instruments and securing a patent on "Improvements in Electro-Telegraphic Apparatus," particularly a new form of induction coil that would later play an important part in the short life of the Atlantic cable. With Brett's backing, he also undertook a series of new experiments on transmission rates along lengths of multiconductor cables being readied for shipment to the Mediterranean. Encouraged by the physicist John Tyndall to focus entirely on his electrical researches, Whitehouse took steps that summer to wrap up his medical practice and, as a Brighton newspaper put it, free himself to "direct his talents to the perfection and extension of the Electric Telegraph."[40] From mid-1856 on, Whitehouse would no longer be a surgeon who dabbled in electricity but a full-time telegraphic experimenter.

By early August 1856, when the British Association met at Cheltenham, Whitehouse was ready to answer Thomson, and he hit back hard. In a paper provocatively titled "The Law of Squares – Is It Applicable or Not to the Transmission of Signals in Submarine Circuits?," published in the *Athenæum* at the end of August, he announced that his experiments pointed to an unequivocal answer: no. Indeed, he declared Thomson's "law" to be no more than "a fiction of the schools," fine in its place but wholly irrelevant to the proper design and operation of submarine cables.[41]

Whitehouse began by defending his use of multiconductor cables, saying that careful measurements showed the effects of mutual induction between the wires to be negligibly small, a point Thomson would soon

---

[40] Whitehouse, *Atlantic Telegraph*, 7–8; Whitehouse's patents are listed and annotated by Allan Green at http://atlantic-cable.com/Books/Whitehouse/Patents/patents.htm. A note in the *Brighton Gazette* (August 21, 1856), 5, reported that Whitehouse, "for many years a distinguished member of the medical profession of Brighton," had "given up the practice of medicine, and left Brighton" to focus on telegraphy. Whitehouse later said Tyndall had encouraged him to turn from medicine to electrical experimentation; see Whitehouse, *Atlantic Telegraph* (1858), 8.

[41] Wildman Whitehouse, "The Law of Squares – Is It Applicable or Not to the Transmission of Signals in Submarine Circuits?," *BA Report* (1856), 21–23, also published in the *Athenæum* (August 30, 1856) 1092.

concede.[42] He next turned to Thomson's telegraphic theory itself, distinguishing three aspects of it, all of which he called the "law of squares." First, there was retardation itself, which Whitehouse defined as the delay before a detectable current first appeared at the far end of a cable. Using sensitive magnetic relays and a paper tape arrangement similar to the one he had employed the year before, he found retardations of 0.08 seconds through 83 miles of wire, 0.79 seconds through 498 miles, and 1.42 seconds through 1020 miles. Far from increasing with the square of the length, he said, the retardation evidently increased "very little beyond the simple arithmetical ratio."[43] Next he took up what he called the *experimentum crucis*, measuring the maximum rate at which distinct signals could be made to pass along a cable. He found that when he doubled the length of an insulated conductor from 500 to just more than 1000 miles, the number of pulses of current it could carry per unit time indeed fell, but only from 350 to 270, not, as the law of squares would imply, to fewer than 90. Long cables were indeed a bit slower than short ones, he said, but not by nearly as much as Thomson's theory had suggested. Finally, Whitehouse reported that increasing the size of the conductor (he again used multiple separately insulated wires rather than a single thicker one) not only did not reduce the retardation but actually increased it: "trebling the size of the conductor augmented the amount of retardation to nearly double that observed in the single wire." As Whitehouse observed, his results were "strikingly opposed" to Thomson's theory; they virtually demanded a response.[44]

Whitehouse later said he had hoped to see Thomson at the Cheltenham meeting so they could hash out their differences in person, but Thomson was then visiting spas on the Continent with his ailing wife and did not learn of Whitehouse's paper until late September, when he happened across the account of it that appeared in the August 30 issue of the *Athenæum*.[45] Concerned that silence might be taken as acquiescence to Whitehouse's criticisms, he dashed off a letter defending his earlier claims and boldly asserting that "all Mr. Whitehouse's experimental results are perfectly consistent with my theory." He admitted, however, that he based this claim simply on his own confidence in his theory, "which, like every *theory*, is merely a combination of established truths"; he could not yet say quite *how* to square it with Whitehouse's seemingly contrary results, just that he was sure there must be a way. Conscious of

[42] William Thomson, "Telegraph to America" (letter), *Athenæum* (November 1, 1856) 1338–39, repr. in Thomson, *MPP 2*: 94–102.

[43] Whitehouse, "Law of Squares," 21–23. [44] Whitehouse, "Law of Squares," 23.

[45] Wildman Whitehouse, "The Atlantic Telegraph" (letter), *Athenæum* (October 11, 1856) 1247; on Thomson's travels from July to September 1856, see Thompson, *Kelvin, 1*: 320–24.

the weakness of his response and with an eye on the plans then brewing to lay a cable across the ocean, Thomson said he hoped to have more to offer soon, for "capitalists ought to require a very 'matter-of-fact' proof of the attainability of a sufficient rapidity of communication of actual messages ... before sinking so large an amount of property in the Atlantic."[46]

Whitehouse replied in the next issue of the *Athenæum*, thanking Thomson for his published response and also for a "long and friendly communication" (now lost) from him. In return, he sent Thomson a detailed account of his experiments and offered to meet him "as far northward as Liverpool any day next week" to explain them more fully. But while confessing that he could not "follow the learned Professor into fine distinctions upon the nature of theory," he said he was at a loss to see how the results of his experiments could possibly be reconciled with Thomson's law of squares. He repeated and extended his claim that "by any considerable increase in the size of a long *submarine conductor*, we positively increase the difficulty of giving telegraphic signals and diminish the speed of their transmission," and declared that if anyone still wished to argue that the maximum rate of signaling attainable on a submarine cable falls off with the square of its length, "I will undertake to prove experimentally the fallacy of that idea."[47]

After closely studying Whitehouse's account of his experiments, Thomson sent the *Athenæum* a long letter in which he sought to explain away the apparent conflicts with his theory. The basic problem, he said, was that Whitehouse had not applied the theory correctly, so that what he had so meticulously measured turned out to be not quite what Thomson had been talking about. In particular, Whitehouse had jumped too quickly from Thomson's full mathematical theory of telegraphic trans-mission to the simple "law of squares" – though it must be said that Thomson had opened the way when he had asserted that "a wire six times the length of the Varna and Balaklava wire, if of the same lateral dimensions, would give thirty-six times the retardation and thirty-six times the slowness of action."[48] He now backed off from such sweeping statements and stressed that "it depends on the nature of the electric operation performed at one extremity of the wire, and on the nature of the test afforded by the indicating instrument at the other extremity, whether or not any *approach to the law of squares is to be expected* in the observed

---

[46] William Thomson, "Telegraphs to America" (letter), *Athenæum* (October 4, 1856) 1219, repr. in Thomson, *MPP 2*: 92–93.
[47] Whitehouse, "Atlantic Telegraph," *Athenæum* (October 11, 1856) 1247. Whitehouse and Thomson did not meet at that time.
[48] Thomson, "Peristaltic Induction," 21–22.

results."[49] According to Thomson, when we take into account the varying power of the batteries, the limited sensitivity of the instruments, and the effects of electromagnetic induction in the coils of the receivers, Whitehouse's careful measurements of retardation and signaling rates were actually well in line with what Thomson's theory predicted; indeed, Thomson later said that "nothing can be more perfect that the agreement of these experimental results with the theory" – when that theory was properly applied.[50]

As for the claim that increasing the size of the conductor led to more, not less, retardation, Thomson said Whitehouse had erred in treating the three separately insulated wires of his cable as equivalent to a single thicker conductor. The key to reducing retardation lay not just in reducing the total resistance of a circuit but also in reducing its associated induction (or capacitance). Since the latter depended crucially on the surface area of the conductor and the thickness and arrangement of the surrounding insulation, the experiments in which Whitehouse sent currents simultaneously along all three wires of his cable were not a proper test of the retardation to be expected along a single thicker conductor. Though he drily said it was "not my part" to explain the increase in retardation Whitehouse had seen when using all three wires, Thomson suggested that the more rapid depletion of the battery when acting through a smaller resistance might have produced the effect.[51]

Amid these defenses of his theory, however, Thomson made a major concession: he had been too hasty, he said, in ascribing the poor performance of the Black Sea cable simply to retardation, and his estimates of the signaling rates achievable on longer cables were therefore mistaken. Drawing on new measurements of the ratio of electromagnetic to electrostatic units by "that most profound and accurate of all experimenters, Wilhelm Weber," Thomson now calculated that the Black Sea line, if pushed to its utmost, should have been able to carry nine letters per second, or about 100 words per minute. That in practice it achieved only a small fraction of that speed was evidently due to the use of sending and receiving instruments that were poorly chosen and badly adjusted. Thomson now calculated that with properly designed apparatus, operated "so as to clear a wire rapidly of residual electricity," a cable of ordinary thickness should be able to carry a distinct letter across the Atlantic every 3.5 seconds. "This, amounting to 17 letters a minute,

---

[49] Thomson, "Telegraphs to America," 1338–39.
[50] William Thomson to Auguste de la Rive, December 17, 1856, in Paul Tunbridge, *Lord Kelvin: His Influence on Electrical Measurements and Units* (London: Peter Peregrinus, 1992), 97.
[51] Thomson, "Telegraphs to America," 1339.

would give 200 messages of 20 words each in the 24 hours," he observed, "and at 30*s*. a message would be not a bad return for 1,000,000 *l* of capital expended."[52] Indeed, it would come to about a 10 percent annual return on a capital of £1 million, and not far short of the 40 percent return Field would soon be touting for the Atlantic Telegraph Company's actual capital of £350,000. The prospects for a successful Atlantic cable had evidently brightened considerably from the dark picture Thomson had drawn just a year earlier.

Whitehouse practically crowed over this response from Thomson. In a letter published in the *Athenæum* on November 8, he said he was happy to let Thomson go on "maintaining the correctness of his theory – as theory," since, as he saw it, "the whole position for which I originally contended is conceded by Prof. Thomson." Whitehouse particularly prized Thomson's admission that a cable not much thicker than those already in use could be expected to carry several words per minute across the Atlantic, and that with carefully contrived techniques and apparatus even higher rates might be possible. This, Whitehouse said, opened up "a field of research, rich, promising, intensely interesting, and practical," and devising such improved signaling methods, based mainly on using pulses of current from induction coils, had already become the focus of his own work.[53]

The whole exchange with Thomson, from Whitehouse's British Association paper in early August to his final letter in the *Athenæum* in early November, played out just as Field, Brett, and Bright were rushing to launch the Atlantic Telegraph Company – and drawing Whitehouse more closely into it. He and Bright had experimented together on Magnetic Telegraph Company lines as early as November 1855 and had even formed a loose partnership at that time.[54] When in late September 1856 Bright joined Field and Brett in pledging to work together to promote an Atlantic telegraph, Whitehouse was not far behind. A key step came on the night of October 2–3, when he and Bright performed a series of transmission tests on a 2000-mile circuit of the Magnetic Telegraph Company's underground lines. Morse came along to observe the experiments and the next morning sent Field an enthusiastic letter – promptly forwarded to the *Daily News* – filled with praise for the "active and agreeable" Bright and "that clear-sighted investigator of electrical phenomena, Dr. Whitehouse." Using Whitehouse's induction coils and magnetic receivers, Morse reported, they had transmitted up to 270 pulses per minute through the full length of the lines, a result that "most satisfactorily resolved all doubts

[52] Thomson, "Telegraphs to America," 1338–39.
[53] Wildman Whitehouse, "Atlantic Telegraph" (letter), *Athenæum* (November 8, 1856) 1371.
[54] Whitehouse, *Atlantic Telegraph* (1858), 8.

of the practicability ... of operating the telegraph from Newfoundland to Ireland."[55]

Morse's letter drew wide attention. Though Whitehouse later said it had been "incontinently published" and that his experiments with Bright should have been carefully verified before being made public, Morse's report had served its purpose, as Field no doubt intended when he forwarded it to the press: by putting the weight of a famous name behind the Atlantic cable project, it had helped calm the fears of investors worried that retardation might render the cable too slow to pay.[56] Field himself was so impressed by Whitehouse, and so intent on securing access to his patents, that in early October he took steps to add him as the fourth "projector" of the nascent company.[57] From then on Whitehouse would occupy a central position in the Atlantic cable project.

The prospectus for the Atlantic Telegraph Company – issued in early November 1856, just as Whitehouse's last reply to Thomson was appearing in the *Athenæum* – went out of its way to praise Whitehouse and Bright's recent "conclusive experiments" on underground circuits and to call their patented instruments "the most perfect mode at present known" for signaling through long submarine cables.[58] To secure the rights to their patents, as well as their future services to the company, the prospectus called for granting Whitehouse and Bright, jointly with Field and Brett, half of all profits the company earned once it had successfully laid its cable and was paying a 10 percent dividend to its shareholders. Whitehouse thus had a substantial financial interest in the success of the venture, as well as in the use of his instruments, and would also earn a handsome salary of £1000 per year as the company's electrician, as Bright would as its chief engineer.[59]

Whitehouse soon went on the road with his fellow projectors, traveling to Liverpool and Glasgow to pitch shares in the company to potential

[55] Samuel Morse to Cyrus Field, October 3, 1856, in Field, *Atlantic Telegraph*, 76–78; also published as "Telegraphic Experiments" in the *Daily News* (October 10, 1856). Morse wrote again to Field on October 10, 1856 to say he believed a cable across the Atlantic could carry at least ten words per minute and was sure to be profitable; see Field, *Atlantic Telegraph*, 78–80.

[56] Whitehouse, *Atlantic Telegraph* (1858), 9–10.

[57] Brett, "Atlantic Telegraph" (letter), *Morning Post* (September 23, 1858), 2.

[58] "Prospectus," in Brett, *Origin and Progress*, 49.

[59] On the appointment of Whitehouse as electrician of the Atlantic Telegraph Company, see ATC Minute Book entry for October 29, 1856, 3; Morse was also named as a company electrician but his title was essentially honorary. On Whitehouse and Bright's salaries, see ATC Minute Book, entry for c. January 10, 1857, 47. In February 1858, after the reverses of the previous summer had tempered their expectations, the four projectors agreed to give up their claim to half of the company's future profits in return for £75,000 in additional shares (just over £28,000 each for Brett and Field, £12,500 for Whitehouse, and £6240 for Bright); see Saward, *Trans-Atlantic Submarine Telegraph*, 11, and ATC Minute Book, entry for February 17, 1858, 293.

investors.[60] On the swing through Glasgow in late November he met with Thomson and explained his experiments to him. Thomson was impressed with Whitehouse's techniques and apparatus, and though the two continued to differ on how to interpret some of Whitehouse's results, Thomson became a warm supporter of the cable project and of Whitehouse himself.[61] After Whitehouse told the assembled group at Glasgow that the proposed cable would be able to carry about seven words per minute, Thomson volunteered that he was now "satisfied that for long distances communication could be much more speedily made than was obtained on the line betwixt Varna and Balaklava, and that the pecuniary return in this speculation would be ample."[62] In early December he joined the project himself when the Glasgow shareholders elected him to the board of directors of the Atlantic Telegraph Company.[63] By January Thomson was publicly declaring that his previous theoretical objections had now been met and that Whitehouse's signaling system promised to give excellent results on the Atlantic cable; by the following September he was praising Whitehouse's patented relays and induction coils in glowing terms at the annual meeting of the British Association; and by November he was asking how he might go about proposing Whitehouse for election to the Royal Society of London.[64] To say that Whitehouse and Thomson had patched up their differences would evidently be an understatement; but that was not quite the end of the story.

Thick or Thin

As the Atlantic Telegraph Company was being formed in the fall of 1856, its leaders faced a crucial decision: what kind of cable should they lay down? What should be the size and composition of its conductor,

[60] On visits by Field, Brett, and Whitehouse to groups of potential investors in Liverpool, see *Times* (November 14, 1856), 12; in Glasgow, see *Glasgow Herald* (November 24, 1856); and in Manchester, though apparently without Whitehouse, see *Times* (November 22, 1856), 6.

[61] On Thomson's meeting with Whitehouse, see Thomson to de la Rive, December 17, 1856, in Tunbridge, *Kelvin*, 97. Thomson later wrote on his copy of Whitehouse's "Experiments on the Retardation of Electric Signals" that it gave "The best account of what is good in Whitehouse's experiments and apparatus. The conclusions, however, are fallacious in almost every point"; see Thompson, *Kelvin*, *1*: 330n.

[62] "The Atlantic Telegraph," *Glasgow Herald* (November 24, 1856).

[63] ATC Minute Book, entries for December 3 and December 9, 1856, 25 and 34.

[64] William Thomson, in discussion of F. R. Window, "On Submarine Electric Telegraphs," *Proc. ICE* (January 1857), *16*: 188–202, discussion 203–25, on 210–11; William Thomson, "On Mr. Whitehouse's Relay and Induction Coils in action on Short Circuit," *BA Report* (1857), 21; William Thomson to G. G. Stokes, November 7, 1857, in David B. Wilson, ed., *Correspondence between Sir George Gabriel Stokes and Sir William Thomson, Baron Kelvin of Largs*, 2 vols. (Cambridge: Cambridge University Press, 1990), *1*: 226.

insulation, and protective outer armoring? In particular, should their cable be thick or thin? It was a question on which Thomson's theory and Whitehouse's experiments pointed in opposite directions, and the answer the company gave would have far-reaching consequences.

In choosing a design for their cable, the company faced several constraints. The first was cost: the company had no more than £350,000 to spend on the entire project. Another was time: Field's promise to lay the cable by the summer of 1857 ruled out adoption of any design that could not be executed quickly. The cable would also have to meet a long list of mechanical requirements for strength, flexibility, and specific gravity, and its total bulk could not exceed the capacity of the two ships that were to carry and lay it. Finally, the cable had to have the requisite electrical qualities to enable it to convey signals across the Atlantic at a commercially viable rate. As Field later observed, this last consideration ought properly to have come first, since solving the mechanical problem of laying a cable across the Atlantic would count for nothing if it proved unable to carry readable signals.[65] In practice, however, more attention was devoted to the mechanical than the electrical needs of the first Atlantic cable.

Immediately after registering their new company on October 20, Field and Brett recruited several business associates to join them on its provisional board of directors, including Samuel Statham of the Gutta Percha Company and George Carr and Charles Tupper, galvanized iron dealers who had worked with Brett on his Mediterranean cables. They soon took up the task of choosing a design for the cable, for as Carr, the chair of the provisional board, told the permanent directors when they took over on December 9, "If it was intended to attempt the enterprise next year, no time was to be lost in deciding upon the cable to be used, and in concluding arrangements for its manufacture."[66]

Whitehouse later said that when he asked for three more months to test sample cables before choosing a final design, Field replied "Pooh, nonsense"; three months of testing, followed by the time needed to manufacture 2500 miles of cable, would have pushed them past the summer laying season recommended by Maury and so put the project back a full year.[67] Unwilling to yield on the 1857 deadline, the provisional board pressed

---

[65] Cyrus Field, in discussion of J. A. Longridge and C. H. Brooks, "On Submerging Telegraphic Cables," *Proc. ICE* (February 23, 1858) *17*: 221–61, discussion 298–366, on 326.

[66] ATC Minute Book, entries for October 23, 1856, 2, and December 9, 1856, 31–32.

[67] Testimony of Wildman Whitehouse, December 15, 1859, in *Joint Committee Report*, 75. On Maury's advice on when best to lay the cable, see his March 28, 1857, letter to Field, in M. F. Maury, *Explanations and Sailing Directions to Accompany the Wind and Current Charts*, 2 vols. (Washington, DC: William A. Harris, 1858), *1*: 182–89, on 189.

ahead with only hurried testing of cable designs. Field had already received samples of several types of cores from Statham and of outer coverings from Richard Glass of Küper and Company. He soon came to favor a core consisting of seven number 22 gauge copper wires (each about 0.028 inches in diameter) twisted together to form a single conductor, all covered with three layers of gutta-percha.[68] Such a stranded conductor had first been used on Field's Cabot Strait cable; it was more flexible than a single wire of the same thickness and also more secure against a loss of continuity, since a break in any one wire would not stop the flow of current.

Field next took up what kind of outer covering to use. Since the Channel cable of 1851, almost all submarine conductors had been protected by an outer armoring of iron wires laid on with a slight spiral, as in a wire rope. While traveling by train in the late summer of 1856, Field happened to encounter Isambard Brunel and showed him some samples of cable designs. Impressed by the stranded conductor, Brunel reportedly suggested that stranded wire be used for the outer covering as well; it would be at least as strong as thicker single wires, he said, and much more flexible.[69] The stranded cable also had the advantage of looking good, and Field later had Glass make thousands of short lengths of it to hand out as marketing props when promoting shares in the new company. The company later paid Glass £1000 to cover the "preliminary expenses" he had incurred in this way.[70]

On October 29, 1856, just nine days after the company was registered and before it had issued a prospectus, the provisional board of the Atlantic Telegraph Company requested tenders from the Gutta Percha Company for the core of the planned cable and from Küper and Company for its outer armoring. Two days later the board appointed Field, Brett, and Tupper to a committee to consider the tenders and recommend which design to adopt. Notably, none of the three were cable engineers or experts on either the mechanical or electrical aspects of cable design. They reported back just ten days later, unanimously recommending that the company adopt essentially the design Field already favored: a stranded copper conductor weighing 107 pounds per nautical mile and "insulated in Gutta Percha of the best quality," applied in three layers

[68] Charles F. Briggs and Augustus Maverick, *The Story of the Telegraph, and a History of the Great Atlantic Cable* (New York: Rudd & Carleton, 1858), 59–60.

[69] Mulally, *The Laying of the Cable*, 29.

[70] Saward, *Trans-Atlantic Submarine Telegraph*, 9. Willoughby Smith, *The Rise and Extension of Submarine Telegraphy* (London: J. S. Virtue, 1891), 45, said the samples "looked very pretty, and were used as decoys to obtain the capital required." On the later payments to Glass, see ATC Minute Book, entries for April 1, 1857, 102, and June 25, 1857, 177.

to bring the diameter of the core up to three-eighths of an inch. This would then be covered with a layer of jute yarn saturated with tar and beeswax, and finally by an outer armoring consisting of "eighteen strands, seven wires each, of No. 22 gauge Best Charcoal Iron bright wire."[71] The report made no mention of the electrical qualities of either the copper or the gutta-percha, nor did it call for any electrical tests on any part of the cable.

Although many later assumed that Whitehouse had chosen this design, he in fact favored a somewhat different one.[72] While assuring the provisional board that he believed the design it had recommended was "well adapted" to its purpose and that he did not at all oppose its adoption, he argued in his "Electrician's Report" that another of the samples the company had been offered, with a slightly thicker covering of gutta-percha and with its outer iron wires embedded in hempen cords, offered several advantages. Such a cable would, he said, be just as strong and nearly as flexible as the design the committee favored, while its lower specific gravity would help buoy it up during laying and so reduce the strain to which it would be subjected. In addition, he said, "in consequence of the greater thickness of the Gutta Percha it will certainly have less induction and retardation, and hence admit of the attainment of higher working speed" than the other design.[73] Whitehouse did not address whether to use a thicker copper conductor, which Thomson had said would also reduce the retardation.[74]

Whitehouse's objections were quickly overruled. Brett flatly rejected the use of hemp-covered armoring, declaring that such a cable would be so weak he "would not have it if it were laid."[75] Bright later said he had favored a much thicker cable, with nearly four times as much copper and half again as much gutta-percha per mile as the core recommended by the committee, but this proposal, if it was ever officially made, was also rejected, largely, he said, because of its greater

---

[71] ATC Minute Book, entries for October 29, October 31, and November 10, 1856, 3–8.

[72] Douglas Galton and George Saward were among those who assumed Whitehouse had designed the cable; see their questioning of him, December 15, 1859, *Joint Committee Report*, 74.

[73] Wildman Whitehouse to Atlantic Telegraph Company, November 5 and November 8, 1856, in ATC Minute Book, entry for November 10, 1856, 8–9.

[74] Thomson, "Telegraphs to America," *Athenæum* (November 1, 1856), 1339. Thomson later said that while a thicker cable would have shown less retardation, he believed the design adopted by the Atlantic Telegraph Company was a good choice for a first attempt, given the cost constraints; see "Banquet to Professor William Thomson," *Glasgow Herald* (January 21, 1859), and Thomson's December 17, 1859, testimony in *Joint Committee Report*, 111.

[75] Whitehouse, *Atlantic Telegraph* (1858), 13, and Brett, "Atlantic Telegraph" (letter), *Morning Post* (September 23, 1858).

expense.[76] Focusing mainly on the mechanical strength of the outer armoring, the provisional board quickly voted to adopt the design recommended by its committee and to invite tenders from manufacturers.[77]

In their choice of conductor, Field and Brett had relied heavily on Whitehouse and Morse's assurances that a wire of ordinary thickness would be able to carry signals at a commercially viable rate across the Atlantic. In fact the Atlantic Telegraph Company soon began to assert not just that a thin conductor would work well enough, but that it would actually be *better* than a thicker one. When Whitehouse had first suggested this in 1856, based on his experiments with three separately insulated wires, Thomson had pointed out that such an arrangement was not at all equivalent to a single thicker conductor, but Whitehouse did not back down, and the idea that thin conductors produce less retardation came to be closely identified with him.[78] In January 1857 it received influential support when Michael Faraday rose during discussion of a paper at the Institution of Civil Engineers to declare that "the larger the wire, the more electricity was required to charge it, and the greater was the retardation of that electric impulse, which should be occupied in sending the charge forward."[79] The leaders of the Atlantic Telegraph Company were delighted to hear the leading electrical authority of the day, and the man who had first brought the problem of retardation to public notice, state that for a submarine cable like theirs, a thin conductor would show less retardation than a thick one. As S. A. Varley and others later pointed out, however, Faraday had erred badly: although the larger surface area of a thick conductor would indeed give rise to more induction, this would be more than offset by its lower resistance, which would result in *less* overall retardation than with a thin wire.[80] Such niceties were lost on most observers, however, and Field later happily quoted Faraday's remark as a direct and authoritative endorsement of the relatively thin conductor the company had already adopted.[81]

---

[76] In a January 1862 discussion at the Institution of Civil Engineers, Bright said that in 1856 he had called for the Atlantic cable to be made with 392 pounds of copper and a like weight of gutta-percha per nautical mile, but "owing to financial and other considerations," this advice was not taken; Bright, in discussion of H. C. Forde, "The Malta and Alexandria Submarine Telegraph Cable," *Proc. ICE* (May 1862) *21*: 493–514, discussion 531–40, on 531.

[77] ATC Minute Book, entry for November 10, 1856, 9.

[78] Whitehouse, "Law of Squares," 23.

[79] Faraday, in discussion of Window, "Submarine Electric Telegraphs," 221.

[80] S. A. Varley, in discussion of Frederick Charles Webb, "On the Practical Operations Connected with Paying Out and Repairing Submarine Telegraph Cables," *Proc. ICE* (February 1858) *17*: 262–297, discussion 298–366, on 330–31.

[81] Field, in discussion of Webb, "Practical Operations," 326.

The Atlantic Telegraph Company committed itself most explicitly to the supposed superiority of thin conductors in an "official manifesto" it issued in the summer of 1857.[82] Stung by criticisms of the project in the press, in April the board of directors ordered the company's engineer, electrician, and secretary (Bright, Whitehouse, and George Saward) to prepare a response.[83] The resulting sixty-nine-page booklet, *The Atlantic Telegraph: A History of Preliminary Experimental Proceedings, and a Descriptive Account of the Present State & Prospects of the Undertaking*, appeared in mid-July, just weeks before the ships carrying the cable were to begin laying it across the ocean.[84] Widely advertised and distributed, the little book heaped so much praise on Whitehouse that many believed he had written it himself.[85] In fact it issued from the pen of R. J. Mann, a physician turned popular scientific writer, though Whitehouse later said that Mann had based it on materials "chiefly furnished by me." Some surviving copies bear the initials "R. J. M." at the end, but Mann later said that after the printing was completed the Atlantic Telegraph Company insisted on paying him fifty guineas and blotting out his initials with a "black lozenge" (as seen in most surviving copies) so that the publication would appear "with the authority of an official document."[86] Similar praise for Whitehouse also appeared in an anonymous piece Mann wrote for the July number of the *Edinburgh Review*, as well as in an article on "The Atlantic Telegraph" in the June 27 issue of *Chambers's Journal*; the latter went so far as to dub the "indefatigable and sagacious" Whitehouse the "lightning-king."[87]

---

[82] The phrase "official manifesto" appears in Briggs and Maverick, *Story of the Telegraph*, 72.

[83] ATC Minute Book, entry for April 22, 1857, 119, evidently in response to "The Great Atlantic Submarine Telegraph Cable," *Times* (April 22, 1857), 9.

[84] [R. J. Mann], *The Atlantic Telegraph: A History of Preliminary Experimental Proceedings, and a Descriptive Account of the Present State & Prospects of the Undertaking, Published by Order of the Directors of the Company* (London: Jarrold and Sons, 1857).

[85] See, for example, "The Atlantic Telegraph" (letter), *Engineer* (November 5, 1858) 6: 355, which says it was widely understood that the booklet "was written by Mr. Whitehouse himself." Signed "A Telegraph Engineer and Practical Electrician," this letter had been written by W. H. Preece; see the manuscript of it, NAEST 17/12.8, IET Archives, London. Whitehouse responded in "The Atlantic Telegraph" (letter), *Engineer* (November 26, 1858) 6: 410, disclaiming authorship and saying the booklet had instead been written by "a talented popular writer."

[86] R. J. Mann to Latimer Clark, November 12, 1880, WC-NYPL; see also Wildman Whitehouse to J. J. Fahie, September 18, 1879, SC/MSS 009/2/151/1, IET Archives, London. No "black lozenge" covers the initials "R. J. M." at the end of the copy held by the University of California at Berkeley, a scan of which is available through Google Books. On the payment of fifty guineas to Mann, see ATC Minute Book, entry for June 25, 1857, 175.

[87] [R. J. Mann], "De la Rive on Electrical Science," *Edinburgh Review* (July 1857) *106*: 26–62; on Whitehouse, see 40–44. Mann is identified as the author in Walter E. Houghton et al., eds., *Wellesley Index to Victorian Periodicals, 1824–1900*, 5 vols. (Toronto: University of Toronto Press, 1966–1989). Mann almost certainly also wrote

In his booklet, Mann gave a full and adulatory account of Whitehouse's various devices and experiments, and after briefly noting the "counsel" provided by "Professor W. Thompson [*sic*] of Glasgow," devoted several pages to recounting Whitehouse's supposed refutations of the law of squares.[88] He then boiled Whitehouse's results down into a set of succinct principles, including the unequivocal statement "that large coated wires used beneath the water or the earth are worse conductors, so far as velocity of transmission is concerned, than small ones, and therefore are not so well suited as small ones for the purposes of submarine transmission of telegraphic signals."[89] Widely quoted in the press, this passage was universally taken to express the official view of the Atlantic Telegraph Company on the question of thick versus thin conductors. It was also widely criticized by experienced cable engineers, including S. A. Varley, F. C. Webb, and Latimer Clark.[90] By the time Mann's booklet appeared in July 1857, however, such criticisms were beside the point: the cable had already been made and would soon be laid, and the size of its conductor was a fait accompli. Thomson, himself by then a director of the company and deeply engaged in its activities, knew that a thicker conductor would give a faster working speed, but he recognized that this would cost more than it might be possible to raise at the time, and in any case he thought he saw a way to make even a relatively thin cable work well enough to pay.[91] Although in the end the signaling method by which he proposed to accomplish this was not carried into practice, it would have important consequences, as we shall see.

"The Atlantic Telegraph," *Chambers's Journal* (June 27, 1857) 7: 401–4, which includes many of the same passages about Whitehouse.

[88] [Mann], *Atlantic Telegraph*, 21.   [89] [Mann], *Atlantic Telegraph*, 26.

[90] Among the many quotations of this passage in the press, see "The Atlantic Telegraph," *Morning Chronicle* (August 22, 1857); "Secrets of the Atlantic Cable," *Morning Chronicle* (September 10, 1858), where it is cited as the reason the Atlantic Telegraph Company adopted a thin conductor for its cable; and [David Brewster], "The Atlantic Telegraph," *North British Review* (November 1858) 29: 519–55, on 533. Brewster is identified as the author in Houghton et al., eds., *Wellesley Index*. The passage also appears in Briggs and Maverick, *Story of the Telegraph*, 81, and in Tal. P. Shaffner, *The Telegraph Manual* (New York: Pudney and Russell, 1859), 624, but was strongly criticized by S. A. Varley in "On the Electrical Qualifications Requisite in Long Submarine Telegraph Cables," *Proc. ICE* (February 1858) 17: 368–85, on 369, and by both Varley and Webb in the discussion of Webb, "Practical Operations," 330–31 and 357. Latimer Clark made penciled exclamation marks beside this passage in his own copy of Mann's pamphlet, WC-NYPL. Note, however, that C. V. Walker endorsed using very thin conductors; *Joint Committee Report*, December 16, 1859, 103.

[91] William Thomson, "On Practical Methods for Rapid Signalling by the Electric Telegraph," *Proc. RS* (November 1856) 8: 299–303, and (December 1856) 8: 303–7, repr. in Thomson, *MPP 2*: 103; see also Thomson's remarks in "Banquet to Professor William Thomson," *Glasgow Herald* (January 21, 1859), and his December 17, 1859, testimony in *Joint Committee Report*, 111 and 124.

Making and Laying the Cable – the First Attempt

As Cyrus Field rushed to launch the Atlantic Telegraph Company in the fall of 1856, he dealt directly with Richard Glass of Küper and Company as the prospective manufacturer of the cable. As word got out that a large contract might be in the offing, however, Glass's rival R. S. Newall began angling for a piece of it. Known as a bold and aggressive businessman, Newall had made and laid several important early cables but had also presided over some embarrassing failures. He was also known to be litigious, particularly about his patent rights, and in November 1856, as the competition to secure the order to armor the cable became, in Saward's later words, "the subject of intrigues," the provisional board of the Atlantic Telegraph Company decided "as an act of policy rather than prudence" to split the contract between the two manufacturers.[92] The Gutta Percha Company of London was to produce 2500 miles of insulated core; half would then be armored by Glass at the Küper works in East Greenwich and half by Newall at his works in Birkenhead near Liverpool, all to be delivered, coiled, stowed, and "ready for sea" by June 1857.[93]

The splitting of the cable contract led to several problems. One came to be seen as emblematic of bungling on the project: after the armoring was complete, it was found that Glass had given the outer wires of his half of the cable a left-handed "lay," or twist, while Newall had given his half a right-handed one. Had the two halves been spliced directly together, they would have unwound each other, exposing the underlying core. This could be prevented only by inserting a heavy frame at the splice, but while the mismatched lays were embarrassing, they had little real effect on the integrity of the cable.[94] More serious was the fact that quality control measures had to be split between two sites more than 200 miles apart, and that no tests could be conducted through the full length of the cable until both halves had been completed and shipped out for laying, a failure that Whitehouse later said he particularly regretted.[95]

While the cable was being manufactured, Whitehouse set about assembling the instruments and apparatus he would use to test and operate it. The expense was substantial; in April 1857 the board authorized spending more than £3200 to acquire sets of Whitehouse's patented batteries, induction

---

[92] Saward, *Trans-Atlantic Submarine Telegraph*, 9; on Newall's litigiousness, see Smith, *Rise*, 15–16.

[93] ATC Minute Book, entry for November 18, 1856, 16–18.

[94] Bright, *Submarine Telegraphs*, 35n, 44–45; Cookson, *Cable*, 73.

[95] ATC Minute Book, entries for January 13, 1857, 50; c. March 1857, 91, 96; Whitehouse, "Electrician's Report," January 4, 1858, WC-NYPL; Whitehouse, *Atlantic Telegraph* (1858), 11.

coils, and relays, and Saward later reported that in all the company spent some £13,000 on Whitehouse's electrical department.[96] The "perpetual maintenance batteries" Whitehouse used to power his huge induction coils were a particularly lavish item; the company reportedly spent £2000 just on their enormous silver plates, though these were later scrapped when graphite plates proved more effective. Perhaps not surprisingly, the board of directors repeatedly grumbled about Whitehouse's spending, leading to tensions that occasionally threatened to boil over.[97]

No ship afloat was big enough to hold the entire cable (the *Great Eastern* was then still under construction), so half was to be carried by the *Niagara*, a steam frigate on loan from the US Navy, and half by the Royal Navy's HMS *Agamemnon* (Figure 2.5). After various complications and delays, Newall's half of the cable was loaded onto the *Niagara* and Glass's onto the *Agamemnon*, a task that took up much of July. While waiting to be loaded, Glass's half of the cable sat in direct sun on some unusually hot days at Greenwich – topping out at 88°F on Sunday, June 28, the hottest day in eleven years – "melting out the gutta percha in many miles of cable," which had to be cut out and replaced.[98]

Once the loading was completed, the two ships, accompanied by several smaller vessels, rendezvoused on July 30 at Queenstown (now Cobh) in southern Ireland, where for the first time the two halves of the cable could be linked together. Conditions were far from ideal, but Whitehouse and Thomson – aided, according to one report, by Whitehouse's wife, Emma – managed to rig connecting wires between the two ships and successfully passed currents through the entire 2500 miles of cable.[99] The link was broken at the turn of the tide, but after reconnecting the wires the next day, Whitehouse began using his induction coils and relays to check signaling rates. He told the press the tests were "most satisfactory," but they in fact raised serious concerns.

---

[96] ATC Minute Book, entries for April 16, 1857, p. 116, and April 22, 1857, p. 119; "The Atlantic Telegraph," *Engineer* (October 8, 1858) 6: 268.

[97] J. N. Hearder, "On the Atlantic Cable," *Phil.Mag.* (January 1859) 17: 27–42, on 40. Most of the cost of the silver plates could be recovered when they were scrapped; see C. V. Walker in *Joint Committee Report*, 102. On Whitehouse's patented battery, see [Mann], *Atlantic Telegraph*, 58–62. For attempts to rein in Whitehouse's spending, see ATC Minute Book, entries for c. June 1857, 168; September 17, 1857, 231; and October 8, 1857, 242.

[98] "Atlantic Submarine Telegraph," *Times* (July 24, 1857), 5; Whitehouse, *Atlantic Telegraph* (1858) 12 and 16; Whitehouse, December 15, 1859, *Joint Committee Report*, 76. On the temperature, see James Glaisher, "On the Meteorology of England, during the Quarter ended June 30th, 1857," *Journal of the Statistical Society of London* (December 1857) 20: 441–42.

[99] Whitehouse, "Electrician's Report," January 4, 1858, WC-NYPL; on Mrs. Whitehouse, "a lady of great skill and experience in electric telegraphic operation," see "The Atlantic Cable – Arrival of the *Agamemnon*," *The Constitution, or Cork Advertiser* (August 1, 1857).

Figure 2.5 Crewmen coiling the first Atlantic cable on the US warship *Niagara* in 1857 or 1858, from a series of stereoscopic views produced by the London Stereoscopic Company.
(Photograph courtesy of and copyright © 2007 Page and Bryan Ginns, www.stereographica.com.)

Morse, who had come over on the *Niagara*, observed the tests and wrote to his wife that while he still believed the project would eventually succeed, he was troubled by how slowly readable signals could be sent: "twenty words in sixteen minutes is now the rate," he said, far below the ten words per minute he had promised when the company was launched.[100] As Whitehouse euphemistically later put it, the Queenstown tests "rendered

---

[100] "The Atlantic Telegraph," *Daily News* (August 6, 1857), 5, particularly the section marked "(Official Intelligence)," the style and content of which suggest it was written by Whitehouse; "The Atlantic Telegraph," *Times* (August 3, 1857), 7; Whitehouse, *Atlantic Telegraph* (1858), 17; Morse to his wife, August 4, 1857, in Prime, *Morse*, 657.

it sufficiently evident that much time and attention might judiciously be bestowed on ... the details and peculiar arrangements required for signaling through so vast and untried a distance, in order to attain a thoroughly certain and commercially satisfactory rate of communication."[101] In other words, his system did not yet work at all well. As the ships carrying the cable steamed out of Queenstown, outward confidence masked nagging doubts.

The original plan had called for the two ships to meet at mid-ocean, splice their cables together, and steam off in opposite directions, the *Niagara* laying its half to Valentia Island on the southwest coast of Ireland and the *Agamemnon* its half to Trinity Bay in Newfoundland. Starting in mid-ocean had the advantage of ensuring that a calm day could be chosen for the risky splicing operation, while also cutting the total laying time in half and so reducing the chance the ships would encounter bad weather. Whitehouse, however, had been forbidden by his physician from sailing beyond Ireland, and wishing to be able to monitor the progress of the expedition from his base at Valentia, he proposed that the *Niagara* instead start directly from there and, after laying its half of the cable, hand over to the *Agamemnon* to complete the route to Newfoundland. The engineers and naval officers strongly opposed this plan – Henry Woodhouse, Bright's assistant, later called it "suicidal" – but the board of directors, perhaps intrigued by the prospect of being able to "speak" through their cable with ships at sea, backed Whitehouse. Thomson, at that point one of Whitehouse's strongest supporters on the board, seconded the motion; he also agreed to take Whitehouse's place on the *Agamemnon*, and in fact would sail on all of the Atlantic cable-laying expeditions.[102]

On August 5, amid speeches and celebrations, the *Niagara* landed its end of the cable at Valentia and began steaming west. In a foretaste of troubles to come, the cable jammed and snapped before the flotilla had left sight of land. Little time or cable was lost, however, and the *Niagara* set out again three days later. Bright's paying-out machinery was of a novel design and not well tested; Woodhouse later said that in the rush to start the laying, "the machines were literally being put together" as the ships made their way to Valentia.[103] The heavy machinery required constant and careful supervision to keep it from putting too much strain on the cable as it was being payed out, and three days out, after the *Niagara* had laid 335 miles and entered deeper waters, the inevitable

[101] Whitehouse, "Electrician's Report" (January 4, 1858), 26, WC-NYPL.
[102] Henry Woodhouse, *Joint Committee Report*, December 9, 1859, 39; ATC Minute Book, entry for c. July 1857, 190; Thompson, *Kelvin*, *1*: 343.
[103] Woodhouse, *Joint Committee Report*, December 9, 1859, 42.

occurred: a workman did not ease the brake at a crucial moment, and the cable parted.[104]

By then it was August 11, too late in the laying season to start another attempt, and too much cable had been lost to be confident that the remaining length would suffice to reach Newfoundland. After gathering reports from Bright, Whitehouse, and the commanders of the *Niagara* and the *Agamemnon*, the board of directors concluded that the failure had resulted from a series of avoidable accidents, and at its September 9 meeting resolved to try again a year later.[105] There was talk for a time of selling the cable stowed in the *Agamemnon* and the *Niagara* to a company that proposed to lay it down the Red Sea to speed communications with India, then in the throes of the Indian Rebellion, with the Atlantic company then using the proceeds to buy a fresh length of cable for its own use the next summer. But the deal fell through, in part because of fears the Atlantic cable would not stand up to the heat of the tropics.[106] The directors decided instead to store their remaining 2000 miles of cable over the winter at the Keyham naval dockyards at Devonport near Plymouth and, after raising some additional capital and securing renewed promises of support from the British and American governments, ordered 900 miles of new cable to replace the length lost and cover any future exigencies.[107] No longer driven by the mad rush to meet Field's original deadline, Whitehouse settled in at Devonport to perform the tests on the full cable that he had long called for.

Credible Measures

From his first involvement with cable telegraphy, Whitehouse had been an active though unorthodox measurer, meticulously recording data produced by instruments of his own design. Thomson later observed that Whitehouse "had his own system" of electrical measurement that he "considered satisfactory," but it remained unique to him; his measurement practices did not link up with those then becoming standard among both laboratory scientists and practical telegraph engineers.[108] As a result, when things began to go wrong, Whitehouse would find himself

---

[104] Bright's report is reproduced in Bright, *Atlantic Cable*, 68–73.

[105] ATC Minute Book, entry for September 9, 1857, 220; see also the entries for August 19 and August 20, 1857, 207–12, and Saward, *Trans-Atlantic Submarine Telegraph*, 21.

[106] "Money-Market and City Intelligence," *Times* (August 22, 1857), 11, and *Times* (September 4, 1857), 8.

[107] Saward, *Trans-Atlantic Submarine Telegraph*, 24. The company raised £31,000 by selling new cheaper shares; see "Money-Market and City Intelligence," *Times* (August 6, 1858), 7.

[108] Thomson, *Joint Committee*, December 17, 1859, 115.

isolated and vulnerable to attacks on the credibility of his methods, instruments, and results.

As the Atlantic Telegraph Company regrouped in the wake of its 1857 failure, Whitehouse and his assistants began an extensive series of tests and measurements on the now land-bound cable. The company had spent more than £3000 to build tanks at Devonport, planning to store the coiled cable underwater to protect its iron armoring and gutta-percha insulation from oxidation. But the tanks leaked, so the company instead coated the cable with tar to stave off rusting and coiled it in open sheds.[109] The time the cable had spent sitting in the sun at Greenwich had already compromised the soundness of its gutta-percha; now, as it was subjected to repeated handling and exposed to heat and air, its insulation no doubt suffered further. In particular, it was later found that where heat had softened the gutta-percha, the copper conductor often settled into an off-center position, sometimes leaving only a thin layer of insulation separating it from the outer covering. In addition, Whitehouse's experiments required frequent cutting and splicing of the cable, risking the introduction of additional faults, and there was no opportunity to test the cable under water, where such faults could be more readily detected.[110]

Whitehouse used several measuring instruments in his work at Plymouth, but the most significant and distinctive was his "magneto-electrometer," or electromagnetic steelyard (Figure 2.6). Whitehouse was very proud of this little device; James Burn Russell, a former student of Thomson's who worked as an assistant on the project, said it was known as Whitehouse's "pet child" – a phrase that also appeared in an article in *Chambers's Journal* in June 1857 – and it served as the touchstone for many of his most important measurements.[111] Whitehouse described it briefly at the 1855 British Association meeting and more fully the

---

[109] Whitehouse, December 15, 1859, *Joint Committee Report*, 77–78; Saward, *Trans-Atlantic Submarine Telegraph*, 24.

[110] Whitehouse, December 15, 1859, *Joint Committee Report*, 76–77; Donard de Cogan, "Dr. E. O. W. Whitehouse and the 1858 Trans-Atlantic Telegraph Cable," *History of Technology* (1985) 10: 1–15; Saward, January 12, 1860, *Joint Committee Report*, 175.

[111] [James Burn Russell], "Paying-Out the Atlantic Cable," *Sydney Morning Herald* (February 8, 1859), 5; excerpt in Thompson, *Kelvin*, 1: 360–64. The texts of both Russell's *Sydney Morning Herald* article and a fuller three-part account ("Atlantic Cable: Leaves from the Journal of an Amateur Telegrapher") he wrote for the *West of Scotland Magazine and Review* in 1859 can be found at http://atlantic-cable.com/Article/1858JBRussell/index.htm, along with a transcription of much of the detailed journal, "Notes of my connection with the Atlantic Telegraph," he kept during his time on the cable project, the original of which is now in the Glasgow City Archives. This transcription is hereinafter cited as Russell, "Notes." See also Donard de Cogan, *They Talk Along the Deep: A Global History of the Valentia Island Telegraph Cables* (Norwich: Dosanda Publications, 2016), 55–59. Russell later became a physician and a leading public health figure in Glasgow; see Edna Robertson, *Glasgow's Doctor: James Burn Russell, MOH, 1837–1904* (East Linton:

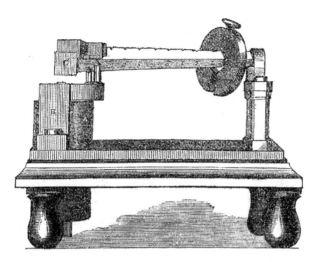

Figure 2.6 Wildman Whitehouse used his "magneto-electrometer" or electromagnetic steelyard to weigh the "value" of pulses of electric current. The little instrument stood about five inches high.
(From *Engineer*, Vol. 2: 523, September 26, 1856; courtesy University of Texas Libraries.)

following year.[112] Designed to measure pulses of current too brief to produce a steady deflection on a galvanometer, the magneto-electrometer consisted of an electromagnet set in a sturdy frame on which was poised a lever, with a soft iron keeper on one end and a movable weight on the other. When a pulse of current entered the coils of the electromagnet, the resulting magnetic force pulled down on the iron keeper and, if strong enough, lifted the weight at the other end of the lever. Whitehouse regarded the maximum weight a pulse of current could lift as the best measure of what he called "its 'value' in telegraphy." He claimed that his magnetic steelyard was so delicate that it could measure the value of a current "too feeble in its energy, too brief in its

Tuckwell, 1998). For an earlier description of the magneto-electrometer as Whitehouse's "pet child," see [Mann], "Atlantic Telegraph" (*Chambers's Journal*), 401.

[112] Whitehouse, "Experimental Observations" (1855), 22; Wildman Whitehouse, "On the Construction and Use of an Instrument for determining the Value of Intermittent or Alternating Electric Currents for purposes of Practical Telegraphy," *BA Report* (1856), 19–21; a fuller account appeared under the same title in *Engineer* (September 26, 1856) 2: 523. Allan Green emphasizes the merits of Whitehouse's device in "Dr. Wildman Whitehouse and his 'Iron Oscillograph'; Electrical Measurements Relating to the First Atlantic Cable," *International Journal for the History of Engineering and Technology* (2012) 82: 68–92.

duration, to give the slightest indication" on a sensitive galvanometer, yet so robust that, with proper adjustments, it could accurately "weigh" currents over a range from less than one grain to more than half a million (i.e., from just over a thousandth of an ounce to nearly one hundred pounds).[113]

Whitehouse's instrument closely resembled one J. N. Hearder had introduced in 1842, though Whitehouse evidently hit on the idea independently.[114] Known as "the blind electrician of Plymouth," Hearder had lost his sight in a chemical accident as a young man but continued to perform experiments, devising ingenious ways to detect electric and magnetic effects by touch or with his tongue. He took a strong interest in the Atlantic cable project and, after Whitehouse arrived in Plymouth, met with him for "frequent friendly discussions," though they found they differed on many points.[115]

One of Hearder's most serious criticisms concerned the operation of Whitehouse's magneto-electrometer. Whitehouse had said that each pulse of current produced a magnetic force "strictly proportioned to its own proper energy," registered by his device as its "value" in grains.[116] He presented no evidence to back up this claim, however, and Hearder said his own long experience with similar devices had convinced him it was not true, or at least depended on what one meant by "its own proper energy." The force generated in the iron core of the electromagnet was not "strictly proportioned" to the strength of the applied current, Hearder said, but instead varied in "a most extraordinary and apparently incongruous" way. Thus, a current that registered a value of ten grains might not really be twice as strong as one that showed a value of five grains. The magnetic steelyard had its uses, Hearder said, but its "relative indications cannot be compared with each other, as expressing corresponding variations in the exciting currents"; one simply could not reliably compare measurements made with it under different conditions.[117] Whitehouse's magneto-electrometer might be his "pet child," but others had serious doubts about adopting it as a tool for electrical measurement.

Whitehouse nonetheless tried hard to convince other electricians of the merits of his device, contrasting its "definiteness and accuracy" with the wandering readings of the "far from reliable" galvanometer.[118] Taking his

---

[113] Whitehouse, "Construction and Use" (*Engineer*), 523.
[114] Hearder, "Atlantic Cable," 29; J. N. Hearder, "Description of a Magnetometer and Appendages," *Annual Report of the Cornwall Polytechnic Society*, (1844) *12*: 98–100.
[115] Hearder, "Atlantic Cable," 33; see also Ian G. Hearder, "Jonathan Nash Hearder," *ODNB*.
[116] Whitehouse, "Construction and Use" (*Engineer*), 523.
[117] Hearder, "The Atlantic Cable" (letter), *Engineer* (April 1, 1859) *7*: 224.
[118] Whitehouse, "Construction and Use" (*Engineer*), 523.

lead from Whitehouse, Mann painted the contrast even more strongly –
and in strikingly gendered terms – in the manifesto the Atlantic Telegraph
Company issued in 1857. A galvanometer was "of no value whatever" for
measuring brief pulses of current, Mann said; "the needle . . . commonly
turns somersets, and jerks backwards and forwards in the most hysterical
and passionate way, instead of maintaining the steady divergence which
alone could be accepted by the eye of science as a satisfactory indication of
strength." Whitehouse's magneto-electrometer, on the other hand, was
simple, direct, and reliable:

> It is only necessary to see this staid and business-like instrument at work by the
> side of the old ecstatic, as well as astatic needle, to comprehend its superiority at
> a glance. The one piece of apparatus tossing so wildly and crazily about, that for
> minutes at a time the most patient and skilful observer can make neither head nor
> tail of its bewildering movements; the other piece quietly tilting up its weight on
> the end of the steelyard, and refusing in the most self-possessed way to lift another
> grain under any inducement that can be brought to bear, and then sending in its
> refusal as the exact estimate of the force it has been commissioned to
> determine.[119]

Where the galvanometer was "hysterical," Whitehouse's magneto-
electrometer was "staid"; where the galvanometer was "ecstatic," his little
steelyard was "business-like." Mann presented Whitehouse's device as
embodying all of the virtues – manly ones, at that – likely to appeal to
either a scientist or a practical man, and as the perfect instrument for
producing solid experimental facts. Others were not so convinced.

Besides being more "business-like" than a galvanometer, Whitehouse's
magneto-electrometer had the advantage, as he saw it, of operating in
a way that more closely resembled the workings of actual telegraph
receivers, including his patented relays. "Unlike the degrees upon the
galvanometer," he declared, the grains lifted on his steelyard were "units
of real value and of practical utility." He boasted that by measuring the
"value" of pulses of current passing through different lengths of cable, he
could "ascertain with certainty and minute accuracy the loss due to the
combined effect of resistance, induction, and defective insulation." He
could not, however, readily disentangle those different effects, nor could
he directly measure resistance, current ("quantity"), or electromotive
force ("intensity") as those were ordinarily defined. Whitehouse knew
that to build up the credibility of his magneto-electrometer and the
"values" it measured, he would need to extend it beyond the limits of
his own testing room and make it the shared property of the broader
electrical community; in particular, it would be "necessary to have one

[119] [Mann], *Atlantic Telegraph*, 18.

common standard of comparison, to which all instruments so made may be adjusted." Toward that end, he offered "to set aside for this special purpose the most accurately-finished and perfect instrument I can obtain, with which I will be most happy at all times to compare those of any of my fellow-labourers in the field."[120] But the offer was not taken up, and Whitehouse's instruments and measuring practices would remain his alone.

At the same time Whitehouse was devising his magnetic steelyard, Thomson was conducting an extensive series of electrical measurements in his Glasgow laboratory, as detailed in the Bakerian Lecture "On the Electro-dynamic Qualities of Metals" he delivered to the Royal Society of London in February 1856.[121] This project grew out of his pioneering work in thermodynamics, particularly on the transformations of energy in thermoelectric phenomena, and it required enormous numbers of meticulous measurements. To carry these out, Thomson began to recruit student volunteers from his natural philosophy classes. The work in the 1850s of this Glasgow "experimental corps" marked the beginnings of the laboratory teaching of physics in Britain, and its focus on precision electrical measurement set the pattern for most of the other physics laboratories that were to spring up in British universities over the next two decades.[122]

Many of Thomson's experiments concerned the way magnetization or mechanical strain affected the electrical resistance of iron or copper wires or plates. To measure these effects, he (or his assistant Donald MacFarlane, or one of the students) typically used the differential arrangement known as a Wheatstone bridge to compare the resistance of a "reference conductor" to that of the wire or plate being tested. Telegraph engineers had been using similar techniques for years, but in the mid-1850s these were all new to Thomson. He was in fact so unfamiliar with the usual practices of electrical measurement when he started this

---

[120] Whitehouse, "Construction and Use" (*Engineer*), 523. On standardization, replication, and networks of credibility, see Bruno Latour, *Science in Action: How to Follow Scientists and Engineers through Society* (Cambridge, MA: Harvard University Press, 1987), 247–57, and Joseph O'Connell, "Metrology: The Creation of Universality by the Circulation of Particulars," *Social Studies of Science* (1993) 23: 129–73.

[121] William Thomson, "On the Electro-dynamic Qualities of Metals," *Phil. Trans.* (1856) *146*: 649–751, repr. with additions in Thomson, *MPP 2*: 189–407.

[122] Crosbie Smith, "'Nowhere But in a Great Town': William Thomson's Spiral of Classroom Credibility," in Crosbie Smith and John Agar, eds., *Making Space for Science: Territorial Themes in the Shaping of Knowledge* (London: Macmillan, 1998), 118–46, on 131–36; Graeme Gooday, "Precision Measurement and the Genesis of Physics Teaching Laboratories in Victorian Britain," *British Journal for the History of Science* (1990) 23: 25–51. James Clerk Maxwell praised Thomson's "experimental corps" in a 15 February 1871 draft letter to E. W. Blore, in Maxwell, *SLP 2*: 613.

project that he had to work out the principle of the Wheatstone bridge for himself, not realizing until just before his lecture to the Royal Society that Charles Wheatstone had described it in his own Bakerian Lecture in 1843, or that Samuel Hunter Christie had in fact first devised the technique as long ago as 1833.[123]

"On the Electro-dynamic Qualities of Metals" was by far the longest paper Thomson would ever write. The original version ran to more than one hundred pages in the *Philosophical Transactions*, and Thomson kept adding new sections to it as late as 1878; when reprinted in his *Mathematical and Physical Papers* in 1884, the paper filled most of the second volume. Together with the brief preliminary reports that led up to it, "Electro-dynamic Qualities" marked Thomson's initiation into the practice of precision electrical measurement, a field in which he would take a leading role for the rest of his long career. It also marked an important step in his exploration of the *theory* of electrical measurement. Drawing on Wilhelm Weber's recent demonstration that electrical resistance could be expressed in "absolute" units of length, mass, and time, Thomson had shown in 1851 how to extend this approach to include "mechanical effect," or energy, as seen, for instance, in the heat generated when a current flows through a resistance. Using others' published measurements, he calculated (in British units of feet, grains, and seconds) the absolute resistance of several wires, as well as the "specific resistance," or resistivity, of copper, silver, and mercury.[124] Around this time he also sent a piece of wire to Weber at Göttingen, asking him to determine its resistance in absolute units so that Thomson might use it as a standard. Weber finally returned the wire in 1855 (apparently after some prodding from Hermann Helmholtz) along with a note stating its measured resistance; thereafter Thomson used it to link together his growing chain of electrical measurements.[125]

Thomson made an especially important series of measurements in early 1857, not long after he joined the board of the Atlantic Telegraph Company. "In measuring the resistances of wires manufactured for

[123] Thomson, "Electro-dynamic Qualities," 732n; Charles Wheatstone, "An Account of Several New Instruments and Processes for Determining the Constants of a Voltaic Circuit," *Phil. Transa.* (1843) *133*: 303–27, on 325; Samuel Hunter Christie, "Experimental Determination of the Laws of Magneto-Electric Induction in Different Masses of the Same Metal, and of its Intensity in Different Metals" *Phil. Trans.* (1833) *123*: 95–142, on 99.

[124] William Thomson, "Applications of the Principle of Mechanical Effect to the Measurement of Electro-motive Forces, and of Galvanic Resistances, in Absolute Units," *Phil. Mag.* (December 1851) *2*: 551–62, repr. in Thomson, *MPP 1*: 490–502.

[125] William Thomson to Hermann Helmholtz, July 30, 1856, in Thompson, *Kelvin, 1*: 321–22, on 322; see also Helmholtz to Thomson, August 11, 1855, H14, Kelvin Papers, GB 247 Kelvin, Glasgow.

submarine telegraphs," he said, "I was surprised to find differences between different specimens so great as most materially to affect their value in the electrical operations for which they are designed."[126] Like almost everyone else, he had assumed that copper was copper and that any reasonably pure wire would conduct a current about as well as any other.[127] He knew that mechanical strains of the kind he had investigated for his Bakerian Lecture could reduce the conductivity of a wire only to a slight degree, and he thought the larger variations in conductivity Weber had found in different copper wires probably just reflected differences in purity.[128] Thomson now found, however, that even ostensibly very pure samples of copper from different suppliers could differ markedly in their conductivity: wires from the manufacturer he labeled "A" conducted nearly twice as well as those from the one he labeled "D." Whatever the source of this difference, it had serious practical implications. As Thomson pointed out in a paper he presented to the Royal Society in June (the italics are his),

It has only to be remarked, that *a submarine telegraph constructed with copper wire of the quality of the manufactory* A *of only $^1/_{21}$ of an inch in diameter, covered with gutta-percha to a diameter of a quarter of an inch, would, with the same electrical power, and the same instruments, do more telegraphic work than one constructed with copper wire of the quality* D, *of $^1/_{16}$ of an inch diameter, covered with gutta-percha to a diameter of a third of an inch,* to see how important it is to shareholders in submarine telegraph companies that only the best copper wire should be admitted for their use.[129]

In short, unknowingly using copper of low conductivity could cost a company dearly.

Two months before he published his findings, Thomson wrote to alert the other directors of the Atlantic Telegraph Company to the problem. They referred the matter to Whitehouse and Bright "for their information," and in July authorized Thomson to spend up to £15 to test more samples of wire.[130] By then, of course, it was too late to do anything about the 2500 miles of cable that had already been manufactured. When the company prepared to order more cable to replace the length lost in the August failure, however, Thomson launched a determined campaign to ensure that it would be made with wire of high conductivity. He wrote to

---

[126] William Thomson, "On the Electrical Conductivity of Commercial Copper of Various Kinds," *Proc. RS* (June 1857) 8: 550–55, on 550, repr. in Thomson, *MPP 2*: 112–17.

[127] As Willoughby Smith later noted, in the early days of cable telegraphy "no attention was given" to the electrical qualities of the conductors "for the simple reason that all copper wire was credited with equal value in these respects"; Smith, *Rise and Extension*, 2.

[128] Thomson, "Commercial Copper," 550; Thomson, "Applications," 561.

[129] Thomson, "Commercial Copper," 551–52.

[130] ATC Minute Book, entries for c. April 1857, 125, and c. July 1857, 194.

the board stressing the "great economical and scientific advantage" of using only the best copper, and after what he later said was "much perseverance" managed to secure passage of a resolution requiring that in the specification for the new length of cable, "provision is made for the chemical purity and conductivity of the copper core."[131] This marked a milestone in submarine telegraphy: for the first time a contract would specify the *electrical* qualities of a cable, not just its mechanical properties and material composition.

At first the Gutta Percha Company balked, claiming it could not possibly test the conductivity of so much copper wire in the short time available. On further inquiry, however, the company said it could do the job for £42 per mile, rather than the previous £40, and it soon began testing conductivity at its north London works.[132] In his January 1858 "Electrician's Report," Whitehouse said that though the process was "somewhat tedious and obstructive," now "every hank of wire to be used for our conductor is tested, and all whose conducting power falls below a certain standard is rejected." The result was "a conductor of the highest value, ranging in conductivity from 28 to 30 percent. above the average standard of unselected copper wire."[133] Of course, the more than 2000 miles of old cable then being stored at Devonport had all been made with "unselected" wire, some no doubt of poor conductivity, but the company would simply have to live with it.

Through the fall and winter, Whitehouse continued to tinker with his coils and relays, hoping to improve on the slow transmission speeds he had achieved in his tests at Queenstown.[134] In the meantime, Thomson was working on a new signaling system of his own that, while it never really panned out, led him to the most important device to emerge from the Atlantic cable project: his mirror galvanometer. In his exchange with Whitehouse in the *Athenæum* the previous fall, Thomson had remarked

---

[131] ATC Minute Book, entries for August 27, 1857, 218, and September 4, 1857, 221–22; Thomson, *MPP* 2: 125n, noted added 1883; Thompson, *Kelvin*, 1: 350–51.

[132] William Thomson, "Analytical and Synthetical Attempts to Ascertain the Cause of the Differences of Electrical Conductivity Discovered in Wires of Nearly Pure Copper," *Proc. RS* (February 1860), *10*: 300–9, repr. in Thomson, *MPP* 2: 118–28, note added June 27, 1883, 125n; see also James M. Curley to William Thomson, September 30, 1857, and George Saward to William Thomson, October 6, 1857, Thomson Family Papers, Glasgow.

[133] Whitehouse, "Electrician's Report," January 4, 1858, WC-NYPL; see also Whitehouse, *Atlantic Telegraph* (1858), 12–13.

[134] In September 1857 the directors voted to offer a premium of £500 "for the best and most suitable form of telegraphic instrument for working through the Atlantic Cable," terms of the competition to be set by Whitehouse and Thomson, but Thomson declined to take part and the board soon dropped the idea; ATC Minute Book, entries for September 10, 1857, 225–26; October 8, 1857, 243; and December 5, 1857, 262.

that signaling rates depended on exactly how the current was applied at one end of the conductor and detected at the other. It soon occurred to him that by carefully calibrating the current entering a cable, one could contrive to make the first term of its Fourier expansion be zero; as the pulse decayed in transit, the remaining induced charges would then largely cancel each other out, reducing the retardation and speeding the transmission rate. Whitehouse's induction coils accomplished this to some extent by accident: the negative pulse that immediately followed each positive one helped clear the induced charge from the line and so prepared the way for the next signal. Thomson now proposed to speed this up further by using his transmission theory to guide exactly how much "curbing" to apply after each pulse, and by receiving the resulting signal on a refined version of a Helmholtz galvanometer. By carefully controlling the size and shape of each pulse, it would be possible, he said, to make each swing of the receiving galvanometer indicate a distinct letter: "The observer will watch through a telescope the image of a scale reflected from the polished side of the magnet, or from a small mirror carried by the magnet, and will note the letter or number which each maximum deflection brings into the middle of his field of view."[135] Instead of needing several dots and dashes to form each letter, a single swing would do the job, speeding up signaling rates by a factor of four or five.

Thomson presented this plan to the Royal Society on November 14, 1856, in a paper entitled "On Practical Methods for Rapid Signalling by the Electric Telegraph," but despite some promising early results, it eventually became clear that his method, at least in its strong form, was not really practical.[136] Too many extraneous factors – earth currents, atmospheric electricity, and other vagaries – came into play for one to be able to control the swing of the galvanometer precisely enough to distinguish different letters in a reliable way. Even as Thomson was forced to abandon his original plan, however, it led him to devise an extraordinarily sensitive galvanometer, able to respond nimbly to very small currents. The key step, foreshadowed in his remark about viewing a scale reflected from the polished side of the magnet, was to replace the needle of an ordinary galvanometer with a beam of light reflected from a tiny mirror. This beam would provide a weightless pointer of enormous length, greatly reducing the mass and moment of the moving parts of the galvanometer and so increasing its sensitivity. Working closely with

---

[135] Thomson, "Practical Methods," 301.
[136] Thomson, "Practical Methods," followed by a "Second Communication," *Proc. RS* (December 1856) *8*: 303–7, repr. in Thomson, *MPP 2*: 107–11, in which he proposed further refinements for use on both cables and overhead landlines.

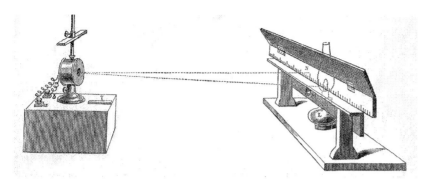

Figure 2.7 William Thomson's mirror galvanometer. Light from the lamp on the right passed through a hole in the screen and was reflected from a tiny mirror within the galvanometer on the left; the operator then read deflections by following the moving spot of light on the screen. (From Fleeming Jenkin, *Electricity and Magnetism*, p. 64, 1873.)

the Glasgow instrument maker James White, later his partner in a substantial business, Thomson refined his design and filed for a patent on his mirror galvanometer in February 1858[137] (Figure 2.7). He also designed a more robust "marine" version suitable for use on the ships that would try again to lay the cable that summer. The costs of this work were substantial, and in April he asked the board of the Atlantic Telegraph Company to grant him £2000 to complete the work. Having already spent thousands on Whitehouse's instruments, the board turned Thomson down, though it later granted him £500.[138] Thomson covered the difference from his own pocket, an investment that would later pay off handsomely.

Whitehouse and Thomson remained on good terms through the spring and summer of 1858, but their work was increasingly headed in opposite directions. Whitehouse continued to develop his huge induction coils and magnetic relays while relying on methods of electrical measurement, founded on his magneto-electrometer, that remained peculiarly his own. Thomson, on the other hand, focused on refining his sensitive

---

[137] Giuliano Pancaldi, "The Web of Knowing, Doing, and Patenting: William Thomson's Apparatus Room and the History of Electricity," in Mario Biagioli and Jessica Riskin, eds., *Nature Engaged: Science in Practice from the Renaissance to the Present* (New York: Palgrave Macmillan, 2012), 263–85. In a note added April 3, 1883, to the reprint of his "Practical Methods" paper (Thomson, *MPP 2*: 105n), Thomson said that the plan he described there, "modified and simplified, became developed a year later into the method of reading telegraphic signals by my form of mirror galvanometer."

[138] Smith, "Great Town," 135–36; Thompson, *Kelvin, 1*: 353–54.

mirror galvanometer and on building up a network of electrical standards and practices that could be widely shared and so become widely credible, and that meshed well with the growing body of established electrical theory. Whitehouse was still the head of the Atlantic Telegraph Company electrical department, but Thomson's was the approach that was destined to prevail.

Success . . .

As the Atlantic Telegraph Company prepared to try again to lay its cable, frictions between Whitehouse and the board of directors repeatedly threatened to erupt into open warfare. Whitehouse's spending was a particular sore point. In his January 1858 "Electrician's Report," he admitted that the work of his electrical department had "naturally involved a somewhat considerable outlay" but insisted that none "had been entered into without the most careful consideration" and asserted that his expenditures on equipment and personnel had been "fully justified" by the positive results they had enabled him to achieve.[139] Nonetheless, the directors repeatedly reminded Whitehouse not to spend so much, and in late January they refused his request to hire additional skilled assistants. Evidently looking to keep him focused on the task at hand, the board also ordered him not to publish his findings in scientific journals or show off his workshop at Devonport to outside experts.[140] Incensed, Whitehouse threatened to resign. He was soon talked out of it, but serious concerns remained, including about the performance of his apparatus.[141] In his January report he had claimed that, by using his improved instruments and adopting "such an amount of abbreviation or code signals as we find safe to use," he and his operators had been able to transmit four words per minute through the full length of the cable. Hearder, who witnessed many of these trials, later said this was an exaggeration, and that the real rate was closer to one word per minute.[142] Moreover, though the board had instructed Whitehouse to focus on training the operators in the use of the signaling apparatus, he reportedly instead kept tinkering with the instruments and directing his staff to conduct new experiments.[143]

---

[139] Whitehouse, "Electrician's Report," January 4, 1858, 25, WC-NYPL.
[140] ATC Minute Book, entry for January 29, 1858, 289.
[141] ATC Minute Book, entry for February 18, 1858, 297–98; see also Whitehouse, *Atlantic Telegraph* (1858), 20.
[142] Whitehouse, "Electrician's Report," January 4, 1858, 28, WC-NYPL; Hearder, "Atlantic Cable," 36.
[143] Saward, *Trans-Atlantic Submarine Telegraph*, 28.

Work at Devonport ramped up in the spring, and Whitehouse got permission from the directors to invite the "eminent electricians" C. V. Walker and W. T. Henley to visit and see the progress being made.[144] Thomson, too, spent much of April and May there, working mainly on his "size of swing" signaling technique. As the difficulty of controlling the swing became evident, he shifted toward using his mirror receiver as simply a very sensitive galvanometer, a role at which it excelled. Looking ahead to the upcoming laying expedition, he asked White, his instrument maker in Glasgow, to deliver one of the new "marine" versions to Devonport as soon as possible.[145]

In late May, shortly before the *Agamemnon* and the *Niagara* were to sail to the Bay of Biscay to test new paying-out machinery, the board of directors gathered at Plymouth to see for themselves how the work was coming along. They were not reassured. In particular, Saward later said, "the condition of the Electrical Department was found to be such as to cause great anxiety to the Directors": not only had the operators not been properly drilled in the use of the equipment, but Whitehouse's "instruments were not in a state nor of a nature calculated to work the cable to a commercial profit." There were also troubling reports of poor insulation in some of the cable already stowed on the ships, and concerns that Whitehouse's testing methods were inadequate.[146] Whitehouse further irritated the board by announcing at the last minute that his physician had forbidden him from sailing on the trial run. As in 1857, Thomson stepped in to take his place on the *Agamemnon*. Thomson was in fact almost the last man to board, as his assistant Donald MacFarlane rushed to the quay to hand over the very first marine galvanometer, brought by express from Glasgow. Looking like "a small brass pot sitting on four legs," it perched amid Whitehouse's array of apparatus in the ship's instrument room, where it would soon take over much of the electrical work of the expedition[147] (Figure 2.8).

The trial run went well, and on June 10 the flotilla set out from Plymouth to lay the cable. Unlike the year before, this time they would

---

[144] ATC Minute Book, entries for March 13 and March 17, 1858, 332 and 352; Whitehouse, *Atlantic Telegraph* (1858), 20. The board also invited Faraday, Wheatstone, William Robert Grove, and J. P. Gassiot to visit Devonport, but there is no evidence they made the trip. Walker was at Devonport from March 27 to April 14, 1858, and Henley apparently a somewhat shorter time; see *Joint Committee Report*, 101, 105.

[145] William Thomson to James Thomson, April 19, 1858, in Thompson, *Kelvin*, 1: 352. William Thomson was then in Glasgow, having just come from Devonport, to which he would return in May. See also Thomson to J. D. Forbes, April 24, 1858, in Smith, "Great Town," 135.

[146] Saward, *Trans-Atlantic Submarine Telegraph*, 28; ATC Minute Book, entry for April 5, 1858, 365; Whitehouse, *Joint Committee Report*, December 15, 1859, 78.

[147] Thompson, *Kelvin*, 1: 354–55; [Russell], "Paying-Out the Atlantic Cable," 5.

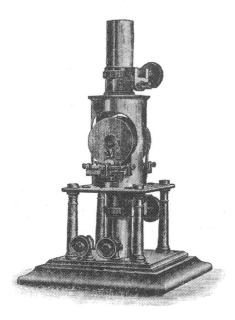

Figure 2.8 A more robust version of Thomson's mirror galvanometer, the marine galvanometer was designed for use aboard ships. (From Silvanus P. Thompson, *Kelvin*, Vol. 1: 355, 1910.)

start from mid-ocean, as Whitehouse now admitted that sending actual messages through the cable from ship to shore had proved unworkable. The most that could be reliably exchanged on shipboard were simple battery currents to check the continuity of the conductor and reveal any faults. The board strongly urged Whitehouse to join the voyage, but he again pleaded "indisposition" and went instead to Valentia to await the landing of the completed cable. Equipment for the Newfoundland end, including Whitehouse's coils and relays, was loaded onto the *Niagara* and its escort, HMS *Gorgon*. This apparatus – some of which, according to Hearder, had never been properly tested or adjusted – was entrusted to Whitehouse's assistant C. V. de Sauty, who would serve as chief electrician on the *Niagara*, while Thomson once again filled in for Whitehouse on the *Agamemnon*.[148]

---

[148] On de Sauty, see his obituary in *Elec.* (April 14, 1893) *30*: 685; on the lack of testing of Whitehouse's coils and relays, see Hearder, "Atlantic Cable," 39. The only known image of one of Whitehouse's five-foot induction coils appears in a painting by Theodor Linde, an operator at the Newfoundland station; see Figure 2.10.

The health concerns that kept Whitehouse ashore may have been legitimate, but his absence from the voyage inevitably undercut his scientific credibility and personal standing: he was not the man on the spot, seeing and doing things for himself, and he was not out there braving the dangers of the North Atlantic.[149] The latter point became especially salient after the ships were hit by a severe storm three days out of Plymouth. Experienced sailors said it was the worst gale they had ever encountered, and the *Agamemnon*, with more than a thousand tons of cable stowed awkwardly in her hold and on her foredeck, suffered especially badly. Ten men were badly injured as the ship rolled violently and nearly capsized; tons of coal broke loose and tumbled across the deck, cabins were flooded, and the cable itself was tossed about and badly kinked.[150] Some of the electrical equipment was also damaged, but Thomson himself bore the ordeal well, winning the respect of the officers and engineers. The experience did not seem to put him off life at sea: he went on to become an avid sailor and later used his earnings from the cable business to buy a much-loved sailing yacht, the *Lalla Rookh*.[151]

After a perilous week the storm finally blew itself out and the ships were able to rendezvous at their appointed spot. They made their splice on June 26 and began paying out, but the cable almost immediately jammed and snapped. After another splice, they managed to pay out eighty miles before Thomson's galvanometer showed a break in the conductor; once again, the ships had to backtrack and make another splice. They made a third attempt but on June 29, after the ships had laid about 250 miles, the cable snapped yet again. Disheartened, the *Niagara* and *Agamemnon* made their way to Queenstown and Field headed to London to meet with the directors. William Brown, the hard-headed Liverpool banker who has served as the first chairman of the permanent board, said they should admit defeat, sell the remaining cable for whatever they could get, and wind up the company. Others called for waiting to make another attempt the next summer. But Field, backed by Brett, Thomson, and the American-born board member Curtis Lampson, pushed to try again at once. It was now or never, they said: the cable could not survive another winter in storage, nor were the British and American navies likely to agree to lend their ships for a third summer.[152] After heated debates, Field and Lampson carried the day and the board voted to make one last try to lay its

[149] On the value of being the scientific man on the spot, see Bruce W. Hevly, "The Heroic Science of Glacier Motion," *Osiris* (1996) *11*: 66–86.

[150] [Nicholas Woods], "The Atlantic Telegraph Expedition," *Times* (July 15, 1858), 10. This vivid account of the gale was reprinted in Bright, *Atlantic Cable*, 91–105.

[151] On Thomson and the *Lalla Rookh*, see Smith and Wise, *Energy and Empire*, 733–40.

[152] Saward, *Trans-Atlantic Telegraph*, 30–31.

cable. Success seemed so little assured, however, that to save the expense of chartering a vessel, the company held off on shipping its heavy shore end cable from Devonport to Valentia until after the main cable had been laid.[153] It was a decision that would lead to sharp controversies.

With none of the pomp that had marked the launching of the earlier attempts and with little real expectation of success, the ships slipped out of Queenstown harbor on July 17. After making their mid-ocean splice on July 29, the *Agamemnon* steamed again toward Ireland and the *Niagara* toward Newfoundland, with Thomson and de Sauty closely watching their galvanometers to gauge the state of the cable. Apart from a mysterious loss of current the first night, and signs of a possible fault in the insulation a couple of days later, things went more smoothly this time and the *Niagara* arrived at Trinity Bay, and the *Agamemnon* at Valentia, on August 5. To the surprise of all, this last forlorn hope had succeeded. Word quickly went out over the wires that the Old and New Worlds were now connected by telegraph, and the celebrations began.

The success was widely acclaimed in the British press and arrangements were soon made to award Bright a knighthood. The hoopla was far greater in the United States, where to the amusement of those who knew the real role British capital and expertise had played in the project, the laying of the cable was widely depicted as a purely American achievement. Field was hailed as the hero of the age, the man who had single-handedly spanned the Atlantic in what a Philadelphia paper called "the greatest triumph of scientific and mechanical *genius* that has been achieved for centuries."[154] Amid an outpouring of press speculation about how the cable would transform global trade and international relations, markets were flooded with ephemera – prints, broadsides, sheet music, and gee-gaws of all kinds. Once the *Niagara* arrived in New York, Field turned its miles of surplus cable to account by having Tiffany and Company make it up into four-inch lengths to sell as souvenirs[155] (Figure 2.9).

Shares in the Atlantic Telegraph Company had risen and fallen with the fortunes of the laying attempts, dipping to a low of £300 in the wake of the June failure. They now jumped overnight from £340 to over £900, though

---

[153] "The Atlantic Telegraph," *Times* (September 10, 1858), 7.

[154] "Newspaper Comments," *New York Times* (September 6, 1858), 1, citing the *Philadelphia Evening Journal*. Bright was knighted in Dublin on September 5, 1858; see "The Engineer of the Atlantic Telegraph," *Freeman's Journal* (September 6, 1858).

[155] For a collection of cable-related ephemera, see Robert Dalton Harris and Diane DeBlois, *An Atlantic Telegraph: The Transcendental Cable* (Cazenovia, NY: Ephemera Society of America, 1994); on Tiffany's sale of short lengths of the cable, see John Steele Gordon, *A Thread Across the Ocean: The Heroic Story of the Transatlantic Cable* (New York: Walker and Co., 2002), 137.

Figure 2.9 After the first Atlantic cable was completed in August 1858, Cyrus Field had the surplus length cut into short pieces, which Tiffany & Co. then sold as souvenirs. The band around this one reads "Atlantic Telegraph Cable – Guaranteed by Tiffany & Co. – Broadway • New York • 1858."

few shareholders appeared willing to sell.[156] Merchants and the public waited eagerly for market orders and news reports to begin flashing back and forth across the Atlantic, and for a new age of oceanic telegraphy to begin.

... and Failure

On arriving in the virtual wilderness of Trinity Bay, de Sauty and his assistants set about assembling Whitehouse's collection of batteries, induction coils, and relays, a task that would take them several days. In the meantime they kept up a steady battery current, with slow reversals, to show Valentia they had arrived and that the conductor was intact. In Ireland, Thomson handed his end of the cable over to Whitehouse and stayed on for several days to help get the station up and running in borrowed rooms at the Knightstown slate works. The prospects looked promising, but they would soon darken.

The plan, touted even before the Atlantic Telegraph Company had been formed, called for using Whitehouse's patented induction coils to send the signals and his relays to receive and record them on paper tapes. The relays, however, never worked as intended; when the Newfoundland operators connected theirs to the cable, they found the coil currents from

---

[156] "Money-Market and City Intelligence," *Times* (August 6, 1858), 7.

Valentia were too weak to trip it, while Thomson later reported that the longest complete word ever received at the Irish end entirely on a relay was "be."[157] After a few frustrating days the operators at Valentia put Whitehouse's relay aside and tried to receive signals on Thomson's mirror galvanometer instead. They were soon rewarded: at 1:45 a.m. on August 10, the swinging spot of light spelled out the first readable words from Trinity Bay: "Repeat, please." It was not the most profound of messages, but Whitehouse and Thomson were delighted; the latter reportedly skipped around the instrument room with joy and treated the staff to a round of porter from the nearby hotel.[158] The Valentia station soon adopted a system in which one operator would watch the spot of light and call out its motions to another, who would then use a local battery and Morse key to record the message on a paper tape. Whitehouse mailed some of these slips of paper to the directors in London, describing them as the "signals first transmitted and received across the Atlantic by the Company's instruments."[159] The directors took these to be just what they looked like: messages recorded directly by Whitehouse's relays. Whitehouse did not mention that they had in fact been received on Thomson's galvanometer.[160]

By then Thomson had already left Valentia. Thinking operations were on track and that Whitehouse would soon have his relays working properly, he had departed on the morning of August 10 for London, Glasgow, and, as he thought, a well-deserved rest. But events at Valentia soon took a turn, and within a week and a half he would be back. Those events centered on three interlocking issues: the possible existence of faults in the insulation of the cable; the effects the use of Whitehouse's induction coils might have on those faults; and Whitehouse's treatment of J. R. France, an experienced telegrapher the directors had sent to assist and, as

---

[157] Newfoundland sent the message "coil signals too weak work relay" on August 12, 1858; see *Joint Committee Report*, 230. For Thomson's December 17, 1859 testimony on the length of words received at Valentia on relays, see *Joint Committee Report*, 121.

[158] Russell, "Notes," August 10, 1858, 7–9. Thomson later testified that "a little single needle instrument of Mr. Henley's" was also in the circuit and received the first message simultaneously with his mirror galvanometer, but the more sensitive mirror device soon came to be used exclusively; Thomson, December 17, 1859, *Joint Committee Report*, 119.

[159] Whitehouse to Directors Atlantic, London, August 10, 1858, in Wildman Whitehouse, *Recent Correspondence between Mr. Wildman Whitehouse and the Atlantic Telegraph Company* (London: Bradbury & Evans, 1858), 17.

[160] On the directors in London assuming that the paper tapes Whitehouse sent from Valentia had been marked directly by his relays, see William Thomson to Board of the Atlantic Telegraph Company, August 21, 1858, Alcatel Archive, Porthcurno, doc. ref. 74/1. This important letter was found in the archive by Allan Green; for his transcription of it, see: http://atlantic-cable.com/Books/Whitehouse/AG/WTLetter.htm.

Whitehouse thought, report on him. All combined to make Whitehouse's relations with the directors in London increasingly fraught.

From early on, Whitehouse suspected the insulation of the cable might have suffered an injury near its Valentia end. It was a plausible idea: since the company had left the heavily armored shore end cable, weighing eight tons per mile, at Devonport to save on immediate costs, the *Agamemnon* had been forced to run its light cable, designed for use only in the deep sea and weighing just one ton per mile, right up to the shore at Knightstown, where it was continually buffeted by waves and risked being snagged by ships' anchors. As early as August 9, Whitehouse had wired London to say it was "absolutely essential" that the company act immediately to protect the light cable in the harbor or risk its destruction. He did not mention, however, that he was already hatching a plan to fix the problem himself.[161]

In the meantime, both Whitehouse and de Sauty were using their induction coils to send huge pulses of current into the cable. Whitehouse's coils were five feet long, with solid iron cores wrapped with miles of wire. Fed by his enormous batteries, such coils could, Whitehouse later said, produce sparks able to jump nearly a quarter inch air gap, implying potentials of thousands of volts. None of Whitehouse five-foot coils appear to have survived and we do not know the exact characteristics of the ones actually used at Valentia and Trinity Bay, but it is clear they could deliver very powerful jolts, enough to give an operator a nasty shock – or worse – if he touched one the wrong way[162] (Figure 2.10).

Whitehouse recognized that if there was indeed a fault in the cable near the Valentia end, shocks from his induction coils could worsen it. As he wrote to de Sauty on August 11 (in a letter that could have reached Newfoundland only much later), with most of the intense current produced by a coil discharge "forcing its way to earth" through such a fault, little would be left to continue on to Newfoundland. This was echoed in the journal kept by James Burn Russell, who had assisted Thomson on the *Agamemnon* and then joined the staff at the Valentia station. Remarking

---

[161] Whitehouse to Directors Atlantic Telegraph Company, London, August 9, 1858, *Recent Correspondence*, 16; Russell, "Notes," August 9, 1858, 7.

[162] Many secondary sources (e.g., Thompson, *Kelvin, 1*: 385) say Whitehouse's induction coils produced tensions of 2000 volts, but cite no basis for this estimate. In his January 5, 1860 testimony (*Joint Committee Report*, 159), C. F. Varley, who had tested the coils at Valentia, noted that their quarter-inch sparking distance implied a tension equivalent to 10,000 to 15,000 Daniell's cells, or in modern units, just over 10,000 to 15,000 volts. Hearder said that while he believed defects of design or workmanship made Whitehouse's coils less powerful than their size would suggest, they could nonetheless produce jolts sufficient "to destroy life in an instant"; Hearder, "Atlantic Cable," 36, 39–40.

Figure 2.10 Theodor Linde, an operator at the Newfoundland end of the first Atlantic cable, painted this watercolor of the station's telegraph room in 1858. On the floor to the right is the only known depiction of a pair of Whitehouse's five-foot induction coils.
(Courtesy Bill Burns.)

on the evidence of "leakage at this end," he wrote that "coils therefore which we have used hitherto are very unsuitable since they give great intensity" and so "force through" the fault.[163] Around this time Whitehouse set his coils aside, except for occasional tests, and shifted to using much less intense currents from a set of Daniell's "quantity" batteries Thomson had provided. At the Trinity Bay end, however, de Sauty continued to use induction coils; indeed, Whitehouse later said that for the first few days after the cable was landed, the coil currents reaching Valentia were so strong "they made the relay speak out loud, so you could hear it across the room."[164] It is not clear if the company officials in London understood quite what was being done, or grasped the difference between the intense currents from the coils and milder ones from batteries, but they cautioned Whitehouse that "application of too much battery power" might injure the cable and asked him to send a full report.[165]

---

[163] Whitehouse to de Sauty and Laws, August 11, 1858, *Recent Correspondence*, 17; Russell, "Notes," August 11, 1858, 9–10.

[164] Whitehouse, December 15, 1859, *Joint Committee Report*, 78.

[165] Saward to Whitehouse, August 10, 1858, *Recent Correspondence*, 17. Saward, who was no electrician, sometimes used "battery power" as a synonym for electromotive force, whether its actual source was batteries or induction coils like Whitehouse's; see, for

Whitehouse used his induction coils steadily for only a few days at Valentia, but the intense jolts of current they produced most likely further damaged an already faulty cable; indeed, Whitehouse himself later said that, on encountering an existing flaw in the insulation, strong currents like those from his coils would tend to "augment the mischief."[166]

Concerned by delays in opening the cable to traffic and by a lack of reports from Whitehouse, the directors arranged to send J. R. France, a well-regarded telegrapher with Brett's Mediterranean Telegraph Company, to Valentia to help get the instruments there working properly. He left London on August 13, carrying a letter directing Whitehouse to give him "every encouragement and assistance," while Saward wired ahead to let Whitehouse know that France was on his way.[167] Whitehouse was incensed: "Advise the Directors to recall France," he wired back. "They have made a great mistake."[168] France might be an able operator, Whitehouse said, but it would not be right to put him above more senior staff already at Valentia. Nor would France's experience with other cable apparatus be of any help at Valentia, where he would encounter "instruments which he has never seen, and of whose nature and construction he can have, from their novelty, the most superficial knowledge."[169] Of course, by the time Whitehouse wrote this, the staff at Valentia had already stopped using his coils and relays, and while France might not have had any earlier experience with Thomson's mirror galvanometer, neither did Whitehouse or anyone else. Whitehouse real objection was that France had evidently been sent to "advise, direct, and, I presume, report on" his work at Valentia, and he resented what he took to be an effort by the directors to check up on him. He said he could use another instrument clerk, however, and told Saward he would be willing to take France on in that capacity, adding archly "I cannot

---

example, George Saward, "The Atlantic Telegraph," *Times* (September 24, 1858), 7, and Saward's questioning of J. W. Brett, December 10, 1859, *Joint Committee Report*, 58.

[166] Whitehouse, December 15, 1859, *Joint Committee Report*, 79. See also Whitehouse's comments in *Letter from a Shareholder to Mr. Whitehouse, and His Reply* (London: Bradbury & Evans, 1858), published in December 1858, about the damage either strong battery power or currents from his "gigantic induction coils" could do to an already faulty cable, though he denied that his own use of the coils had caused any problems. A transcription of the pamphlet can be found at http://atlantic-cable.com/Books/White house/1858-FM/index.htm.

[167] Samuel Gurney and George Saward, Atlantic Telegraph Company, London, to Whitehouse, August 12, 1858, *Recent Correspondence*, 20. France presented this letter to Whitehouse on arriving at Valentia on August 15, 1858.

[168] Saward to Whitehouse, August 13, 1858, and Whitehouse to Saward, August 13, 1858, *Recent Correspondence*, 21.

[169] Whitehouse to Saward, August 16, 1858, *Recent Correspondence*, 26–29, on 29.

recognise him in any other."[170] Tensions with London continued to mount, along with suspicions that Whitehouse was trying to hide something.

Amid all this, a hopeful sign emerged on August 13: the Newfoundland station, having put a mirror galvanometer in circuit for some tests, managed to receive its first full word: "Atlantic." But the Trinity Bay operators then immediately went back to their relay – and were unable to read anything more. The staff at Valentia regarded this an another example of "an extra proportion of obtuseness" at the Newfoundland end, but the Trinity Bay operators were simply following established protocols while waiting for their Valentia counterparts to sort out their equipment and procedures.[171] Fruitless efforts to get de Sauty and his staff to give up their relay continued for another day until they finally caught on and switched to using an ordinary needle galvanometer. The result was not great – the needle only moved about half a degree with each pulse – but it was enough to enable the Newfoundland station to receive messages, albeit very slowly.[172]

Why were the signals so much weaker at Newfoundland than at Valentia? As experienced electricians knew, a fault has its greatest effect on the end of the cable farthest from it. Consider a fault with a resistance equal to ten miles of cable and located ten miles from the Irish end. A current coming from Newfoundland will split at the fault; in this case, half will go directly to earth through the fault and half will continue on to Valentia, where a signal will arrive weakened but still quite readable. A current starting from Valentia will also split at the fault, but since the resistance through the fault is equal to just ten miles of cable, while that on to Newfoundland is equal to 2000 miles, almost all of the current will escape through the fault, leaving only 1/200th of it to go on to Trinity Bay – far too little to work a relay. One could, of course, try to compensate by sending more intense currents into the Valentia end, but only at the risk of damaging the insulation and worsening the fault.

Whitehouse did not frame the problem of locating the suspected fault in terms of measurable resistances, nor did he possess a set of standard resistance coils or any other reliable way to gauge how far away the fault was. While he recognized that the relative weakness of the signals at Trinity Bay pointed toward a fault lying nearer the Valentia end, he could do no more than guess at its actual location, nor did he seem to grasp that even a fault 200 or 300 miles from Valentia could greatly

[170] Whitehouse to Saward, August 15, 1858, *Recent Correspondence*, 23.
[171] Russell, "Notes," August 31, 1858, 54.
[172] Whitehouse to Saward, August 16, 1858, *Recent Correspondence*, 26–29, on 27.

weaken the currents received at Trinity Bay. In any case, Whitehouse soon convinced himself that the fault must lie very near the Valentia end, and he set about making plans to repair it.[173]

Company procedures in such a case called for Whitehouse to notify the directors in London and refer any repairs to Bright and his engineering department. But Bright had left Valentia for England soon after the cable was landed, and Whitehouse was concerned that an inquiry to London about repairing even a minor fault in the harbor might set off a "panic" that would damage the company. Fixing a fault so near the shore would, he thought, be a quick and easy task, and once completed, he could presumably go back to using his coils and relays as originally planned.[174] No one need ever know about these embarrassing teething problems, which Whitehouse was convinced all stemmed from the company's failure to install the heavy shore end in a timely way.

On August 13, Whitehouse wired Samuel Canning, an engineer on Bright's staff who was then in Dublin, and asked him to come to Valentia and underrun a few miles of cable in the harbor. London soon got wind of this plan – "by mere accident, and not from you," as Saward told Whitehouse – and the directors were not pleased.[175] Thomson and others in London did not think there could be a serious fault very near the Valentia end; as Thomson later explained, if such a fault existed there, no currents strong enough to work Whitehouse's relays, which had substantial internal resistance, would have been able to get past it – and the paper tapes Whitehouse had mailed to London gave every sign of having been marked by his relays.[176] Of course, the tapes notwithstanding, Whitehouse had not really received the messages on his relays, but here his coyness about using Thomson's mirror galvanometer had backfired on him.

The board's real objection, however, was not to Whitehouse's chasing a possibly imaginary fault but to his acting without its authorization and infringing on Bright's proper responsibilities. On August 14 the chairman, deputy chairman, and Saward sent Whitehouse an urgent telegram ordering him "not to underrun or otherwise interfere with the submerged

[173] Russell, "Notes," August 9, 1858, 7; Whitehouse to Directors, August 13, 1858, *Recent Correspondence*, 20.
[174] Whitehouse to Saward, August 16, 1858, *Recent Correspondence*, 26–29, on 28.
[175] Russell, "Notes," August 14, 1858, 12; Saward to Whitehouse, August 14, 1858, *Recent Correspondence*, 22, 24.
[176] Saward to Whitehouse, August 14, 1858, *Recent Correspondence*, 22; Thomson to Board, August 21, 1858, Alcatel Archive, Porthcurno; Saward, "Atlantic Telegraph," *Times* (September 24, 1858), 7. After the first few days, the currents from Trinity Bay became too weak to work Whitehouse's relays; see Whitehouse, "Professor Whitehouse and the Atlantic Telegraph," *Daily News* (September 29, 1858).

cable" until he had received permission to do so, and scolding him for not consulting them before even considering such a step.[177] By then Canning had in fact already underrun the cable in the harbor and found several kinks in it, though rough seas kept him from being able to cut and repair the suspect spots at that time. Whitehouse suspended the work after the order came in from London, but he remained convinced there was a fault in the harbor and was determined to do something about it[178] (Figure 2.11).

When France arrived at Valentia the next day, Whitehouse turned him away. Feeling that his own experience and expertise had been under-valued and ignored, and no doubt frustrated that the elaborate instru-ments in which he had invested so much time and effort had proven useless, Whitehouse sent Saward and the directors two long letters in which he tried to explain himself. He had been left on his own, he said, to face a myriad of urgent problems at Valentia, and "in the absence of any one at hand to whom I can instantly refer, it becomes my duty to act; in doing so I assume responsibility, and may herein be blamed by the Directors."[179] He denied he had violated their order not to underrun the cable, saying he had begun that operation before the directors' tele-gram arrived and had ordered Canning to stop as soon as it was received – but he did not mention that he had sent Canning out again on August 16, the very day he wrote his second letter. This time the underrunning interrupted the transmission from Valentia of the Queen's congratulatory message, leading to embarrassing reports that it had taken more than sixteen hours to send just ninety-nine words, though when allowance was made for the interruption, the time actually spent in sending the message was far less.[180] Rough seas again kept Canning from actually cutting and repairing the cable that day, but Whitehouse could hardly claim this time that he had not violated the explicit order from London that he not meddle with the cable.

Fed up with what it saw as his rank insubordination in turning away France and underrunning the cable, on August 17 the board of directors voted to dismiss Whitehouse as the electrician of the com-pany he had helped to found and to summon him to London to

[177] Chairman (Samuel Gurney), Deputy Chairman (Curtis Lampson), and George Saward to Whitehouse, August 14, 1858, *Recent Correspondence*, 22.
[178] Russell, "Notes," August 14, 1858, 12–13; Whitehouse to Directors, August 14, 1858, and Whitehouse to Saward, August 14, 1858, *Recent Correspondence*, 21.
[179] Whitehouse to Directors, August 15, 1858, *Recent Correspondence*, 24–25, on 24.
[180] Russell, "Notes," August 16–August 17, 1858, 18–22. Because of the interruption, the banal first sentence of the Queen's message was distributed in America before the rest arrived, prompting questions about its authenticity; see "America," *Times* (August 30, 1858), 7.

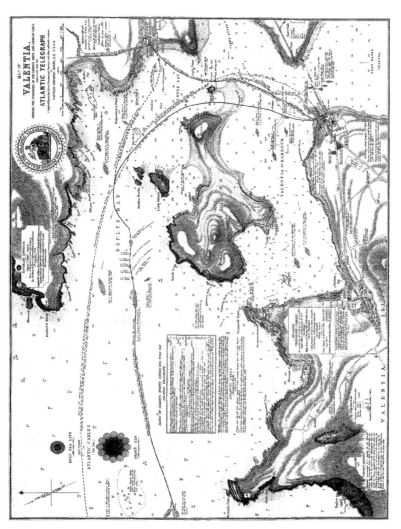

Figure 2.11 In early 1859 Captain Frederic Brine published an extraordinarily detailed map of Valentia harbor, showing the routes and landing places of the 1857 and 1858 cables, the positions of the ships involved in laying them, and even the price of rooms at the hotel in Knightstown.
(From Frederic Brine, *Map of Valentia, Shewing the positions of the various ships and lines of cable connected with the Atlantic Telegraph*, 1859; courtesy Bill Burns.)

explain himself.[181] By then Whitehouse had concluded that if the cable failed, he would bear the blame anyway, so in flagrant violation of the board's direct orders, he decided to make one last try to repair the fault he was sure was the source of all the trouble and, as he later said, "either to complete the operation and resign, or still more nobly to succeed, and rescue, the undertaking."[182] As he was leaving for London early on August 18, he therefore sent Canning and his crew out yet again. This time they underran the cable for about three miles, as far as Doulas Head, cut it, and replaced the suspect section with a surplus length that had been left by the *Agamemnon*. Whitehouse received reports on this work while he was en route to London, and he wired the directors from Dublin that he was confident Canning's efforts had fixed the problem, or at least most of it.[183] The staff at Valentia initially agreed, reporting that they were again using Whitehouse's coils to signal to Newfoundland, but on reviewing the instrument room logs a month later, Russell concluded that replacing the cable in the harbor had not really improved the quality of signaling, but in some ways had made it worse. "Mr. Whitehouse plainly I think must be astray in his testing," Russell wrote in his journal; while there may indeed have been some leakage near the Valentia end of the cable, he said, the evidence pointed toward the existence of another and far more serious fault much farther out to sea.[184]

Thomson remained perhaps Whitehouse's last supporter on the board of directors. He reluctantly went along with the August 17 vote to dismiss Whitehouse and then agreed to take over direction of the station at Valentia, but on arriving there four days later found things to be in better shape than he had been led to expect. In particular, Thomson found that the relays had been set aside in favor of his mirror galvanometer as early as August 10, rendering moot his earlier argument that there could not be a serious fault near the Valentia end – though he would soon find other evidence that no such fault existed. While admitting that Whitehouse had been wrong to turn away France and to disobey direct orders from

---

[181] Saward to Whitehouse, August 17, 1858, *Recent Correspondence*, 30, enclosing an extract from that day's minutes of the board of the Atlantic Telegraph Company informing him that "his engagement and authority as an officer of the Company have now ceased." In formal terms, Whitehouse's salaried appointment as electrician (and Bright's as engineer) were "to continue until the cable be laid down or until the Board shall think fit to dispense with their services," but both Whitehouse and the board had assumed he would stay on at least until routine operations at Valentia and Trinity Bay had been established; see ATC Minute Book, c. January 10, 1857, 47.

[182] Whitehouse, "Professor Whitehouse," *Daily News* (September 29, 1858).

[183] Telegrams from "Valentia" to Whitehouse at Killarney, Mallow, and Dublin, August 18, 1858, *Recent Correspondence*, 30–31; Whitehouse (at Dublin) to Directors, August 19, 1858, *Recent Correspondence*, 31; Russell, "Notes," August 18, 1858, 25–28.

[184] Russell, "Notes," September 17, 1858, 77.

London not to underrun the cable, Thomson now said that looking for a fault in the harbor had not been unreasonable, and he urged the board to reconsider its dismissal of Whitehouse.[185] He wired Whitehouse to express his support and even told the board he would like to have Whitehouse back to help him at Valentia. But Whitehouse, his pride injured and honor impugned, told Thomson he would not accept reinstatement without "ample honourable amende."[186] In any case, there was no prospect of any such reinstatement; as the board made clear in its reply to Thomson, the issue was no longer a technical one of the location of faults or the use of instruments, but one of Whitehouse's insubordination and insolence.[187]

The quality of signaling on the cable continued to fluctuate, as it had before Whitehouse's departure. Thomson pressed the operators at Trinity Bay to use his mirror galvanometer, and when they finally began doing so on August 22, they reported "signals beautiful."[188] This clarity did not last, however, as the currents gradually weakened and were often overwhelmed by earth currents. Thomson managed to nurse the cable along for a bit longer, sending and receiving several dozen messages over the next ten days, but it was increasingly clear that the insulation was badly compromised.

Perhaps the most important of the messages the cable carried in its last days were two the British government sent to Newfoundland on August 31. The army had earlier ordered two regiments to sail from Canada to aid in putting down the Indian Rebellion, but as the uprising was quelled, the authorities concluded that the troops were no longer needed. By using the cable to cancel the earlier orders before the regiments had sailed, the government saved itself about £50,000. These messages led Russell to reflect on

what important services the cable may perform for our Govt. – both in saving money; and in knitting the limbs of empire into one gigantic frame. When we have extended these wonderful wires to India & Australia, Great Britain and her Colonies will resemble in economy the human body. London the seat of supreme intellect, whence the electric lines, the nerves, ramify and distribute themselves, the medium by which her behests are made known and executed in the remotest parts of the huge structure.[189]

[185] Thomson to Directors, August 21, 1858, *Recent Correspondence*, 35; Thomson to Board, August 21, 1858, Alcatel Archive, Porthcurno.
[186] Thomson to Whitehouse, August 21, 1858, and Whitehouse to Thomson, August 21, 1858, *Recent Correspondence*, 35.
[187] Atlantic Telegraph Company directors to William Thomson, August 25, 1858, in Thompson, *Kelvin*, 1: 370–72.
[188] Register of messages received at Valentia, August 21, 1858, *Joint Committee Report*, 234.
[189] Russell, "Notes," August 30, 1858, 52.

It would be more than a decade before the "nerves of empire" would extend as far as Russell envisaged, and in the meantime, the link from Ireland to Newfoundland was rapidly failing. Thomson's tests indicated a dead earth between 200 and 400 miles west of Valentia, probably in waters too deep for the cable to be lifted and repaired with the means then available. After September 1, the cable spoke only in a few fitful and isolated words, and it would soon breathe its last. As Thomson remarked to Russell on September 5, "Is this not an unhappy termination to our labours?"[190]

Of course, Thomson's labors were not really over. Joined by Cromwell Fleetwood Varley of the Electric and International Telegraph Company, he would spend weeks at Valentia in what he called "the dull and heartless business of investigating the pathology of 'faults' in submerged conductors," as they tried every expedient they could think of to bring the cable back to life.[191] There were a few flickers, but nothing really worked, and when in November the company finally laid the heavy shore end that Whitehouse had so long called for, it did nothing to cure the problem; the fault, as Thomson and Varley's careful resistance measurements had established, lay in inaccessible waters hundreds of miles out to sea.[192] In the end the Atlantic Telegraph Company had to admit defeat, at least for this round. Recriminations followed, of course, with Whitehouse bearing the brunt, but the real question would be what lessons ought to be drawn from this grand failure, not just about how future cables should be made and laid, but about how electrical phenomena should best be measured and understood.

---

[190] Russell, "Notes," September 3 and September 5, 1858, 58–60.

[191] Thomson to James Joule, September 25, 1858, in Thompson, *Kelvin*, 1: 378–79, on 379.

[192] On the laying of the heavy shore end in November 1858, see Frederic Brine, *Map of Valentia, Shewing the Positions of the Various Ships and Lines of Cable Connected with the Atlantic Telegraph* (London: Edward Stanford, 1859).

3   Redeeming Failure
The Joint Committee Investigation

In the wake of the failure of the Atlantic cable in 1858, followed two years later by the even costlier collapse of the Red Sea and India line, oceanic submarine telegraphy came to be widely seen as a great *failed* technology. Of over 1000 miles of submarine cables that had been laid up to the spring of 1861, only about 3000 miles were actually working, and of those, almost all were relatively short and lay in shallow waters.[1] If the dream of a global network of undersea cables knitting together the vast British Empire and quickening the pace of world commerce was ever to be realized, proponents of submarine telegraphy needed to find a way to explain why so many of their previous efforts had failed and to lay out means and methods by which the success of future cables could be assured.

This was an enormous task, but it was substantially accomplished in the three years following the 1858 failure. First came a wave of heated polemics over who and what to blame for the earlier failures, aimed at depicting them as resulting not from any intrinsic impossibility of deep sea submarine telegraphy but simply from a series of personal failings and correctable missteps. This was followed by a careful government-funded investigation, led by the prestigious "Joint Committee to Inquire into the Construction of Submarine Telegraph Cables," into just what those missteps had been and how they might be avoided in the future. The result was a massive report, issued in the summer of 1861, that pulled together expert testimony and extensive experimental evidence to establish best practices in the field. The *Joint Committee Report* was widely praised as thorough and judicious; Sir Charles Tilston Bright later called it "the most valuable collection of facts, warnings, and evidence ever compiled concerning submarine telegraphs."[2] Deeply rooted in government concern to secure rapid and reliable communications with India and other imperial possessions, the investigation and subsequent report

---

[1] *Joint Committee Report*, vii.
[2] Sir Charles T. Bright, "Inaugural Address of the New President," *Journal of the Society of Telegraph-Engineers and Electricians* (1887), *16*: 7–40.

would help set the course for British work not just in submarine telegraphy but in much of electrical science for decades to come. While no committee report, however worthy, could in itself fully restore public confidence in oceanic cable telegraphy, the clear practical successes that would eventually do so – above all the new Atlantic cables laid in 1865–1866 – owed much to the work of the Joint Committee. It cleared away the rubble of past failures and helped lay the scientific and engineering foundations on which the future success of submarine telegraphy could be built.

Disposing of Whitehouse

The first order of business for proponents of oceanic submarine telegraphy in the fall of 1858 was to find a way to explain, or explain away, the humiliating failure of the Atlantic cable. This was an especially thorny problem for the leaders of the Atlantic Telegraph Company, as they strove to do so in a way that not only did not discredit the larger project of cable telegraphy but also spared the principal figures of their own company – particularly Cyrus Field – from direct blame. Even before the Atlantic cable had fallen completely silent, defenders of the company had found a convenient figure on whom to cast the bulk of the blame: Wildman Whitehouse. It was in many ways an easy choice. Whitehouse had certainly contributed substantially to the failure of the cable, both by his overall handling of the electrical department of the company from its founding in the fall of 1856 to his dismissal two years later and by his subjection of the already faulty cable to jolts from his huge induction coils after it was laid. Moreover, the directors of the company had already dismissed him for insubordination before the inquest into the death of the cable had really begun, and they were happy to point out that they had not freely chosen him to serve as their chief electrician but had been saddled with him as one of the original "projectors" of the company.[3] Besides his conflicts with the board, Whitehouse had also managed to alienate many leading British telegraph engineers and electricians, who regarded him as an interloper whose views and methods lay well outside the norms of their community. They, like the leaders of the Atlantic Telegraph Company, sought to distance themselves from Whitehouse and from responsibility for the failure of the cable, and so to preserve their own claims to competence in carrying out future projects.

---

[3] George Saward, "The Atlantic Telegraph" (letter), *Morning Chronicle* (September 22, 1858), and in *Times* (September 24, 1858), 7.

In retrospect, it is clear that the first Atlantic cable failed because of a combination of excessive haste, poor design, hurried manufacture, inadequate testing, rough handling, improper storage, and generally sloppy management. Given all of this, one might well agree with the veteran telegraph engineer F. C. Webb, who told the Joint Committee that rather than being surprised that the cable had failed, they should regard it as "a miracle that it was even possible to get a few messages through the cable at all."[4] Had Field not been in such a rush to launch the company in the fall of 1856, and had he not promised to try to lay the cable the following summer, he and his associates might have taken the time to perform proper tests and been led to choose a better design – one that, while perhaps not as pretty as the stranded iron rope they selected, might have had a thicker conductor and a more durable outer covering.

Even given the deficiencies of its design, had the cable been more carefully made and tested, and had Bright and his team of engineers used better paying-out machinery from the first, so that they could have laid the cable successfully on their first try instead of needing to unload it, coil it, and store it for the winter, it seems likely it could have withstood the shocks from Whitehouse's induction coils and managed to carry at least a few words per minute across the Atlantic for some months or even years. But given the flaws that existed in the insulation of the cable by the time it was laid, the strong jolts from Whitehouse's coils could not help but hasten its demise. Perhaps, as William Thomson suggested in 1860, if his mirror galvanometer had been used as the sole receiver from the first and if "no induction coils and no battery power ... exceeding 20 cells of Daniell's ... had ever been applied to the cable since the landing of its ends, imperfect as it then was, it would now be in full work day and night, with no prospect or probability of failure."[5] We will never really know. But even if the jolts from Whitehouse's coils only precipitated the failure of the cable rather than being its underlying cause, it served the interests of both the Atlantic Telegraph Company and the community of cable engineers to focus public attention on Whitehouse's manifest failings – to make him the "fall guy." By tagging Whitehouse as the principal culprit in the debacle, and then asserting that his views, instruments, and proced-ures were far out of step with their own, cable engineers and electricians

<hr />

[4] F. C. Webb, August 24, 1860, *Joint Committee Report*, 267.
[5] William Thomson, "Telegraph, Electric," *Encyclopedia Britannica*, 8th ed. (Edinburgh: Adam and Charles Black, 1860), *21*: 94–116, on 99n. On the causes of the failure of the 1858 cable, and its likely longer life had the operators used only small batteries and Thomson's galvanometer, see Donard de Cogan, *They Talk Along the Deep: A Global History of the Valentia Island Telegraph Cables* (Norwich: Dosanda Publications, 2016), 70–72.

could seek to escape blame for the failure of the cable and reassure the public that oceanic submarine telegraphy could, if properly pursued, be reliably accomplished.

The Atlantic Telegraph Company did not issue any public announcement when it dismissed Whitehouse on August 20, 1858, and as Thomson nursed the cable along for the next two weeks, the company kept up an optimistic front. In the first days of September, members of its board of directors even gathered at Valentia to prepare for what they hoped would be the gala opening of the line to public traffic. At that point Thomson was still defending Whitehouse to his fellow directors, but they would have none of it, answering him with a long letter detailing Whitehouse's many offenses. In the meantime, Whitehouse stewed in London over what he saw as the incompetence of the board and the injustice of his dismissal.[6]

After the cable had been silent for several days and as prospects for its quick recovery dimmed, the secretary of the company, George Saward, sent the *Times* a brief report, published on September 6, stating that because of a fault "at a point hitherto undiscovered," no intelligible signals were being received at Valentia. "Various scientific and practical electricians" were investigating the problem, he said, but "under these circumstances no time can at present be named for opening the wire to the public."[7]

Saward's letter cast a pall over planned celebrations and caused shares in the company to plummet, while the *Times* noted obliquely that "Some disagreements between the electricians and the Board of Directors have latterly existed, and these it may be presumed tend to embarrass the general proceedings."[8] Whitehouse took Saward's letter as his cue to break his public silence and initiate what became a month-long series of increasingly bitter exchanges in the press. Writing to the *Times*, he complained that the company had "studiously suppressed" any mention of his own contributions to the success of the project and now by its bungling was putting that success in jeopardy. The directors had ignored his repeated warnings about what he said was a worsening fault in the harbor at Valentia, and when he had taken the initiative to order the cable underrun and the fault repaired, they had rewarded him with summary dismissal. Yet Whitehouse claimed that only after this repair had been effected had it been possible for President Buchanan's reply to the Queen

---

[6] Atlantic Telegraph Company to Thomson, c. August 24, 1858, in Thompson, *Kelvin*, 1: 370–72; Wildman Whitehouse, *Recent Correspondence between Mr. Wildman Whitehouse and the Atlantic Telegraph Company* (London: Bradbury & Evans, 1858).

[7] George Saward, "The Atlantic Cable" (letter), *Times* (September 6, 1858), 7.

[8] "Money-Market and City Intelligence," *Times* (September 7, 1858), 5.

to be "received at Valentia, and recorded under my own patent."[9] Now the fault had evidently reappeared, Whitehouse said, and he warned that if the company did not mend its ways, as well as the injured cable in the harbor, all might soon be lost.

Thomson read Whitehouse's letter with consternation. His own careful tests at Valentia had convinced him there was no serious fault within the harbor and that the source of the problem must lie some 200 or 300 miles west, at or near the drop-off into the deep waters of the Atlantic. He knew Whitehouse did not agree, but put it down to an honest difference of opinion, reflecting Whitehouse's idiosyncratic methods of electrical measurement. More troubling was Whitehouse's claim that the president's message had been received and recorded on his patented relays. After making careful inquiries among the clerks at Valentia and examining the log of incoming signals, Thomson had ascertained that the president's message, and virtually all others, had been received at Valentia on his own mirror galvanometer, and that the clerks had employed Whitehouse's relays only to record the dots and dashes on paper tapes, using a finger key and a local battery.[10] Moreover, Whitehouse must have known this, as Thomson reminded him in an anguished response to his letter to the *Times*. "How could you possibly have allowed yourself so far to forget the facts," Thomson asked, "as to say that the President's reply had been received and recorded under your patent?"[11]

Prompted in part by Thomson, Saward wrote to the *Times* from Valentia to say that Whitehouse's letter had been "read with much astonishment at this place." The greater part of it was "grossly untrue," he said, while the rest was "so disingenuously garbled" as to further justify Whitehouse's dismissal. Saward promised to give a fuller response once he was able to consult company records in London.[12] The fight was clearly getting nasty.

One of Whitehouse's critics later observed that he had a facile pen and "much power of language," but was also "imbued with a fatal passion of a love of rushing wildly into print."[13] Whitehouse certainly did not hold

---

[9] Edward Orange Wildman Whitehouse, "The Atlantic Telegraph" (letter), *Times* (September 7, 1858), 7. Whitehouse addressed this from the Royal Institution in London, as he did most of the letters he sent to the press that fall.

[10] Thomson to Directors, in Thompson, *Kelvin*, *1*: 374; Thomson, December 17, 1859, *Joint Committee Report*, 122. As Thomson noted, 119, some early messages, including the very first one, were also received on a sensitive needle galvanometer made by W. T. Henley.

[11] Thomson to Whitehouse (draft), c. September 9, 1858, in Thompson, *Kelvin*, *1*: 375.

[12] George Saward, "The Atlantic Telegraph" (letter), *Times* (September 15, 1858), 7.

[13] "A Telegraph Engineer and Practical Electrician" (letter), *Engineer* (November 5, 1858), 6: 355. This was in fact written by W. H. Preece; see the manuscript in NAEST 17/12.8, IET.

back now: in the space of just over three weeks he sent the London newspapers a series of letters totaling nearly 20,000 words, most of which he also had printed separately as pamphlets. The *Mechanics' Magazine* jibed that it was "a great misfortune for Mr. Whitehouse that his health, which was not sufficiently good to enable him to do his duty by accompanying the late telegraph expeditions, has proved good enough to enable him to spend the whole of his time since in writing tremendously long and laborious documents in his own honour, and to the dishonour of everyone then associated with him."[14]

On September 15 Whitehouse sent the *Times* a relatively short letter defending himself against Saward's charges and calling for an impartial investigation of the events at Valentia. He also offered Thomson a partial apology for having "unintentionally deprived a friend of credit due to him," saying now that the president's message had been "received and recorded at Valentia" not solely on his patented relay but "by the combined use of instruments of Professor Thomson and my own."[15] This was technically true, in that the message had been received from across the Atlantic on Thomson's mirror galvanometer and then recorded from across the room on Whitehouse's relay, but also deeply misleading. Thomson wrote privately to Whitehouse to express his dismay; what separated them now, he said, was not "a question of a difference of opinion" but of "truth and honour."[16]

Whitehouse followed with a letter that filled five columns of the September 20 edition of the *Morning Chronicle* and also appeared as a separate pamphlet.[17] "Dismissing every other feeling than the interests of Science," he here offered "an outline of the glorious enterprise as it has developed itself under the eyes of one of its earliest promoters." Not surprisingly, he praised his own contributions to the project while attacking the leaders of the Atlantic Telegraph Company for their haste and negligence, particularly in not having immediately laid the heavy shore end of the cable at Valentia. "Having a success which startled the whole world laid at their feet," he said, "they know not how to use it, but by

---

[14] "Mr. Whitehouse and His Injuries," *Mechanics' Magazine* (October 9, 1858), 69: 342–43, on 342.

[15] Edward Orange Wildman Whitehouse, "The Atlantic Telegraph" (letter), *Times* (September 16, 1858), 7.

[16] Thomson to Whitehouse (draft), September 23, 1858, in Thompson, *Kelvin, 1*: 376–77, on 376.

[17] Edward Orange Wildman Whitehouse, "The Atlantic Telegraph Cable" (letter), *Morning Chronicle* (September 20, 1858), also published as Whitehouse, *The Atlantic Telegraph: The Rise, Progress, and Development of Its Electrical Department* (London: Bradbury & Evans, 1858).

apathy and incompetence suffer it to elude their grasp, and the grand enterprise of the day to fall into collapse."[18]

By this time the directors had clearly decided to throw as much of the blame for the failure as they could onto their erstwhile electrician, and at their behest Saward issued a lengthy statement, addressed to the share-holders and published in the *Morning Chronicle* on September 22 and in the *Times* two days later, in which he took Whitehouse apart limb by limb.[19] A sampling of the subheadings inserted by the *Morning Chronicle* aptly conveys the flavor of Saward's statement: "That Mr. Whitehouse Failed in His Duty," "That Professor Thomson Did It for Him," "That Mr. Whitehouse Mistook the Locality of the Defect," "That He Kept the Board in Ignorance," "That He Misled Them," "That He Was in Error," "That He Ought to Have Known Better," "That His Apparatus Was Faulty," and "That Much Money was Lost Through Him." In short, Whitehouse was incompetent, insubordinate, vainglorious, and dishon-est. The real hero of the story, on Saward's telling, was Thomson, who had stepped in to sail on the expeditions when Whitehouse was "indis-posed" and had on his own initiative developed the mirror galvanometer that had served so well when Whitehouse's costly and cumbersome instruments had proved unworkable. Whitehouse lacked the skill and knowledge even properly to locate the fault that had silenced the cable, Saward said, or the good sense to leave the matter to abler hands.

Saward did not indict Whitehouse's induction coils as culprits in the failure of the cable; that was left to Cromwell Fleetwood Varley in a report that the Atlantic Telegraph Company issued alongside Saward's statement.[20] The chief electrician for the Electric and International Telegraph Company and a pioneer in the development of electrical methods for locating faults, Varley had been called to Valentia in early September to assist in efforts to resuscitate the cable. His resistance measurements and other tests showed that the main fault must lie between 240 and 300 miles west of Valentia, likely in waters too deep for the cable to be hauled up and repaired. He also stated that "It is not at all improbable that the powerful currents from the large induction coils have impaired the insulation, and that had more moderate power been used the cable would still have been capable of transmitting messages." To satisfy himself on this point, Varley made a small nick in the insulation of a gutta-percha-covered wire, immersed it in a jug of seawater, and then

[18] Whitehouse, *Rise*, 28.
[19] George Saward, "The Atlantic Telegraph" (letter), *Morning Chronicle* (September 22, 1858), and in *Times* (September 24, 1858), 7.
[20] C. F. Varley, "Report on the State of the Atlantic Telegraph Cable," *Morning Chronicle* (September 22, 1858).

subjected the wire to a series of jolts from Whitehouse's coils. The results were dramatic: the discharges "burnt a hole in the gutta percha under the water, half an inch in length, and the burnt gutta percha came floating up to the surface." It was evident, he said, that "when there are imperfections in the insulating covering, there is a very great danger arising from using such intense currents."[21] Varley's account of his simple but striking experiment was widely reprinted and did much to convince the public that whatever else may have contributed to weakening the cable, it was jolts from Whitehouse's coils that finally did it in.[22]

Whitehouse responded to Saward in a long, florid, and remarkably vehement "Reply to the Statement of the Directors" that appeared in the *Morning Chronicle* on September 28 and in the *Daily News* the next day.[23] "Unconscious of the blow which has been secretly endeavoured to be struck against my character and conduct – engaged in peaceful and philosophical pursuits – unwilling to enter the arena of hostile controversy, and desirous only to deprecate an obloquy undeserved," he now came forward, he said, "openly and boldly to answer all the accusations that have been untruly and unjustly brought against me, and to declare that if science (as has often been the case) must have its victims, I will not fall the butt of unrefuted slander and detraction." He had been charged, he said, with incompetency, duplicity, and disobedience to the directors of the Atlantic Telegraph Company. "These charges I retort upon themselves: the incompetency was theirs – the duplicity was theirs," and his supposed disobedience had, he said, been in service to a higher duty. In bitter and injured tones, and presenting himself very much as a disinterested man of science, he attacked not just Field's "frantic fooleries" and the lax leadership of the board of directors but also the methods of electrical measurement that had led Thomson and Varley to conclude – quite wrongly, Whitehouse said – that the fatal fault must lie far beyond Valentia harbor. He also continued to insist that readable signals had been received at Valentia directly on his relays and strongly

---

[21] Varley, "Report," *Morning Chronicle* (September 22, 1858).

[22] Varley's report was also published in the *Times* (September 22, 1858), 7; *Daily News* (September 22, 1858); *Birmingham Daily Post* (September 23, 1858); *Lloyd's Weekly* (September 26, 1858); and *Aberdeen Journal* (September 29, 1858); excerpts highlighting the damaging effects of Whitehouse's coils appeared in *Liverpool Mercury* (September 23, 1858), *Newcastle Courant* (September 24, 1858), and *Manchester Times* (September 25, 1858).

[23] Edward Orange Wildman Whitehouse, "The Atlantic Cable. Mr. Whitehouse's Reply to the Statement of the Directors of the Atlantic Telegraph Company," *Morning Chronicle* (September 28, 1858); also in *Daily News* (September 29, 1858) and *Engineer* (October 15, 1858), 6: 287–88, and as Edward Orange Wildman Whitehouse, *Reply to the Statement of the Directors of the Atlantic Telegraph Company* (London: Bradbury & Evans, 1858), a transcription of which can be found on the Atlantic-cable.com website.

implied, though he did not quite state, that complete messages had also been received in that way.[24]

Though somewhat delayed, Thomson's response to Whitehouse was direct and forceful. Exhausted by his struggles with the cable, he had left Valentia in late September to spend a few weeks recuperating on the Isle of Arran. London newspapers were slow to reach him there, but when on October 14 Thomson finally saw Whitehouse's long reply to the directors, it touched a nerve. He immediately dashed off a letter to the *Morning Chronicle* stating that Whitehouse's characterization of Thomson's reasoning and conclusions about the location of the fault was "entirely incorrect," that Whitehouse's own treatment of the question was "absurd," and that his whole approach to such matters consisted simply of "conjecture unsupported by experiment." Moreover, Whitehouse's statements about the instruments used to receive messages at Valentia were, he said, filled with "gross perversions and misrepresentations."[25] Thomson had been for a time one of Whitehouse's closest friends and strongest allies in the scientific world; now, within the space of a few weeks, the breach between them had become complete.

Whitehouse's swan song as an active electrical experimenter came in late September at the Leeds meeting of the British Association for the Advancement of Science. There, just as his "Reply" was appearing in the newspapers, he presented to the mathematical and physical section of the association no fewer than six papers on topics ranging from techniques for locating faults to "A Few Thoughts on the Size of Conductors for Submarine Circuits."[26] In the latter, he said that "from having let fall an opinion that I thought it possible that a submarine conductor might be too large for practical use, it has been erroneously supposed and stated that I held the converse opinion." He now said that rather than arguing that thinner conductors were better, he had merely suggested, based on his experiments on retardation on multiconductor cables, that the case for using thicker conductors was not as clear-cut as was sometimes claimed, and that the whole question of the combined effects of resistance and induction on telegraph signals deserved further study. He closed with the wistful remark that "Present leisure from engrossing details of official duties will, I hope, afford me the opportunity of examining this subject experimentally."[27]

---

[24] Whitehouse, "Reply," *Morning Chronicle* (September 28, 1858).

[25] William Thomson, "The Atlantic Cable. Professor Thomson in Reply to Mr. Whitehouse" (letter), *Morning Chronicle* (October 18, 1858).

[26] On Whitehouse's presence at the British Association meeting, see "The British Association at Leeds," *Morning Chronicle* (September 30, 1858).

[27] Wildman Whitehouse, "A Few Thoughts on the Size of Conductors for Submarine Circuits," *Engineer* (October 29, 1858), 6: 329.

Whitehouse sent copies of his British Association papers to the *Mechanics' Magazine*, the *Engineer*, the *Athenæum*, and the *Civil Engineer and Architect's Journal*, where they were published over the next few months.[28] They would be the last of his papers on electrical topics ever to appear in print. Significantly, when the full *British Association Report* was published the next year, it did not include even abstracts of Whitehouse's papers but listed only an overall title: "Contributions on the Submarine Telegraph."[29]

Whitehouse's claim at Leeds that he had never said thin cable conductors were better than thick ones was not allowed to pass unchallenged, and it set off a revealing exchange in the *Engineer* with someone who signed himself "A Telegraph Engineer and Practical Electrician." This was in fact W. H. Preece, then a young engineer with the Electric and International Telegraph Company and later the chief engineer of the Post Office telegraph system.[30] Like most others, Preece assumed that Whitehouse had written the "manifesto" that the Atlantic Telegraph Company had issued in July 1857, and he quoted its categorical statement "that large coated wires used beneath the water or the earth are worse conductors, so far as velocity of transmission is concerned, than small ones" as proof that Whitehouse had indeed held "what is considered by all electricians to be an absurd and ridiculous theory."[31] Drawing no immediate response, Preece extended his attack two weeks later, arguing that Whitehouse had misunderstood his own experiments on retardation and that both proper theory and actual experience proved – and "our leading practical electricians" all knew – that thicker conductors produce less retardation.[32]

[28] All six of Whitehouse's British Association papers were published in *Engineer* (October 29, 1858), 6: 329; in *Mechanics' Magazine* (October 16, October 23, and October 30, 1858), 69: 366–67, 395–96, and 417–18); and in *Civil Engineer and Architect's Journal* (December 1858), 21: 408–10. "On Some of the Difficulties in Testing Submarine Cables" also appeared in the *Athenæum* (October 30, 1858), 553.

[29] Wildman Whitehouse, "Contributions on the Submarine Telegraph" (title only), *BA Report* (1858), 25.

[30] [W. H. Preece], "The Atlantic Telegraph" (letter), *Engineer* (November 5, 1858), 6: 355; signed "A Telegraph Engineer and Practical Electrician." The manuscript of this and other letters from "A Telegraph Engineer" are among Preece's papers in NAEST 17/ 12.1–12.15, IET. On Preece, see E. C. Baker, *Sir William Preece, F.R.S., Victorian Engineer Extraordinary* (London: Hutchinson, 1976). Preece had been drawn into telegraphy by a family connection: his sister Margaret married Latimer Clark in 1854; after they divorced in 1861, she married another cable engineer, F. R. Window.

[31] [R. J. Mann], *The Atlantic Telegraph: A History of Preliminary Experimental Proceedings, and a Descriptive Account of the Present State & Prospects of the Undertaking, Published by Order of the Directors of the Company*, (London: Jarrold and Sons, 1857), 26; [Preece], "Atlantic Telegraph," *Engineer* (November 5, 1858). On Mann's authorship, see R. J. Mann to Latimer Clark, November 12, 1880, WC-NYPL.

[32] [W. H. Preece], "The Atlantic Telegraph" (letter), *Engineer* (November 19, 1858), 6: 391–92, signed "A Telegraph Engineer and Practical Electrician"

Whitehouse finally responded the next week, denying he had written the 1857 pamphlet or had ever said that thicker cable conductors produce more retardation – a claim that was hard to square with his statement in the *Athenæum* in 1856 that "it is now proved that by any considerable increase in the size of a long *submarine conductor*, we positively increase the difficulty of giving telegraphic signals and diminish the speed of their transmission."[33] He renewed his old attack on the law of squares, citing again Thomson's 1855 remark that an Atlantic cable no thicker than the one laid across the Black Sea would show thirty-six times the retardation and said this theory-based claim had been roundly refuted not only by his own experiments but by the signaling rates actually achieved on the Atlantic cable. "In this manner," he said, "does its brief and barely utilised existence triumphantly confute and set at nought the unnatural constraint to which mere rigid theorists would subject the operations of nature's laws"; "men of figures" were thus again taught, "if they would but receive it in a genial manner, that 'there are more things in heaven and earth than are dreamt of in their philosophy.'"[34] Whitehouse's flirtation with presenting himself as a man of science engaged simply in "philo-sophical pursuits" had passed, and he again positioned himself as an opponent of theory-led science and as a follower simply of experiment and practical experience.

Telegraph engineers, or at least "A Telegraph Engineer," would have none of it; they were intent on ruling Whitehouse firmly out of the practical camp as well as the scientific one. "When Mr. Whitehouse was high in the counsels of the Atlantic Telegraph Company," Preece wrote, "his errone-ous views of the laws of induction and conduction were a source of much concern to practical electricians, because it was [feared] that his ignorance of practical electricity would damage, if not destroy, the grand experiment of ocean telegraphy. Now, however, that he no longer sways the electrical rod in Broad-street, the engineering profession can afford to smile at his fallacious opinions."[35] Britain's telegraph engineers had a duty to go beyond such private smiles, however; to protect their professional reputa-tions, they must, Preece declared, expose Whitehouse's errors before the public and make it clear that "the profession to which I belong by no means coincide with the wild theories and opinions that have been circulated by

---

[33] Wildman Whitehouse, "The Atlantic Telegraph" (letter), *Athenæum* (October 11, 1856), 1247.

[34] Wildman Whitehouse, "The Atlantic Telegraph" (letter), *Engineer* (November 26, 1858), 6: 410.

[35] [Preece], "Atlantic Telegraph," *Engineer* (November 5, 1858). The word rendered here as "feared" was published as "found" in the *Engineer*, but comparison with the original manuscript, NAEST 17/12.8, IET, shows this to be a mistranscription. "Broad-street" was the location of the offices of the Atlantic Telegraph Company.

him."[36] Whitehouse had, in effect, to be publicly cast out. Denying that he was "actuated by any personal animosity" to Whitehouse, Preece said that he simply wished, "for the credit of the profession to which I have the honour to belong, to let the public, at home and abroad, know that Mr. Whitehouse is not our mouthpiece; that his theories are considered fallacious, and that he is in many respects opposed to the received opinions of our leading practical electricians."[37]

In an especially telling observation – one that points up how important Whitehouse's idiosyncratic measuring practices were in marking him as out of step with both scientists and practical men – Preece noted that "in the perusal of Mr. Whitehouse's papers, those most essential and important terms, 'quantity,' 'intensity,' and 'resistance,' are almost ... lost sight of. It cannot be that Mr. Whitehouse is ignorant of these properties; but it certainly leads to the opinion that he does not attach sufficient importance to what are considered by all practical electricians as the bone and sinew of the electric current."[38] As the measuring practices of electrical scientists and practical men converged in the late 1850s and early 1860s toward a focus on "quantity," "intensity" (or "tension"), and resistance, as defined in Ohm's law and reified by embodiment in appropriate standards and measuring instruments, those like Whitehouse who eschewed such quantities and instruments and insisted on instead defining their own electrical "values" found themselves squeezed out. They were excluded, even expelled, from the realms of both science and practice, and left, as Whitehouse soon discovered, with no place to stand in the new electrical world then emerging.

Toward the end of his "Reply to the Statement of the Directors," Whitehouse said he was "a supplanted man."[39] He had never been more than a peripheral member of either the scientific or engineering communities, and by October 1858 he had effectively been pushed out of both. By disposing of Whitehouse, proponents of oceanic submarine telegraphy sought to draw a line under the failure of the Atlantic cable and open a new chapter for their enterprise. They would, however, find this more difficult than they had hoped.

The Red Sea Debacle

While the fight over the failure of the Atlantic cable was still playing out, an even costlier telegraphic debacle was taking shape. When the Indian

---

[36] [Preece], "Atlantic Telegraph," *Engineer* (November 5, 1858).

[37] [W. H. Preece], "The Atlantic Cable" (letter), *Engineer* (December 3, 1858), 6: 431, signed "A Telegraph Engineer and Practical Electrician."

[38] [Preece], "The Atlantic Telegraph," *Engineer* (November 19, 1858).

[39] Whitehouse, "Reply," *Morning Chronicle* (September 28, 1858).

Rebellion broke out in May 1857, it took six weeks for word of it to reach London. This was scarcely acceptable in the new age of electrical communications, and critics soon declared it to be a national and imperial imperative that Britain establish a telegraphic line to India as quickly as possible.[40] After heated debates over what route such a line should take and how it might be financed, the British government agreed in July 1858 to provide generous backing for a project to lay a chain of cables down the Red Sea and on to India. Even before the final splice was made in February 1860, however, faults began to appear and communication was interrupted. Sections of the line proceeded to fail one by one, but because of the way the guarantee had been written, the British government remained on the hook to pay the shareholders of the cable company tens of thousands of pounds a year for decades to come. It was hardly a result calculated to build public trust in submarine telegraphy.

The Red Sea cable project got its start in 1854 when Lionel Gisborne, a well-connected young British engineer, approached Ottoman officials in Constantinople about obtaining a concession to allow him to lay cables connecting various points in the Ottoman Empire.[41] The sultan and his advisers mainly wished to secure a link to Egypt, believing it would give them more control over the pasha there, and in return for a promise from Gisborne to lay a cable from the Dardanelles to Alexandria, they were willing to grant him permission to lay cables and establish telegraph stations down the Red Sea and along the coast of the Arabian Peninsula. With support from the British ambassador, Gisborne secured the last of the required concessions in December 1856 and soon arranged to transfer them, reportedly for £15,000, to a group headed by John C. Marshman, contingent on Marshman's Red Sea Telegraph Company raising enough money actually to make and lay the cables. Financing the project was complicated by a proposal from a rival British company to erect and operate an overland telegraph line down the Euphrates Valley. Both the Treasury and the East India Company favored this overland route, and intense rivalry ensued as partisans of each scheme pointed out the other's weaknesses: "wild Arabs" were sure to attack the Euphrates Valley line, while coral outcrops in the Red Sea would cut to pieces any cable laid there. Marshman insisted there would

---

[40] "Telegraphic Communication with India," *Daily News* (July 20, 1857); "A Telegraphic Line to India an Immediate Want," *Leeds Mercury* (August 13, 1857).

[41] Lionel Gisborne was not related to Frederic Gisborne, the engineer who had first interested Cyrus Field in extending a cable to Newfoundland. On the origins of Lionel Gisborne's Red Sea and India project, see an essay by one of his brothers: [Francis Gisborne], "Telegraphic Communication with India," *Cambridge Essays* (London: J. W. Parker, 1857), 106–24, esp. 110.

surely be enough traffic to support both routes, but the endless sniping no doubt delayed the adoption of either one.[42]

After the Atlantic cable snapped during the first attempt to lay it in August 1857, at the height of the Indian Rebellion, the Red Sea company offered to buy the unused portion of the Atlantic cable and lay it down the Red Sea just as soon as the *Agamemnon* could get there.[43] The Atlantic Telegraph Company professed to be open to the idea as long as it could be assured of getting enough new cable to try again to span the Atlantic the next summer, but the deal soon fell apart amid reports that heat had damaged the gutta-percha insulation of the Atlantic cable, raising doubts about how well it would stand up to the hot conditions of the Red Sea.[44]

At this juncture the Ottoman government announced that instead of letting a British company build and operate a landline down the Euphrates Valley, it would take the job into its own hands.[45] The East India Company was willing to rely on such a Turkish line, along with a planned short cable down the Persian Gulf and along the coast to Karachi (finally completed in 1864), but amid public calls for a route that would be wholly under British control, Marshman renewed his pitch for the Red Sea cables.[46] In the spring of 1858, his group made a series of deals with Gisborne and the cable maker R. S. Newall. The key to the concessions Gisborne had been granted by the sultan, and that were now held by the Red Sea Telegraph Company, was his promise to lay a cable from the Dardanelles to Alexandria. He and Newall had already worked

---

[42] On sniping between advocates of the alternative routes delaying the adoption of either, see "Money-Market and City Intelligence," *Times* (May 4, 1858), 10; see also John Marshman, "The Red Sea Telegraph" (letter), *Times* (September 3, 1857), 8.

[43] John C. Marshman to R. D. Mangles, August 11, 1857, and to H. J. Baillie, March 6, 1858, in J. C. Marshman, *Telegraph to India, Extracts from Correspondence Relative to the Establishment of Telegraphic Communication between Great Britain and India, China and the Colonies* (London: J. E. Adlard, 1858), 4–6, 16–17.

[44] On the scheme to purchase the unused length of the Atlantic cable and lay it in the Red Sea, and on the reports of heat damage that helped scuttle the plan, see "Money-Market and City Intelligence," *Times* (August 31, 1857), 4, and (September 4, 1857), 8. Field quickly denied that there had been any such heat damage and said the Atlantic Telegraph Company had considered selling its unused length of cable to the Red Sea company only on the understanding that a new length of cable could be manufactured in time to lay it in the Atlantic the following spring; see his letter in "Money-Market and City Intelligence," *Times* (September 10, 1858), 9.

[45] Yakup Bektas, "The Sultan's Messenger: Cultural Constructions of Ottoman Telegraphy, 1847–1880," *Technology and Culture* (2000) 41: 669–96.

[46] Christina Phelps Harris, "The Persian Gulf Submarine Telegraph of 1864," *Geographical Journal* (1969) 135: 169–90. Latimer Clark and Charles Tilston Bright served as engineers for the project. The cover of this book depicts Bright directing the perilous landing of this cable in April 1864 across the mudflats of the Shatt al-Arab at the head of the Persian Gulf; see "The Indo-European Telegraph," *Illustrated London News* (July 8, 1865) 47: 21–22.

together to lay several short cables in the eastern Mediterranean, including ones connecting the Dardanelles to Candia (Heraklion) on Crete. Now Newall undertook to lay the final link from Candia to Alexandria, thus preserving the all-important concession. In return, Marshman's company agreed to give Newall the contract to make and lay its cables down the Red Sea and on to Karachi, as well as to install Gisborne as its chief engineer on the project.[47]

With these agreements in place, Marshman and other proponents of the Red Sea route (including Gisborne and Newall) met with the prime minister, Lord Derby, in early June 1858 to make their case. They must have been persuasive, for the government soon agreed to provide the backing the Red Sea company said it would need in order to attract investors and build the line: a 4½ percent return on a capital of £800,000, guaranteed for fifty years, and conditioned on satisfactory completion of each section of the cable.[48] Backed by this guarantee, and amid the optimism generated by the apparent success of the Atlantic cable, the shares sold quickly. Marshman was happy to report to the Treasury on August 27 that his group, now reorganized as the Red Sea and India Telegraph Company, had succeeded in securing the required capital and would soon sign a contract with Newall to lay the first lengths of cable.[49]

Not everyone was happy with the deal. Lord Stanley, the Secretary of State for India, questioned why the Red Sea company proposed to award its contract to Newall without seeking other bids, especially since Glass, Elliot and Company, Newall's chief competitor in the business, had said it could do the job for £100,000 less.[50] Marshman was forced to explain that the crucial Turkish concession would lapse unless the long-promised link from the Dardanelles to Alexandria was completed by the end of the year; since Newall's firm had undertaken to do so from its own resources in time to save the concession, the Red Sea company felt bound to give it the contract.[51] As for Glass, Elliot's complaint that it had been unfairly shut out of the project, the Treasury Office replied blandly that it would leave all such matters to the discretion of the Red Sea company.[52] The British government in fact chose to exercise remarkably

---

[47] J. C. Marshman to W. H. Stephenson, September 28, 1858, in *Red Sea Contract*, 12.
[48] "Court Circular," *Times* (June 8, 1858), 9; "Red Sea Telegraph Company," *Times* (August 2, 1858), 6; *Joint Committee Report*, x.
[49] J. C. Marshman to the Secretary of the Treasury, August 27, 1858, *Red Sea Contract*, 5–6.
[50] J. Cosmo Melvill to George Hamilton, September 23, 1858, *Red Sea Contract*, 11; Glass, Elliot to Lords Commissioners of the Treasury, June 20, 1858, in *Further Corr. 269*, 5–6.
[51] J. C. Marshman to W. H. Stephenson, September 28, 1858, *Red Sea Contract*, 12
[52] Treasury Minute, August 4, 1858, *Further Corr. 269*, 6.

little direct supervision over a project on which it was putting such large sums at risk.

The terms of the contract gave Newall a very free hand; Willoughby Smith of the Gutta Percha Company later said "it was purely a contractor's 'go as you please' sort of arrangement," of which "full advantage was taken."[53] As chief engineer for the Red Sea and India Telegraph Company, Gisborne was allowed to inspect and test the cable during its manufacture and laying but "not to interfere with the progress of the work"; the company's electrician, Werner Siemens of the Berlin firm of Siemens and Halske (Newall's partners in other ventures), operated under similar restrictions.[54] There were consequently few checks on Newall's handling of the project, and Smith for one found much to criticize, from badly made joints to the mistaken use of lengths of core with substandard insulation.[55] Moreover, an unusual clause in the contract allowed Newall to keep "any surplus cable remaining after the final completion of the contract."[56] This gave him an incentive to lay the cable as tautly as possible, potentially straining its joints and stretching both its conducting and armoring wires, while leaving little slack to fill the hollows in any uneven stretches of seabed.

Before he began laying cables down the Red Sea, Newall turned to a piece of unfinished business: connecting Candia to Alexandria and so completing the link from Constantinople on which Gisborne's original concession had been predicated. Newall tried repeatedly in 1858 and 1859 to lay this cable, starting with a light unarmored core covered only with tarred hemp. Whether this was a legitimate experiment in cable design or simply an attempt to do the job on the cheap is unclear, but in any case the line snapped during laying.[57] Newall tried twice more with cables armored in various ways but had no better luck and in fact never managed to deliver on the promise that had been the ostensible reason for

---

[53] Willoughby Smith, *The Rise and Extension of Submarine Telegraphy* (London: J. S. Virtue, 1891), 67.

[54] "Contract with Messrs. R. S. Newall & Co.," October 23, 1858, *Red Sea Contract*, 14. On Siemens and Halske's partnership with Newall, see Wilfried Feldenkirchen, *Werner Siemens: Inventor and International Entrepreneur* (Columbus: Ohio State University Press, 1994), 70–72. In his August 24, 1860, testimony to the Joint Committee, F. C. Webb noted the extreme restrictions Newall put on the engineers who represented the Red Sea company; *Joint Committee Report*, 268, 270.

[55] Smith, *Rise and Extension*, 67–68.

[56] "Contract with Messrs. R. S. Newall and Co.," October 23, 1858, *Red Sea Contract*, 13–15, on 14.

[57] On the attempts to lay a cable from Alexandria to Candia, see Smith, *Rise and Extension*, 63–64, and Fleeming Jenkin's letters to his wife from May and June 1859, in R. L. Stevenson, "Memoir of Fleeming Jenkin," in Sidney Colvin and J. A. Ewing, eds., *Papers Literary, Scientific, &c., by the late Fleeming Jenkin*, 2 vols. (London: Longmans, Green, and Co., 1887), *1*: xi–cliv, on xcv–xcviii.

awarding him the Red Sea contract. The Ottoman authorities repeatedly granted Newall extensions to complete the line, but in the fall of 1859, under pressure from the British government, they withdrew the clause of their concession that had given him the exclusive right to land cables on the coast of Egypt. Amid spluttering protests, Newall gave up the effort.[58] By then, however, he was facing much bigger problems.

To avoid the retardation effects that slowed signaling on very long cables, the Red Sea company originally planned to break its line from Suez to Karachi into eight segments, none longer than 470 nautical miles.[59] Surveys of coastal conditions and concerns about poor security ashore later led the company to drop two intermediate stations and shift two others, resulting in stages as long as 629 and 718 nautical miles.[60] Nonetheless, Newall remained confident that, with proper sending and receiving instruments, all of the sections would still be able to carry the ten words per minute specified in his contract.

As the Red Sea company was completing its landline from Alexandria to Suez in April 1859, Newall was preparing to lay the first lengths of cable from Suez to Aden, with intermediate stations at Kosseir (El Qosier) and Suakin. The laying went smoothly, watched over by Gisborne and Siemens as the engineer and electrician, respectively, for the Red Sea company, though they had little power to intervene in Newall's operations. After landing the cable at Aden on May 28, Newall and the rest of the cable-laying crew boarded the *Alma*, a P & O passenger liner, for the return voyage to Suez. On its first night out from Aden, however, the ship ran aground on a coral reef, stranding its passengers and crew on a tiny islet. While Gisborne took charge of doling out the meager supply of water (and bottled beer) to the sun-baked survivors, Newall set off in an open boat to seek help. It was a harrowing experience, but after three days all

---

[58] See the extended exchanges in *Correspondence Respecting the Dardanelles and Alexandria Telegraph*, British Parliamentary Papers, 1863, LXXIII.601 (London: Her Majesty's Stationery Office, 1863), 1–13; on the grant to Newall of the exclusive right to land cables on the Mediterranean coast of Egypt or any other Ottoman territory, see *Further Corr. 269*, 116. On the cancellation of those rights and Newall's abandonment of the project, see Newall's testimony to the Joint Committee, August 10, 1860, *Joint Committee Report*, 256.

[59] "Contract with Messrs. R. S. Newall & Co., dated 23 October 1858," *Red Sea Contract*, 13–16.

[60] The length of the cables as finally laid is given in *Joint Committee Report*, 518. On the route, see Captain W. J. J. Pullen, "The Red Sea Telegraph Line," *Times* (May 21, 1858), 12; on concerns about security ashore at Maculla, see F. C. Webb, "Old Cable Stories Retold–VII," *Elec.* (June 5, 1885) *14*: 65–66, on 66. In June 1858 about twenty Europeans, including the French consul and British vice-consul, were killed at Jeddah, contributing to the abandonment of plans to establish a cable station there; "Massacre at Jeddah," *Times* (July 15, 1858), 9.

were saved, and Newall was hailed as a hero.[61] It was in some ways the high point of his Red Sea experience.

Apart from some minor repairs at Kosseir in July, the Red Sea cables initially worked well. They greatly sped communications between Aden and Alexandria and earned a modest profit, though the lack of cable connections at either end meant that messages bound for Europe or India still had to complete their journeys by ship[62] (Figure 3.1).

Newall's contract had called for him to complete the remaining links to Karachi by the end of 1859, but in the event he did not begin laying them until January 1860. Starting from Karachi, he first spanned the Gulf of Oman to Muscat, where the expedition was greeted by the local imam. Their next stop was Hallani (Al-Hallaniyah) in the Khuriya Muriya Islands, a bleak little archipelago the imam had ceded to Queen Victoria in 1854 for use as a cable station. It was so lonely a spot that the captain of one of the ships on the laying expedition reportedly said that any clerk who had served a year there would deserve a pension of £1000 a year for life.[63] Hallani was among the first of the many isolated postings that would lead later cable clerks to dub their fraternal organization the "Exiles Club."[64]

Newall and the laying expedition continued down the coast of the Arabian Peninsula until just past the port of Maculla (Mukalla). There they fished up a length of cable they had laid eastward from Aden some months earlier and on February 12 made their final splice, thus for the first time putting India in direct touch with Egypt – or so they thought. In the same letter to the *Times* in which it confirmed the completion of the chain of cables from Karachi to Aden, the Red Sea and India Telegraph Company had to report "an interruption between Suakin and Aden (on the Red Sea line), which prevents the directors from announcing the establishment of a complete communication between Alexandria and India, which, however, they hope to be able to do in the course of a few days."[65] On arriving in Aden after making his final splice, Newall had answered a toast to his health by declaring that "the cable would last for

[61] "Loss of the *Alma*," *Daily News* (July 7, 1859). Werner Siemens, who was aboard the *Alma*, recounted the shipwreck and rescue in *Inventor and Entrepreneur: Recollections of Werner von Siemens* (London: Lund Humphries, 1966), 138–44.
[62] "Alpha," "Red Sea Telegraph" (letter), *Times* (August 24, 1860), 8.
[63] Webb, "Old Cable Stories Retold–VII," *Elec.* (October 16, 1885) *15*: 428–29, on 428.
[64] On the Exiles Club, established by staff of the Eastern Telegraph Company in the late nineteenth century, see Jenny Rose Lee, "Empire, Modernity, and Design: Visual Culture and Cable & Wireless' Corporate Identities, 1924–1955," PhD dissertation, University of Exeter, 2014; and Hugh Barty-King, *Girdle Round the Earth: The Story of Cable and Wireless* (London: Heinemann, 1979), 120.
[65] C. L. Peel, "Submarine Cables" (letter), *Times* (February 29, 1860), 9.

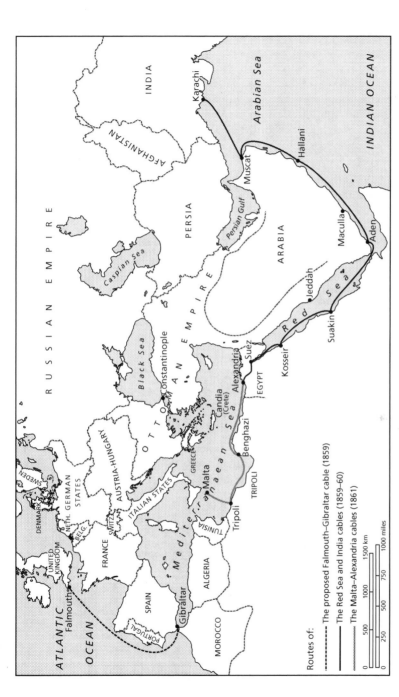

Figure 3.1 This map shows the route of the ill-fated chain of cables R. S. Newall laid in the Red Sea from Suez to Aden in 1859 and the continuation he laid in 1860 to connect Aden to Karachi. It also shows as a dashed line the route of the cable the British government proposed to lay from Falmouth to Gibraltar in 1859, as well as the route from Malta to Alexandria, via Tripoli and Benghazi, along which that cable was eventually laid in 1861. (Based on a map in Frederic John Goldsmid, *Telegraph and Travel*, 1874.)

one hundred years!"[66] Perhaps, but first it would need to be put in working order. Since he was on the spot at Aden with extra cable (he had laid the cables from Karachi so tautly that he had 400 miles' worth left over), the Red Sea company authorized Newall to try to fix the faulty stretch. No sooner was one fault repaired, however, than another appeared. A few messages made it all the way through from India to Egypt in March 1860, but the line soon broke down again.[67] After Newall returned to Britain, Gisborne and a crew of engineers and electricians spent another five months tracking down and patching faults, but the line remained "continually out of order"; when they pulled lengths of cable up from the seabed, they found "the outside iron wires ... so corroded as to leave the gutta percha almost bare," while the gutta-percha itself was riddled with small holes.[68]

By August the British government was forced to acknowledge that the Red Sea cable had failed, with "little hope that it will be repaired."[69] The only section that consistently worked well was the one from Muscat to Hallani, but no one wished to send messages to or from that desolate spot.[70] The Red Sea company prepared to sue Newall for nonperformance, but in August 1860 he struck first, suing the company for nonpayment of £25,000 it had held back pending satisfactory completion of the cables and of over £106,000 for the repair work he had undertaken after the first faults appeared. The public looked forward to seeing the case hashed out in court, but after much back and forth, Newall and the company agreed early in 1861 to settle out of court for £65,000, half of Newall's original claim.[71] With that, Newall wrapped up his unhappy dealings and largely withdrew from the cable industry he had helped to pioneer.

The real losers in all of this, of course, were the British taxpayers. The guarantee the government had given the shareholders of the Red Sea company was predicated only on each *section* of the cable being shown

---

[66] Webb, "Old Cable Stories Retold–VII," 428.

[67] According to F. C. Webb, in late March "a message from me at Kurrachee to my wife in London passed right through, and was one of the few messages that ever went through from India to England in 1860"; see Webb, "Old Cable Stories Retold–VII," 429. C. L. Peel to Secretary of the Treasury, March 14, 1860, in *Further Corr. 461*, 3.

[68] "The Isthmus of Suez Canal and the Red Sea Telegraph," *Times* (August 16, 1860), 11; "Alpha," "Red Sea Telegraph" (letter), *Times* (August 24, 1860), 8. "Alpha," who had recently returned from the Red Sea, said it was generally believed that the holes were produced by the expansion of air bubbles left during the manufacture of the gutta-percha.

[69] "Money-Market and City Intelligence," *Times* (August 16, 1860), 10.

[70] The cable from Kosseir (Al-Qoseir) to Suakin also worked adequately after some early problems with its insulation, but there was little demand for communication between those two stations either; *Joint Committee Report*, 519.

[71] "Red Sea and India Telegraph," *Times* (February 1, 1861), 7.

to work satisfactorily for thirty days after it was laid down. Since each separate section had met this requirement and been accepted by the Red Sea company, the Treasury was obligated to pay 4½ percent per year for fifty years on the company's capital of £800,000, even though the chain of cables had worked all the way through for only for a few days at best. Not surprisingly, the prospect of paying out £36,000 every year for fifty years for a dead cable aroused strong criticism in the press and Parliament, not just against the ill-considered terms of the government guarantee but against the whole enterprise of submarine telegraphy.[72]

John Crawfurd, a distinguished orientalist and the president for 1861 of the geographical section of the British Association, summed up the attitude of many of his fellow countrymen at that year's meeting. "Our oceanic cables," he said, "had been total failures. We had sunk two or three millions of money, which might as well as have been thrown in the form of sovereigns into the Red Sea. Pharaoh had lost his chariots and horsemen; but this country had lost a sum amounting to 50,000 l. per annum for the next fifty years, – a monstrous sum to spend, and to spend for nothing."[73] The collapse of the Red Sea line, coming on top of the spectacular failure of the Atlantic cable, had put oceanic submarine telegraphy into a deep hole. By the time Crawfurd said these words, however, the cable industry had already begun to climb back out.

Forming the Joint Committee

The "Joint Committee to Inquire into the Construction of Submarine Telegraph Cables" had its origins in an offshoot of the Red Sea cable project – one prompted by British concerns about imperial communications and European geopolitics. As preparations went forward in the latter half of 1858 to link Alexandria telegraphically to India, it became a pressing question just how messages would first reach Alexandria. The long-promised line from Constantinople had not yet materialized – nor would it – and in any event a telegram sent from London would first have to pass through France, one or more German or Italian states, and the Austrian Empire before it even got as far as Turkey. By 1858–1859, those routes were looking increasingly insecure: Napoleon III was rattling sabers in France, agitation for independence and unification was convulsing the Italian peninsula, and tensions among the continental powers were nearing the breaking point. With the Rebellion still smoldering in

---

[72] "Money-Market and City Intelligence," *Times* (August 16, 1860), 10; "Parliamentary Intelligence," *Times* (August 20, 1860), 6.
[73] "Section E. – Geography and Ethnography," *Athenæum* (September 14, 1861), 350.

India, the British government sorely felt its lack of a secure and independent line of communication with its empire to the east.

At this juncture several groups stepped forward to offer a solution: a chain of submarine cables running from Falmouth to Gibraltar and on to Malta and Alexandria. The Great Indian Submarine Telegraph Company, recently formed to promote the use of Scottish engineer Thomas Allan's design for light unarmored cables, was the most aggressive in pushing this route. Beginning in August 1858, Allan and other leaders of the company repeatedly pitched their proposed chain of cables to the prime minister and foreign secretary, describing it as "not simply a commercial speculation" but "an engine of Government Administration" and an instrument of imperial power.[74] Emphasizing that their cables would not be subject to the "expense, delays, and espionage of the foreign lines of telegraph across the continent of Europe," advocates of the Falmouth-Gibraltar-Malta-Alexandria route framed the project as vital to Britain's national and imperial interests and as fully deserving of a government guarantee comparable to the one the Red Sea company had received.[75]

The Treasury Office repeatedly demurred, perhaps having developed second thoughts about the wisdom of the Red Sea guarantee. Allan and his allies persisted, however, noting the rising tensions on the continent in early 1859 and urging the value of "a great national line of telegraph, free from the risk of foreign interference in time of peace, and secure from foreign interruption in time of war."[76] The outbreak of the Franco-Austrian War in northern Italy at the end of April gave an added edge to such appeals, and both Whitehouse and the South Atlantic Telegraph Company added their voices to the call for the government to back the establishment of "a direct and independent line of telegraphic communication" to Alexandria via Gibraltar and Malta.[77]

[74] John Chapman (director, Great Indian Submarine Telegraph Company) to the Earl of Derby, August 9, 1858, in *Further Corr. 461*, 179; the company prospectus appeared in an advertisement, *Times* (June 21, 1858), 5.

[75] Thomas Allan to the Earl of Malmesbury, November 18, 1858, *Further Corr. 269*, 180; Edmund Hammond to George Hamilton, November 29, 1858, and "Prospectus of Great Indian Submarine Telegraph Company," *Further Corr. 269*, 180–82. On Allan and his persistent but unsuccessful efforts to promote his light cables, see Steven Roberts, "Distant Writing," distantwriting.co.uk.

[76] Chapman to Northcote, March 8, 1859, *Further Corr. 269*, 186; see also Allan to Derby, April 9, 1859, and Allan to Malmesbury, April 25, 1859, *Further Corr. 269*, 183, and Treasury minutes of February 15, 1859 and May 17, 1859, *Further Corr. 269*, 180 and 184, declining to support the proposal.

[77] William [sic] Whitehouse to Derby, May 16, 1859, *Further Corr. 269*, 186–87, and W. H. Bellamy (of the South Atlantic Telegraph Company) to Northcote, May 19, 1859, *Further Corr. 269*, 187.

On May 20 the India Office weighed in as well, forwarding to its Treasury counterparts a carefully reasoned memorandum that the Indian Railway and Telegraph Department had drawn up a few months before. "The insular position of England is at present made detrimental to her secure and independent communication by telegraph with her European and Eastern possessions," the memorandum observed, "whereas it ought to be made instrumental to that end." Cables would be the key and would free Britain from reliance on foreign powers. With the Red Sea cables set to open soon, and those from Aden to Karachi to follow, Britain would find it "unsatisfactory" to have to rely on overland lines through France, Italy, and Austria, as such links would surely be broken off at any rupture of relations, just when they were most needed. Declaring that "England should, as far as possible, be independent of all European countries for her communication with India and with her Mediterranean possessions," the author urged the government to back the immediate laying of a chain of cables to Gibraltar, Malta, and Alexandria. "The value of such a line, in a political point of view, cannot be overrated"; useful as it would no doubt be in peacetime, "in the event of war, its true value would be learnt," as it would become the backbone of Britain's strategic communications. The recent failure of the Atlantic cable aside, techniques for making and laying cables were improving so rapidly that "there can no longer be any doubt as to the practicability of such a scheme"; the only remaining question was whether it should be undertaken by a private company with government support or "by direct Government agency," perhaps as a first step toward establishing a state-run telegraph system modeled on the Post Office. In the latter case, the memorandum concluded, the government should choose a reliable contractor and "employ a cable which might be fixed upon by a committee of scientific and practical men."[78] Here was a first seed of what would become the Joint Committee, its origins clearly tied to considerations of empire and defense.

By the time this memorandum reached the Treasury Office, the wheels were already turning there. Amid the anxieties stirred up by the Franco-Austrian War, and in evident response to the proposals from Allan and others, Sir Stafford Northcote of the Treasury wrote to the eminent engineers Robert Stephenson and Sir Charles Tilston Bright on May 17 to ask what form of cable they would recommend for a line from Falmouth to Gibraltar. On receiving their replies a few days later, Northcote pushed the project ahead quickly, largely along the lines laid

---

[78] "Memorandum" from the Indian Railway and Telegraph Department, January 1859, *Further Corr. 269*, 188–89.

out in the India Office memorandum: on May 25 the Treasury told the South Atlantic company that "Her Majesty's Government has determined, after mature deliberation, to take the construction of the proposed line into it own hands," and that same day it informed the India Office that "my Lords have the satisfaction of stating that arrangements are now nearly completed for the immediate construction of a line of telegraphic communication between this country and Gibraltar."[79]

Drawing on Thomson's theory of induction effects and his "law of squares," Bright and Stephenson both recommended using a core much thicker than that of any existing cable. Gisborne and his partner H. C. Forde, appointed by the Treasury as its engineers on the project, agreed, proposing that to achieve a "fair practical speed" of five to seven words per minute, the core should be made with 400 pounds of copper and a like amount of gutta-percha per nautical mile – nearly four times as much copper, and half again as much gutta-percha, as the Atlantic cable.[80] Acting on Bright's advice that "the manufacture of the conductor and insulating material should not be delayed on account of the outer covering, if time is as important to the execution of this work as your letter leads me to suppose," the Treasury immediately placed a rush order with the Gutta Percha Company for 1200 nautical miles of such core to be ready by August 20, so that the cable might be laid before the stormy season set in on the Bay of Biscay.[81]

As Bright's remark had hinted, the choice of outer covering proved far more contentious than that of the core. Bright proposed armoring the cable with hemp-covered iron wires, while Stephenson strongly favored using tarred hemp alone, saying its lighter specific gravity would spare the cable from excessive strain during laying. Forde sided with Bright, noting that all previous attempts to lay cables covered only with hemp had ended in failure and arguing that "in a work of such importance and moment as a telegraph between this country and Gibraltar, it would be unwise to attempt anything experimental or speculative."[82] Northcote leaned toward accepting Forde's advice, but given the lack of consensus among

---

[79] "Treasury Minutes," May 25, 1859, *Further Corr. 269*, 187 and 189.

[80] "Treasury Minute," May 27, 1859, *Further Corr. 269*, 159; Sir Charles Bright to Northcote, May 21, 1859, *Further Corr. 269*, 160–61; Robert Stephenson to Northcote, May 25, 1859, *Further Corr. 269*, 161–63; and H. C. Forde to W. H. Stephenson, May 25, 1859, *Further Corr. 269*, 163–66; see also *Joint Committee Report*, 441.

[81] See letters between Forde and H. F. Barclay of the Gutta Percha Company, June 2 to June 8, 1859, *Further Corr. 269*, 167–69; on the order placed on June 1, 1859, see Smith, *Rise and Extension*, 71.

[82] Bright to Northcote, May 21, 1859; Robert Stephenson to Northcote, May 25, 1859; and Forde to W. H. Stephenson, May 25, 1859, *Further Corr. 269*, 160–66.

experts, he thought it "desirable that some further experiments should be made before finally deciding this very important point"; he therefore directed Bright and Forde to conduct more tests on prospective cable designs.[83]

After a few hurried weeks of experimenting, Forde sent his specification for the cable to the Treasury Office, which forwarded it, along with a sample length of the proposed design, to Bright, F. C. Webb, and Samuel Canning for their opinions. These turned out to differ substantially. Canning and Bright expressed only fairly minor reservations, but Webb questioned the adequacy and accuracy of the tests Forde had made on the tensile strength of the cable, as well as the advisability of covering its outer armoring wires with hemp. Webb said he could not endorse the design Forde had proposed and was "decidedly of the opinion" that "it would be prudent if Her Majesty's Government provided themselves with such a series of experiments ... as would at once be an answer to the slightest imputation of precipitancy on a work of such magnitude."[84]

If Webb's report was not itself enough to give proponents of the project pause, by the time he submitted it there had been a change of government, as well as of wider circumstances. On June 10 Lord Derby, who had been governing with a minority in Parliament, lost a vote of confidence. He was succeeded by Lord Palmerston, with William Gladstone replacing Benjamin Disraeli as Chancellor of the Exchequer. The Indian Rebellion was by then effectively over, and though the Franco-Austrian War would continue for a few more weeks, French victories at Magenta and Solferino had made its outcome clear. In this changed atmosphere, and with expert engineers disagreeing on how best to proceed, Gladstone decided to pull the plug on the cable project, or at least slow it down. In a July 2 memorandum he stated that

Her Majesty's Government deem it best, after maturely weighing all the circumstances of the case, not to prosecute the manufacture of an electric cable with a view to its being laid down between England and Gibraltar during the present season. In arriving at this decision, they have had regard to the divided state of opinion with regard to the sufficiency of the experiments on the proposed composition of the cable, and to the period of the year, as well as to other matters.[85]

Gladstone still professed to believe that "the laying of an independent line of telegraphic communication from England to Gibraltar, and

[83] "Treasury Minute," May 27, 1859, *Further Corr. 269*, 159.

[84] F. C. Webb to W. H. Stephenson, July 1, 1859, *Further Corr. 269*, 170–71; Bright to W. H. Stephenson, July 1, 1859, *Further Corr. 269*, 171–72; Samuel Canning to W. H. Stephenson, July 1, 1859, *Further Corr. 269*, 172.

[85] "Memorandum by the Chancellor of the Exchequer," July 2, 1859, *Further Corr. 269*, 172–73.

likewise from Gibraltar to Malta, is on public grounds much to be desired"; more concretely, the government had already signed a contract with the Gutta Percha Company for 1200 nautical miles of core, much of which had already been manufactured.[86] The question of what to do with this enormous length of insulated wire – known at the Gutta Percha Company works as the "Gib core" – would preoccupy government cable policy for the next two years.[87] It would also give an important impetus to the development of improved electrical testing techniques and to the formation of the Joint Committee.

In his memorandum, Gladstone turned responsibility for the Gibraltar cable over to the Board of Trade, which evinced little enthusiasm for a project that had, after all, originated in the Treasury Office of a previous government. After affirming the appointment of Gisborne and Forde as engineers for the project and, on Gisborne's recommendation, of the Siemens firm as electricians in charge of testing the core during its manufacture, the Board directed the Gutta Percha Company to moderate the pace of its work, since there was no longer any need to rush to complete the core that summer.[88]

In his initial order, Gladstone had directed the Board of Trade to attend to "the further prosecution of experiments on the composition of the cable," and that was one part of its charge the Board seemed happy to take up.[89] Its chief engineer, Captain Douglas Galton, had no special background in telegraphy, but he was thorough and conscientious and knew where to turn for advice. He soon called on Stephenson and Charles Wheatstone, who had been investigating cable problems for the Electric and International Telegraph Company, of which Stephenson was then chairman. With the recent failure of the Atlantic cable in mind, Stephenson told Galton that he thought "the knowledge at present possessed is not such as to justify the submerging of a second deep-sea cable without further experiments being made."[90] Galton agreed and asked the Board to request £3000 from the Treasury for Stephenson and Wheatstone to conduct "experiments on the composition of submarine electric cables," particularly their outer coverings. By then reports were circulating that the Atlantic Telegraph Company, trying to regroup after

---

[86] "Memorandum," *Further Corr. 269*, 172; on the manufacture of the core, see Smith, *Rise and Extension*, 71–73.

[87] On the "Gib core," see Smith, *Rise and Extension*, 72.

[88] "Board of Trade Minute," July 16, 1859, *Further Corr. 269*, 173–76.

[89] "Memorandum by the Chancellor of the Exchequer," July 2, 1859, *Further Corr. 269*, 172.

[90] "Board of Trade Minute," July 16, 1859, *Further Corr. 269*, 173–76, on 175. On Galton, see R. H. Vetch and David Channell, "Sir Douglas Strutt Galton," *ODNB*; he was a cousin of the eugenicist Francis Galton.

its earlier failures, was assembling a committee of its own, including what the *Daily News* called "those eminent authorities" Stephenson, Wheatstone, and Thomson, to advise on the "external form, specific gravity, and electrical construction" of the new cable it hoped to lay if it could attract enough investors.[91] In its July 21 memo approving the funds Galton had requested, the Treasury Office noted it had "reason to believe that the Atlantic Telegraph Company are about to take measures for making similar experiments" and suggested that the Board and the company work together and share expenses. With that, the Joint Committee was born.[92]

What became of the "Gib core" that had gotten all of this started? It would follow a winding path to its final resting place, which lay nowhere near either Falmouth or Gibraltar. After the Gutta Percha Company had finished manufacturing the core and Siemens's staff had subjected it to searching electrical tests, the enormous length of insulated wire was sent to Glass, Elliot in Greenwich at the end of 1859 to receive its outer armoring, the engineers having finally settled on steel wires wound with tarred hemp as the best covering for a deep sea cable. Before that covering was completed, however, the government came to doubt the wisdom of trying to lay a cable in the depths of the Bay of Biscay, and in hopes of speeding communications in the Far East during the Second Opium War, it decided in March 1860 that the cable should instead be laid in the shallow coastal waters from Rangoon to Singapore.[93] Before the cable was ready to ship out, however, the war in China ended, and amid various misadventures, including Newall suing Glass, Elliot over a supposed infringement of one of his patents and the ship carrying the cable running aground at Plymouth, the government decided in January 1861 to change the destination of the cable yet again and lay it instead from Malta to Alexandria. Still spooked by deep waters, it ordered the cable to be laid in three sections along the North African coast, with stops at Tripoli and Benghazi. Glass, Elliot undertook the job in the summer of 1861, and except for a somewhat deeper stretch between Malta and Tripoli, most of the cable was laid within a few miles of shore and in depths of less than

---

[91] "Money Market," *Daily News* (July 2, 1859); see also "Money-Market and City Intelligence," *Times* (July 2, 1859), 10. C. F. Varley and "other authorities" were reportedly also to serve on the committee; in the end Thomson did not.

[92] "Treasury Minute," July 21, 1859, *Further Corr. 269*, 176.

[93] See the documents in *Papers Explanatory of the Intended Transfer of the Falmouth and Gibraltar Electric Telegraph Cable to a Line from Rangoon to Singapore*, British Parliamentary Papers, LXII.257 (London: Her Majesty's Stationery Office, 1860), especially "Treasury Minute," November 14, 1859, 2. Because the cable was now to be laid entirely in shallow waters, the plan to sheath it with hemp-covered steel was dropped.

a hundred fathoms.[94] (C. V. de Sauty served as Glass, Elliot's electrician during the laying and brought his old boss Whitehouse along as a consultant; it would be Whitehouse's last hurrah in the cable business.) The elaborate efforts in 1859 to devise a cable suited for laying in deep waters were thus rendered moot, but the chain of coastal cables from Malta to Alexandria, though often snagged by wayward anchors, nonetheless worked reasonably well until it was finally replaced by a direct deep-sea cable in 1868.[95] By far its most important service, however, lay in stimulating the formation of the Joint Committee.

Testifying

On August 4, 1859, a brief item appeared at the end of the "Money-Market and City Intelligence" column in the *Times*:

A series of experiments is about to be conducted by Captain Galton on the part of the Board of Trade, Mr. R. Stephenson, M.P., and Professor Wheatstone, to determine the best description of cable for the proposed line of telegraph from Falmouth to Gibraltar. The experiments are to be carried on with the aid of the Atlantic Telegraph Company, and will comprise an investigation into the comparative value of gutta-percha and india-rubber, the best mode of insulation, and other questions connected with the manufacture of deep submarine lines.[96]

It had originally been expected that Stephenson, among the most distinguished engineers in Britain, would lead the committee, and soon after it was formed he reportedly planned out much of its future work.[97] He had long suffered from nephritis, however, and in late summer his health worsened; he died on October 12, leaving the committee in Galton's hands. Besides Galton, who served as chairman, the Board of Trade chose three other members of the committee: Wheatstone, professor of natural philosophy at King's College London and one of the inventors of the telegraph, and the engineers William Fairbairn and George Parker Bidder. The Atlantic Telegraph Company also chose four members: George Saward, the company's secretary; Cromwell Fleetwood Varley, who had succeeded Whitehouse as its chief electrician; Latimer Clark, recently appointed its chief engineer; and Clark's older brother Edwin,

---

[94] H. C. Forde, "The Malta and Alexandria Submarine Telegraph Cable," *Proc. ICE* (May 1862), *21*: 493–514, and discussion, 531–40, in which Bright, Thomson, and William Siemens took part.

[95] On the Malta-Alexandria cable of 1861 and its later replacement, see Charles Bright, *Submarine Telegraphs: Their History, Construction, and Working* (London: Lockwood, 1898), 62–64.

[96] "Money-Market and City Intelligence," *Times* (August 4, 1859), 7.

[97] M. W. Kirby, "Robert Stephenson," *ODNB*; on Stephenson's role in planning the work of the Joint Committee, see *Joint Committee Report*, v.

also a well-known engineer. Varley and the Clarks also held leading positions as electricians and engineers in the Electric and International Telegraph Company, and both of the Clarks had long been close associates of Stephenson's. James Stuart-Wortley, who had become chairman of the board of the Atlantic Telegraph Company late in 1858, took part in several meetings of the committee, but health problems eventually forced him to withdraw and he did not sign its final report in 1861.[98]

The work of the Joint Committee had three main aspects: taking testimony from experts; collecting experimental studies, some directly commissioned by the committee or conducted by its members and some submitted by outside contributors; and drawing up a final report, in which Galton and the other committee members would marshal the available evidence and lay out their collective conclusions. When it appeared in mid-1861 as a large "blue book" issued by Her Majesty's Stationery Office, the *Joint Committee Report* was widely praised for its thoroughness, balance, and good judgment; in the words of William Siemens, "by providing an impartial and complete record of the principal facts" bearing on the troubled history of submarine telegraphy, it "disembarrassed the question of much uncertainty."[99] Bright later said the *Report* was so valuable that "no telegraph-engineer or electrician should be without it.... It is like the boards on ice marked 'Dangerous' as a caution to skaters"[100] (Figure 3.2).

The committee had been instructed to investigate the *construction* of submarine telegraph cables, and it devoted most of its attention to the proper choice and design of their conductors, insulation, and outer coverings. It also considered techniques for laying cables, especially in deep waters, and the electrical conditions for the transmission of signals and the detection of faults. It had not been asked to conduct a formal inquest into the failure of the Atlantic or other cables, and though it has often been said that the committee was formed in response to the failure of the Red Sea line, it is worth noting that those cables did not collapse until some months after the committee had begun its work. Nonetheless, questions about the causes of the Atlantic and Red Sea failures came up repeatedly in the course of the investigation and the committee addressed them briefly in its report, partly in the form of advice on how contracts and

---

[98] The members of the committee are listed in *Joint Committee Report*, xxxvi; on its composition see also Stuart-Wortley's remarks in "The Atlantic Telegraph," *Glasgow Herald* (July 15, 1859).

[99] Charles William Siemens, "On the Electrical Tests Employed during the Construction of the Malta and Alexandria Telegraph, and on Insulating and Protecting Submarine Cables," *Proc. ICE* (May 1862), *21*: 515–30, on 515.

[100] Sir Charles T. Bright, "Inaugural Address of the New President," *Journal of the Society of Telegraph-Engineers and Electricians* (January 1887), *16*: 7–40, on 32.

REPORT

OF

THE JOINT COMMITTEE

APPOINTED BY THE

LORDS OF THE COMMITTEE OF PRIVY COUNCIL FOR TRADE
AND THE ATLANTIC TELEGRAPH COMPANY

TO INQUIRE INTO THE

CONSTRUCTION OF SUBMARINE
TELEGRAPH CABLES:

TOGETHER WITH

THE MINUTES OF EVIDENCE

AND APPENDIX.

Presented to both Houses of Parliament by Command of Her Majesty.

LONDON:
PRINTED BY GEORGE EDWARD EYRE AND WILLIAM SPOTTISWOODE,
PRINTERS TO THE QUEEN'S MOST EXCELLENT MAJESTY.
FOR HER MAJESTY'S STATIONERY OFFICE.

1861.

Figure 3.2 The *Joint Committee Report*, issued in the summer of 1861, had enormous influence on the subsequent development of British cable telegraphy. This copy formerly belonged to the cable engineer and historian Charles Bright (son of Sir Charles Tilston Bright), and later to Rollo Appleyard, the historian of the Institution of Electrical Engineers.
(From *Joint Committee Report*, title page, 1861.)

specifications should be drawn up to lessen the chance of similar debacles in the future.

The committee questioned forty-three witnesses, two of whom (William Siemens and Charles West) appeared twice. It heard from almost everyone who had played a significant role in the first decade of the submarine cable industry (apart from Cyrus Field, who remained in the United States) and from many peripheral figures as well. The committee took testimony in sessions stretching from December 1, 1859 to September 4, 1860, but it held half of its sessions in the first three weeks of December and half of the rest between January 5 and February 2; the remaining few sessions were scattered at long intervals between March and September 1860.[101]

Galton chaired all twenty-two sessions and usually took the lead in questioning witnesses. Wheatstone was the most faithful other member, missing just one session, though he often said relatively little. Saward attended the first fourteen sessions and one later one and was an especially assiduous questioner on anything touching on the fortunes of the Atlantic Telegraph Company. The other members attended less regularly: Bidder and Varley eight times each, Edwin Clark six times, Latimer Clark four times, and Fairbairn just twice, while Stuart-Wortley made it to five sessions when his health allowed. Notably, Saward, Varley, and Latimer Clark also testified at length themselves, as well as submitting supplementary documents and reports. The testimony was all taken down and published, apparently verbatim, with numbered questions and replies. Procedures were not always as scrupulously formal as this might suggest, however; figures who were not members of the committee, including William Siemens, Robert FitzRoy (of *Beagle* fame), and John Marshman and Charles Peel of the Red Sea company, sometimes sat in and asked questions, and some sessions devolved into exchanges among the committee members or with witnesses who had finished their own testimony but were still in the room.[102] All of the testimony was voluntary; the committee had no power to compel anyone to appear before it. There is, however, no sign that anyone refused; even Whitehouse and Newall, who would face some notably sharp questioning, appeared willingly before the committee.

The first witness the committee called was Lionel Gisborne, followed immediately by William Siemens. They were asked mainly about the Red Sea cables and various Mediterranean projects, as well as the relative merits of gutta-percha and India rubber, how heat affected both materials, and the results of various testing procedures. Over the next three weeks the committee would hear from a mix of experienced telegraph

---

[101] A list of witnesses appears in *Joint Committee Report*, iii.

[102] See, for example, the testimony of FitzRoy, February 1, 1860, in which Siemens and two naval officers joined the questioning, and the testimony of William Mayes, September 4, 1860, in which Marshman and Peel also posed questions; *Joint Committee Report*, 195–99, 275–80.

engineers, including Bright, Varley, and Canning; established core and cable makers, including Willoughby Smith and Richard Glass; and proponents of various new schemes and materials, notably Thomas Allan, the ubiquitous promoter of companies committed to using his patented light cables. All drew a respectful hearing, though Allan, apparently unaware of the "joint" nature of what he had assumed was a purely governmental inquiry, professed surprise at finding himself questioned by the secretary of the Atlantic Telegraph Company, which he regarded as a competitor to his own companies.[103] None of the other witnesses seemed to question the propriety of having a private company take part on nearly equal footing in an official inquiry.

The committee delved most deeply into the failure of the Atlantic cable on December 15, when Whitehouse gave testimony that would fill more than twelve double-columned pages in the published *Report*.[104] Whitehouse could be excused for thinking the committee had been stacked against him: half of its members had, after all, been chosen by the company that had dismissed and denounced him a little over a year before, and two of his questioners, Saward and Varley, had been his chief antagonists in the heated exchanges that had filled the London newspapers. He nonetheless seemed happy to have a chance to appear in an official forum and explain again why he should not be blamed for the failure of the cable, though his success on that score was mixed at best. After the *Report* was published, one commentator found it "highly amusing, though melancholy, to read Mr. Whitehouse's account of the Atlantic cable's vicissitudes"; another observed that despite his best efforts to evade blame for the demise of "that unfortunate rope," Whitehouse's own testimony showed clearly that he had "subjected the Atlantic cable, with open eyes and in broad daylight, to the 'rack of martyrdom.'"[105]

After recounting his early investigations of signal transmission and the experiments that he claimed had refuted Thomson's law of squares, Whitehouse made it clear that he did not ascribe the failure of the cable to any missteps of his own, but rather to Field's headlong rush and the resulting cutting of corners. He downplayed the widely held belief (based, it must be said, on his own statements) that he had ever claimed that thin conductors would produce less retardation than thick ones, while also pointedly denying he had chosen the form of cable the company had

[103] Thomas Allan to Douglas Galton, January 2, 1860, in *Joint Committee Report*, 62–63.

[104] Wildman Whitehouse, December 15, 1859, *Joint Committee Report*, 69–82.

[105] "Momus" (letter), *Elec.* (January 17, 1862) *1:* 128–29, on 129, and "P. Q." (letter), *Elec.* (January 24, 1862) *1:* 143–44, on 144.

adopted.[106] He also said he had suspected at the time that the cable had serious flaws even before it was laid. Early on he had found and removed a bad joint made by some "lazy rascal" at the Gutta Percha Company works, he said, but he feared that other defective pieces might have slipped through – as some evidently did, judging from several notably shoddy joints Saward would later display during his own testimony. Whitehouse had also cut out many miles of cable that had been damaged by exposure to heat but said he could not have replaced all of the suspect lengths without scrapping almost the entire cable.[107] Saward questioned Whitehouse sharply as to why he had kept his doubts about the soundness of the cable to himself, pointing out that "Those suspicions were not communicated to the directors, or to myself, were they?" Whitehouse replied that he had hesitated to raise doubts about the quality of the insulation without direct evidence, which given the limits of his testing methods he could not supply.[108] He had thus been left in the awkward position of approving the use of lengths of cable he privately suspected were faulty.

Whitehouse strongly denied that jolts from his induction coils had killed the cable, though the real thrust of his argument was that they would not have damaged the insulation had it been sound to start with. When pressed by Galton and Varley, he conceded that if there were already small flaws in the insulation – as he had already admitted there were – strong currents from any source would "augment the mischief," though even then he insisted that battery currents, because of what he called their greater "heating power," would do more damage than would jolts from his enormous coils.[109] Varley was distinctly unconvinced, but the committee seemed disinclined to make an issue of it. Once the committee had put its questions and Whitehouse's answers on the record, it appeared to be satisfied to leave Whitehouse and his induction coils behind as simply a wrong turn in the story of cable telegraphy.[110]

When Thomson testified two days later, the questioning was much friendlier, even deferential (Figure 3.3). Galton repeatedly asked

[106] Whitehouse, December 15, 1859, *Joint Committee Report*, 74; in "The Atlantic Telegraph" (letter), *Athenæum* (October 11, 1856), 1247, Whitehouse had said directly that thin cables showed less retardation than thick ones.

[107] Whitehouse, December 15, 1859, *Joint Committee Report*, 76. Saward later showed the committee some "fearful" joints found in the unused portion of the cable; *Joint Committee Report*, 175.

[108] Whitehouse, December 15, 1859, *Joint Committee Report*, 78.

[109] Whitehouse, December 15, 1859, *Joint Committee Report*, 79–80.

[110] Charles Bright, "The Atlantic Cable of 1858," *Electrical World* (September 30, 1909) 54: 788–91; see also Latimer Clark, February 2, 1860, *Joint Committee Report*, 205–6.

(W. Thomson) reading a letter or letters from Fleeming
Jenkin, about experiments on [sub-]
marine cables probably about March 1859

Figure 3.3 William Thomson in 1859, reading one or more letters from
Fleeming Jenkin about experiments on submarine cables. (From Agnes
Gardner King, *Kelvin The Man*, frontispiece, 1925.)

Thomson if he "would . . . be so good as to tell us" his thoughts on various
subjects, and then let him speak at length – enough to fill sixteen pages of
the *Report*. Thomson said little against either the design of the cable or
Whitehouse's use of his induction coils but focused instead on what he
saw as deficient testing methods, particularly Whitehouse's failure to use
standard resistance coils or to test the insulation of the completed cable
under water. With such tools and procedures, Thomson said, any flaws in
the conductor or insulation could have been found and corrected early
on, and with proper instruments – that is, small batteries and his mirror
galvanometer – the cable could, he believed, have been made to work
reasonably well. When Saward asked the question that, in one form or
another, he put to almost every witness who had worked on the Atlantic
cable – in light of earlier failures, "has your confidence been shaken in the
future permanent success of a properly constructed and properly laid
Atlantic cable?" – Thomson's reply was unequivocal: "My confidence

has not been in the slightest degree shaken."[111] This was, of course, just what the secretary of the Atlantic Telegraph Company wanted to hear and have put on the public record, and he was clearly pleased when several other witnesses answered him in similarly confident terms.

Although testimony to the committee was not published until its *Report* was issued in mid-1861, some parts evidently circulated earlier, and Galton quoted substantial extracts from it in a piece on "Ocean Telegraphy" that appeared in the *Edinburgh Review* in January 1861.[112] By the time Newall appeared before the committee in August 1860 he had caught wind of the evidence J. W. Brett had given the previous December, and he devoted much of his own testimony to contesting statements Brett had made about cables Newall had laid for him in the Mediterranean.[113] Some of Newall's testimony was challenged in turn. For instance, he claimed to have seen messages transmitted on his Black Sea cable at up to seventeen words per minute, but S. A. Varley, who had supervised the operation of that line, told Galton that its actual rate of working was "very slow, not more than five words per minute."[114] Newall's credibility was by then already at a low ebb; his Red Sea cables had failed ignominiously and all efforts to repair them had proved unavailing. His exit from the industry soon thereafter was not widely mourned.

Newall's firm had, however, made at least one lasting contribution to the scientific side of submarine telegraphy: it had recruited Fleeming Jenkin into the field. The son of a naval officer, Jenkin had been educated in Edinburgh (where he was James Clerk Maxwell's school-mate) and on the continent.[115] After working briefly on railway projects, he was drawn into cable work in 1857 by Newall's business partner Lewis D. B. Gordon, a friend and former colleague of Thomson's at the University of Glasgow. Jenkin served (with mixed success) on several of Newall's cable-laying expeditions in the Mediterranean but found his true métier in performing electrical measurements at the firm's Birkenhead works. He began to collaborate with Thomson early in 1859 and would be closely associated with him for the rest of his life.

---

[111] Thomson, December 17, 1859, *Joint Committee Report*, 125.

[112] [Douglas Galton], "Ocean Telegraphy," *Edinburgh Review* (January 1861) *113*: 113–43. Galton is identified as the author in Walter E. Houghton et al., eds., *Wellesley Index to Victorian Periodicals, 1824–1900*, 5 vols. (Toronto: University of Toronto Press, 1966–1989).

[113] R. S. Newall, August 10, 1860, *Joint Committee Report*, 252–53.

[114] Newall, August 10, 1860, *Joint Committee Report*, 255; S. A. Varley to Galton, May 4, 1861, in *Joint Committee Report*, 506.

[115] Gillian Cookson and Colin A. Hempstead, *A Victorian Scientist and Engineer: Fleeming Jenkin and the Birth of Electrical Engineering* (Aldershot: Ashgate, 2000).

Jenkin was only 26 when he was called to testify before the Joint Committee in December 1859, but he had already established a substantial reputation as both a practical engineer and a scientific investigator. He was also becoming known for his volubility – "the best talker in London," a friend later said – and he held forth to the committee at length.[116] "I talked a good deal the other day at the Committee," he told Thomson. "What stuff it looks when reported by a shorthand writer, at least mine did."[117] Jenkin's "stuff" consisted largely of describing how he had measured the variation with temperature of the conductivity of gutta-percha and copper, as well as experiments that had, he said, "completely verified" Thomson's "very beautiful" theory of signal propagation, including the law of squares.[118] In a bow to his mentor, Jenkin expressed his measurements of resistance in what he called "Thomson's units," part of a system Thomson had introduced in 1851 in parallel to Wilhelm Weber's "absolute" system but based on imperial units of feet, grains, and seconds rather than Weber's metric units of millimeters, milligrams, and seconds. The resistance of a nautical mile of the Red Sea cable came to 25×10^7 Thomson's units, Jenkin reported – about eight ohms, to use a unit Jenkin himself would later help introduce. Jenkin stressed the importance for both electrical science and practical telegraphy of adopting an agreed unit of resistance, whether Thomson's or some other, and of embodying it in standard coils, a point Clark, Varley, and Thomson had all made as well.[119] The call for such units and standards became a leitmotif in their testimony to the committee, and one that would have far-reaching consequences for both cable telegraphy and electrical science.

Jenkin supplemented his testimony with six tables of data and five large graphs of arrival curves. Seemingly unable to keep himself from adding yet more, he followed it months later with a long letter, addressed from Turin, on the causes of faults, the processes that could make them grow worse, and the best ways to detect and locate them.[120] By the time Jenkin's letter reached London, the Joint Committee had heard from its final witness, the sea captain William Mayes, who testified on September 4 about his

[116] "Fleeming Jenkin" (obituary notice), *Proc. ICE* (1885) *82*: 365–77, on 369.

[117] Jenkin to Thomson, January 5, 1860, J26, Kelvin Papers, GB 247 Kelvin, Glasgow.

[118] Jenkin, December 22, 1859, *Joint Committee Report*, 135–36, 142–43.

[119] Jenkin, December 22, 1859, *Joint Committee Report*, 139. On Thomson's "British absolute units," see Thomson, "Applications of the Principle of Mechanical Effect to the Measurement of Electro-motive Forces, and of Galvanic Resistances, in Absolute Units," *Phil. Mag.* (December 1851) *2*: 551–62, repr. in Thomson, *MPP 1*: 490–502.

[120] Jenkin to Galton, September 28, 1860, in *Joint Committee Report*, 145–48. Jenkin was in Turin en route to a cable-laying expedition for Newall in the Mediterranean; see Jenkin to Thomson, September 18, 1860, J42, Kelvin Papers, GB 247 Kelvin, Glasgow.

fruitless attempts to raise and repair the Red Sea cables and the ill effects of Newall having laid them too tautly.[121] It was a fitting end to the committee's extended gathering of testimony about the failures that had plagued submarine telegraphy.

Experimenting

The Joint Committee got its start when the Treasury Office called for a few hand-picked engineers to test different outer coverings for the planned Gibraltar cable, but on taking charge, Galton greatly broadened its program of experimentation and opened it to more participants. Under his leadership, the committee gave advocates of different materials, designs, and procedures a chance to compete on what Willoughby Smith later said was "a fair field with no favour," and with "good umpires and all necessary appliances being provided," it was able to settle quite effectively many previously contentious issues.[122] In his presidential address to the Institution of Civil Engineers in January 1860, Bidder looked forward to when the completed findings of the committee would become "the common property of the country," and when the final *Report* appeared in mid-1861, the submarine cable industry proceeded to use them to reform its procedures and rebuild its credibility.[123]

The published *Report* included eighteen appendices, the last of which listed all significant cables laid since 1851, along with information on who had made each one, its length, conductors, insulation, and outer covering, as well as its repair history and current state as of early 1861.[124] A few other appendices reproduced documents related to the laying and testing of the Atlantic and Red Sea cables and the formation of the committee. The rest – thirteen in all, filling over 200 pages of the published *Report* and including many large plates and foldout graphs – consisted of accounts of experiments and related investigations by members of the committee, its staff, or outside experts. Fairbairn, for example, contributed a detailed report on the permeability of gutta-percha and India rubber to water when subjected to pressures of up to 20,000 pounds per square inch, while Smith and his colleague John Chatterton recounted their own extensive tests on the insulating

[121] William Mayes, May 4, 1860, *Joint Committee Report*, 275–80.

[122] Smith, *Rise and Extension*, 87.

[123] Bidder's remarks appear in "Institution of Civil Engineers," *Morning Chronicle*, January 13, 1860. The published *Report* was advertised for sale in the *Athenæum* (August 4, 1861), 136.

[124] "Table Showing the General Particulars of the Various Submarine Telegraph Cables, Arranged Chronologically," *Joint Committee Report*, 512–19.

properties of gutta-percha.[125] Chemists weighed in as well, with Augustus Matthiessen analyzing how even slight impurities could affect the conductivity of copper and W. A. Miller reporting on how quickly gutta-percha broke down when exposed to light and air – and how well it held up when kept in the dark under water, as in an undersea cable.[126]

In two of the longest appendices, Gisborne, Forde, and William Siemens reported the results of their tests of the strength of various outer coverings, while the committee's two staff electricians, Owen Rowland and E. G. Bartholomew, recounted how heat affected gutta-percha and India rubber. Both reports included extensive tables of data that later cable engineers would draw upon as reference sources, as well as large plates depicting the testing apparatus used and no fewer than twenty-two different forms of cable – serving in effect as an atlas of submarine telegraphy[127] (Figure 3.4).

The most important of the reports submitted to the committee focused on the electrical properties of conducting and insulating materials, on conduction, induction, and retardation phenomena, and on methods of electrical measurement. Wheatstone contributed a long report on the inductive discharge of cables, but while it was, as Brian Bowers has remarked, "clearly the outcome of extensive research," it served mainly to confirm established results and added little that was really new.[128] The appendices that best reflected the state of the field and did most to advance it were those submitted by Werner and William Siemens, Fleeming Jenkin, and Latimer Clark. Together, they helped set the course of much future work by both telegraph engineers and scientific researchers.

[125] William Fairbairn, "On the Permeability of Various Kinds of Insulators of Submarine Electric Cables," *Joint Committee Report*, 342–49; John Chatterton and Willoughby Smith, "Experiments upon Gutta Percha as an Insulator for Submarine Electric Conductors," *Joint Committee Report*, 383–85.

[126] Augustus Matthiessen, "Report of an Investigation Relating to the Causes of the Different Electric Conducting Powers of Commercial Copper," *Joint Committee Report*, 335–37; William Allen Miller, "Report upon the Results of Chemical Investigation into the Causes of the Decay of Gutta-Percha Used for the Insulation of Wires for Conveying Electric Currents," *Joint Committee Report*, 337–40.

[127] "Abstract and Results of Experiments Made by Messrs. Gisborne, Forde, and C. W. Siemens for Determining the Relative Strength of the Outer Covering of Submarine Telegraph Cables," and for "Determining the Strength of Steel and Iron Wires, and Hempen Strands Separate and Combined," *Joint Committee Report*, 391–412 and 413–41; E. G. Bartholomew and Owen Rowland, "Experiments Made by Direction of the Committee to Determine the Influence of Temperature and Pressure on Various Insulating Materials," *Joint Committee Report*, 350–78. On the role of "atlases" in science, see Lorraine Daston and Peter Galison, *Objectivity* (Cambridge, MA: MIT Press, 2007).

[128] Charles Wheatstone, "On the Circumstances Which Influence the Inductive Discharge of Submarine Telegraphic Cables," *Joint Committee Report*, 281–93; Brian Bowers, *Sir Charles Wheatstone* (London: Her Majesty's Stationery Office, 1975), 151.

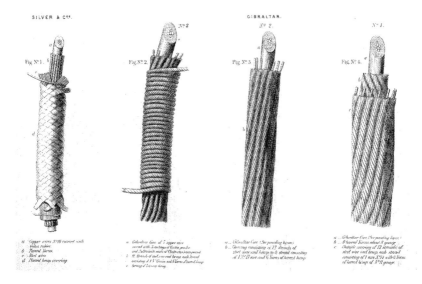

Figure 3.4 The *Joint Committee Report* included depictions of cables of various designs; the three on the right show different armorings applied to the "Gib core" that the British government ordered in 1859 and originally intended to lay from Falmouth to Gibraltar. (From *Joint Committee Report*, appendix 9, plate 2, 1861.)

The Siemens brothers contributed four appendices: one laying out ways to measure specific resistance and inductive capacity, along with tables of the values of these quantities they had found for gutta-percha and other insulating materials; two reporting on the extensive tests they and their staff had made on the "Gib core" and the Red Sea cables; and one giving the full text of a paper on the "principles and practice" of cable testing that the brothers had delivered at the Oxford meeting of the British Association in July 1860.[129] In all of these they emphasized the

---

[129] "Notes and Experiments by Dr. Werner Siemens and Mr. William Siemens," *Joint Committee Report*, 379–82, with four fold-out tables; "Memorandum of Method Adopted in Manufacturing and Testing the Falmouth and Gibraltar Cable by Messrs. Gisborne and Forde, and Messrs. Siemens and Halske," *Joint Committee Report*, 441–48; "Report on the Electrical Condition of the Red Sea Telegraph, by Siemens, Halske, and Co.," "Report on the Electrical Conditions of the Aden–Kurrachee Cable, Shortly Before, during, and for One Month after its Submersion," and "Siemens, Halske, and Co.'s Method of Determining the Distances of Faults," *Joint Committee Report*, 458–63, with six fold-out tables; Werner Siemens and C. W. Siemens, "Outline of the Principles and Practice Involved in Dealing with the Electrical Conditions of Submarine Electric Telegraphs," *Joint Committee Report*, 455–58. A shorter version of the latter paper appeared in the *BA Report* (1860), 32–34.

importance of referring all resistance measurements to a uniform standard, in particular to the "Siemens unit," a unit based on the resistance of
a column of mercury one meter long and one square millimeter in section
that Werner Siemens had recently introduced.[130] Smith later criticized
some of the Siemens brothers' procedures and conclusions, especially the
aspersions they cast on the reliability of gutta-percha as an insulator, but
even he praised their introduction of a standard of resistance and the
emphasis they put on measuring and recording the resistance of each mile
of cable as it was being manufactured.[131]

The Siemens brothers made a point of measuring the resistivities of
both copper and gutta-percha on the same scale – a very wide scale, to be
sure, since their values differed by about twenty orders of magnitude.[132]
Putting the resistivities of both a "conductor" and an "insulator" on
a single scale served to break down any strict line between the two, making
it evident, as Thomson put it, "that the distinction between conductors
and … insulators, is not an absolute distinction, but a distinction of
degree."[133] This insight – that all material substances should ultimately
be regarded as conducting dielectrics – lay at the heart of Faraday's
approach to electrical phenomena and would later be central to
Maxwell's theory of the electromagnetic field.

Jenkin was, if anything, even more intent than the Siemens brothers on
blurring the line between conductors and insulators. He opened his
report "On the Insulating Properties of Gutta-Percha" by declaring that
in it he would treat gutta-percha "as a conductor offering a resistance to
the electric fluid similar to that offered by any so-called good conductor,
such as copper or iron," and went on to recount measurements of the
conductivity of gutta-percha he had made at Newall's Birkenhead works
in August and September 1859.[134] Jenkin had originally intended his
paper for the *Philosophical Transactions*; Thomson communicated it to
the Royal Society in February 1860 and an abstract of it appeared soon
thereafter in the *Proceedings* of the Society.[135] Galton, however, steered
Jenkin away from seeking to publish in "philosophical journals" work that
was slated to be included in the committee's *Report*, and though Thomson

[130] Werner Siemens, "Vorschlag eines reproducirbaren Widerstandsmaasses," *Annalen der Physik* (1860) *110*: 1–20, trans. as "Proposal for a New Reproducible Standard Measure of Resistance to Galvanic Currents," *Phil. Mag* (1861) *21*: 25–38.
[131] Smith, *Rise and Extension*, 81, 89.
[132] Siemens and Siemens, "Outline of the Principles," *Joint Committee Report*, 455.
[133] Thomson, "Telegraph, Electric," 95.
[134] Fleeming Jenkin, "On the Insulating Properties of Gutta-Percha," *Joint Committee Report*, 464–81, on 464.
[135] Fleeming Jenkin, "On the Insulating Properties of Gutta Percha," *Proc. RS* (March 1860) *10*: 409–15.

had said publicly that he hoped Jenkin's paper would soon appear in the *Philosophical Transactions*, it never did.[136]

Thomson had expressed this hope in an unusual place: the pages of the *Encyclopedia Britannica*. Although Jenkin's full paper had not yet been published when Thomson was drafting his *Britannica* article on "Telegraph, Electric," he nonetheless praised it lavishly, writing in the venerable encyclopedia that "by experiments of a very precise character," Jenkin had "put in practice systematically for the first time" methods for the absolute measurement of the specific resistance of insulating substances – measuring them, that is, not just relative to an arbitrary material standard like the Siemens unit but in "absolute" terms based on units of length, mass, and time.[137] Thomson also reproduced Jenkin's tables of the resistivity of both pure gutta-percha and the compound used to insulate the Red Sea cables, though in a gesture of modesty he cited what Jenkin had called "Thomson's units" as "British absolute units."[138] He later praised Jenkin's experiments at Birkenhead even more highly, declaring that they had yielded "the very first true measurement of the specific inductive capacity of a dielectric" since Faraday's original discovery of the property in 1837, and so had helped establish the reality and importance of a property that had been "either unknown, or ignored, or denied, by almost all of the scientific authorities of the day."[139] Jenkin's measurements were not above reproach – Willoughby Smith later said they were marred by confusion over the composition of some of the insulating materials he had tested – but they marked a real advance in electrical technique and formed an important point of contact between the practices of telegraph engineers and laboratory researchers.[140]

The most comprehensive of the electrical reports submitted to the Joint Committee was Latimer Clark's. In the published blue book, it bore the simple heading "Report," but when Clark brought it out under separate covers later in 1861, he gave it a more informative title: "Experimental

---

[136] Jenkin to Thomson, August 27, 1860, J40, Kelvin Papers, GB 247 Kelvin, Glasgow, quoted in Cookson and Hempstead, *Jenkin*, 128.

[137] Thomson, "Telegraph, Electric," 95, 97.

[138] Thomson, "Telegraph, Electric," 98; Jenkin, "Gutta Percha," *Proc. RS*, 415.

[139] William Thomson, "Note on the Contributions of Fleeming Jenkin to Electrical and Engineering Science," in Colvin and Ewing, eds., *Fleeming Jenkin Papers*, 1: clv–clix, on clvii. In Fleeming Jenkin, "Experimental Researches on the Transmission of Electric Signals through Submarine Cables. – Part I. Laws of Transmission through various lengths of one Cable," *Phil. Trans.* (1862), *152*: 987–1017, on 1016, Jenkin said that "the specific inductive capacity of the dielectric could be calculated" from the values he had tabulated, but he did not give the calculation, saying "these points … will be more fully treated of in the second part of the paper," which never appeared.

[140] Smith, *Rise and Extension*, 92–93.

Investigation of the Laws Which Govern the Propagation of the Electric Current in Long Submarine Telegraph Cables."[141] As Charles Bright (son of the engineer Sir Charles Tilston Bright) observed in 1898 in his magisterial *Submarine Telegraphs*, Clark's report was clearly "the result of much thought and experiment" – in fact far more than would ever appear in print, for while the published version ran to over 70,000 words, it had, Clark said, been much condensed from the full report he had originally prepared.[142]

Clark opened his report by clarifying some basic terms, particularly "quantity," "tension" (which he preferred to the more widely used but often misunderstood "intensity"), and "resistance," and by showing how Ohm's law tied them all together. Like Jenkin and the Siemens brothers, he emphasized that no sharp line divided conductors from "so-called non-conducting materials," declaring that "all conduct to a certain extent."[143] Clark put particular emphasis on the accurate measurement of resistance, a term and concept that was, he said, "in constant use in telegraphy," and summarized in tabular form the results of his many tests of how the resistivity of copper, gutta-percha, and other materials varied with temperature and pressure. He took great care with these measurements, most of which he carried out at the London works of the Electric and International Telegraph Company in intervals from his work as its chief engineer, but modestly asked his readers to "judge them in a lenient spirit," as they had been "almost entirely conducted at night, after the labours attendant on the maintenance of the most extensive system of telegraphs in Europe."[144] Clark spared no effort in his work for the Joint Committee; it was, he believed, his best opportunity to contribute to the advancement of science as well as of Britain and its empire.

The central sections of Clark's report concerned a subject whose study he had pioneered: induction and retardation on submarine cables. "There is no phenomenon in electricity that has a more important bearing on the electric telegraph than that of induction," he said, "and none which

---

[141] Latimer Clark, "Report," Joint Committee Report, 293–335, repr. in Latimer Clark, *Experimental Investigation of the Laws Which Govern the Propagation of the Electric Current in Long Submarine Telegraph Cables* (London: George E. Eyre and William Spottiswoode, 1861), 5–47. This slim volume also included an unpaginated preface by Clark; an account of "Milner's Electrometer," 47; Latimer Clark, "On Electrical Quantity and Tension," 47–48, an abstract of a lecture he gave at the Royal Institution on March 15, 1861; and Latimer Clark and Sir Charles Bright, "On the Principles Which Should be Observed in the Formation of Standards of Measurement of Electrical Quantities and Resistance," 49–50, a paper they delivered at the 1861 British Association meeting, which will be discussed in the next chapter.

[142] Bright, *Submarine Telegraphs*, 60n; Clark, "Preface," *Experimental Investigation*.

[143] Clark, *Experimental Investigation*, 6.

[144] Clark, "Preface," *Experimental Investigation*.

interferes more with the commercial success of telegraphic enterprise." In particular, "if it were not for this evil presenting itself in the form known as retardation of the current, any telegraph cable, however long, could be worked at almost any speed."[145] Completely eliminating retardation appeared to be out of reach, he said, but "much may be done to reduce its effects," and finding how best to do so was the chief aim of his investigations. Clark's pursuit of this eminently practical goal would lead him to address some of the deepest and most consequential questions in electrical science.

After briefly recounting how he had first noticed retardation on cables at the Gutta Percha Company works in 1852, Clark described more fully the many experiments he had demonstrated to Faraday and others in October 1853, as well as the more recent investigations he had undertaken for the Joint Committee. In all of these, he focused on measuring both the induction and retardation phenomena themselves and the various conditions that could affect them, particularly the size and resistivity of the conductor and the thickness and specific inductive capacity of the insulation. His examination of how these factors played out in the operation of submarine cables led him to take up what he said was "a new element in these investigations, viz., that of *time*."[146] This was a crucial step; as Clark noted, "the entire commercial value of the telegraph depends on the time occupied in charging and discharging, and the rate at which signals can be distributed through the cable within a given period."[147] By focusing, as the title he gave his published report implied, on how currents *propagate* along submarine cables, Clark turned from simply measuring the strength of electric and magnetic forces at a given moment, as one might in a laboratory, to tracing how those forces varied through time and space. This shift of focus to *time* and to propagation through space was among the most important contributions submarine telegraphy would make to electrical science, as it turned investigators' attention toward the role successive actions in the surrounding field played in electromagnetic phenomena.

Clark closed his report by noting that an understanding of induction and retardation suggested several ways signaling rates on long cables might be increased, particularly by using thicker conductors and adjusting the ways signals were produced and detected. In light of such advances, he said, it could be "safely asserted that a cable of the same length as the Atlantic cable designed at the present day would convey messages at five or six times the speed of that cable, and at very

[145] Clark, *Experimental Investigation*, 15.   [146] Clark, *Experimental Investigation*, 35.
[147] Clark, *Experimental Investigation*, 35.

moderately increased cost, while its electrical perfection," secured through more careful manufacture and testing, "would be incomparably greater." "The value of this gain is, both in its social and commercial aspects, immense," he said, "and the acquisition of this power would alone amply repay the members of this Committee for their labours."[148] Moreover, as the work of Clark and other contributors to the Joint Committee showed, those labors also yielded scientific dividends whose consequences would reach far beyond cable telegraphy.

Reporting

In the spring of 1861 Galton pulled together the mass of testimony, experimental reports, and other evidence the Joint Committee had gathered over the preceding year and a half and drew up his official "Report." Issued that April over his signature and those of the seven other members of the committee, it ran to over 25,000 words and filled thirty-two pages of the printed volume that issued from Her Majesty's Stationery Office that summer. As the *Journal of Gaslighting* observed, the *Joint Committee Report* formed "a Blue Book of the largest size, containing upwards of 500 pages, and costing £1," making it "not, therefore, tempting reading to persons not pecuniarily interested in the subject." Nonetheless, it contained such "a fund of information on a most important subject" that its contents deserved to be known even to those not in the cable business, and the *Journal* said it was "doing a useful service to our readers" by reporting its main points.[149] The *Engineer* and the *Electrician* went further, serializing all of Galton's "Report," though not the lengthy accompanying testimony and appendices.[150]

One of the chief points the report addressed was why the most ambitious cable projects had failed so spectacularly. It traced the failure of the Atlantic cable to "the original design having been faulty owing to the absence of experimental data, to the manufacture having been conducted without proper supervision, and to the cable not having been handled, after manufacture, with sufficient care," while the Red Sea and India line had collapsed because the cable had been "designed without regard for the conditions of the climate or the character of the sea over which it had to be laid" and the contractor had been given too free a hand during manufacture and laying.[151] After recounting the troubles that had beset

---

[148] Clark, *Experimental Investigation*, 47.

[149] "Notes upon Passing Events," *Journal of Gaslighting* (July 16, 1861) *10*: 495–97, on 495.

[150] *Joint Committee Report* was serialized in four parts in *Engineer* (Vol. 12) between September 6 and September 27, 1861 and in fifteen parts in *Elec.* (Vol. 1) between November 9, 1861 and March 7, 1862.

[151] *Joint Committee Report*, ix, x.

various Mediterranean cables, the report stated perhaps its most important conclusion: "It will be observed that the failures of all of these
submarine lines are attributable to defined causes, which might have
been guarded against." That is, the disasters that had befallen cable
telegraphy were not intrinsic to the enterprise but resulted simply from
a series of correctable errors. "We believe," the members of the committee declared, "that there are no difficulties to be encountered in laying
submarine cables, and maintaining them when laid, which skill and
prudence cannot and will not overcome."[152] This official expression of
confidence in the future prospects of oceanic cable telegraphy offered
a much needed lifeline to an industry that by 1861 was badly floundering.

How, the committee asked, had cable telegraphy fallen so low, and how
could it now right itself? The report emphasized that the industry was still
very young – scarcely ten years old – and noted that its first efforts had
been the products of "bold" action rather than "the application of any
well ascertained data." Moreover, it was "a remarkable fact, and, as
regards the science of the subject, probably an unfortunate one, that
complete success attended the laying of the first telegraph cables," notably those across the English Channel.[153] Those lucrative early triumphs
had given the leaders of the industry misplaced confidence that they could
use the same designs and techniques in other seas and oceans, and they
had gone on to lay more cables with little further study. In a pattern
common to many emerging industries, a string of failures ensued.
These had prompted a reexamination of foundations, with the Joint
Committee leading the effort to gather evidence, weigh experience, and
rationalize procedures.[154] It was here that scientific investigation came
into play, not so much to pioneer new breakthroughs as to enable the
reliable operation and incremental improvement of existing technologies.
In turn, science itself benefited as researchers gained access to resources
far beyond those available in any ordinary laboratory.

The reciprocal relationship between electrical science and cable telegraphy was especially clear, and would prove especially fruitful, in the field
of electrical measurement. The Joint Committee had declared accurate
measurement and careful quality control to be the keys to the successful
manufacture and operation of submarine cables, and it urged the industry
to adopt agreed standards, particularly of electrical resistance, so that
measurements could be readily compared and contract specifications
stated in a way that would be clear and binding on all parties.[155] Such

---

[152] *Joint Committee Report*, xii.   [153] *Joint Committee Report*, xiii.
[154] See, for example, John G. Burke, "Bursting Boilers and the Federal Power," *Technology and Culture* (1966) 7: 1–23.
[155] *Joint Committee Report*, xvi, xxxiii.

standards would soon become central to electrical science as well, and the committee noted that in the extensive experiments it had commissioned on electrical measurement, "new methods of testing and new instruments" had been employed that "cannot fail to prove of the highest scientific interest to the theoretical philosopher and to the practical electrician."[156]

The same could be said of the work of Clark, Jenkin, and the Siemens brothers on induction and retardation phenomena. Building on Faraday's conception of the electric field and Thomson's mathematical analysis of signal propagation, they had brought new theoretical ideas and experimental techniques to bear on what the Joint Committee report called "the most important problem to solve in submarine telegraphy."[157] In the process they had not only contributed to solving the practical problem of how to increase signaling rates on long cables but had initiated investigations that, as Thomson later observed, would soon reach into "the loftiest regions and subtlest ether of natural philosophy" and lead to seminal advances in the understanding of the electromagnetic field.[158]

The Joint Committee concluded its report in a way that acknowledged the troubled history of oceanic cable telegraphy while also expressing confidence in its future. "We are convinced," the committee declared, "that if regard be had to the principles we have enunciated in devising, manufacturing, laying, and maintaining submarine cables, this class of enterprise may prove as successful as it has hitherto been disastrous."[159] This judgment would be borne out in the years to come, as the *Joint Committee Report* became the bible of the cable industry. The principles it laid down were not at all radical, and in fact most submarine cables would long remain remarkably similar in design to the first one laid beneath the English Channel in 1851: copper conductors covered with gutta-percha insulation and sheathed in an outer armoring of iron wires. The key to success lay not in revolutionary new designs but in improved procedures, particularly more careful manufacture and handling and better quality control, guided by precise and extensive measurement.

Looking back from the early 1870s, when the network of submarine cables was spreading rapidly and reliably around the globe, Sir James Anderson praised the *Joint Committee Report* as "full and complete" and as a signal contribution to the success of an industry that was rapidly becoming one of the chief bulwarks of the British Empire. The captain of the

[156] *Joint Committee Report*, xvii.   [157] *Joint Committee Report*, xxiv.
[158] William Thomson, "Presidential Address," *BA Report* (1871), lxxxiv–cv, on xciii, repr. in Thomson, *Popular Lectures and Addresses*, 3 vols. (London: Macmillan, 1889–1894), 2: 132–205, on 162.
[159] *Joint Committee Report*, xxvvi.

Great Eastern on its Atlantic cable-laying expeditions of 1865 and 1866 and later a leading figure in the Eastern Telegraph Company, Anderson declared (the italics are his) that the Joint Committee had "*established principles which up to the present time have uniformly guaranteed success, while the neglect of them has as uniformly resulted in partial loss or failure.*"[160] The Joint Committee had been formed in response to anxieties about imperial communications, as the British government had stepped in to address a pressing technological problem. The resulting work of the committee went far toward redeeming the failures of the Atlantic and Red Sea cables; in the process, it also gave an important impetus to new work in electrical science.

[160] Sir James Anderson, *Statistics of Telegraphy* (London: Waterlow and Sons, 1872), 64.

4   Units and Standards
The Ohm Is Where the Art Is

Since the 1860s, the humble resistor has been central to the daily work and crucial to many of the leading achievements of both physicists and electrical engineers. Without the ready availability of accurate and reliable resistance standards, whether in the form of resistance coils or their later digital descendants, the precision electrical measurement that underlies so much of modern science and technology would be virtually impossible. As Fleeming Jenkin, himself a pioneer in the development of the standard ohm, observed in 1865, "Resistance-coils . . . are now as necessary to the electrician as the balance to the chemist."[1] Standard resistances – ohms – have become so ubiquitous as to be almost invisible; they are now simply taken for granted. But how did they achieve this virtually unquestioned authority? Who undertook the immense labor necessary to secure and disseminate the first reliable resistance standards, and why? In particular, what role did the practical concerns of British telegraph engineers play in the origin of the ohm, and to what extent did the early standard coils in fact embody their aims and expertise?

Two quotes from William Thomson and James Clerk Maxwell will shed some light on these questions. In a lecture on "Electrical Units of Measurement" at the Institution of Civil Engineers in 1883, Thomson declared that British cable engineers had been ahead of physicists in the practice of accurate electrical measurement from the late 1850s to the early 1870s. "Resistance coils and ohms," he said, "and standard condensers and microfarads, had been for ten years familiar to the electricians of the submarine-cable factories and testing-stations, before anything that could be called electric measurement had come to be regularly practised in almost any of the scientific laboratories of the world."[2] Along similar

---

[1] Fleeming Jenkin, "Report on the New Unit of Electrical Resistance Proposed and Issued by the Committee on Electrical Standards Appointed in 1861 by the British Association," *Proc. RS* (April 1865) *14*: 154–64, repr. in *Phil. Mag.* (June 1865) *29*: 477–86, and in Smith, *Reports*, 277–90, on 279. Resistance coils are perhaps better compared to a chemist's graduated weights, with the balance itself corresponding to a Wheatstone bridge.

[2] William Thomson, "Electrical Units of Measurement," in *Practical Applications of Electricity. A Series of Lectures Delivered at the Institution of Civil Engineers* (London:

lines, Maxwell wrote in a review of Jenkin's 1873 textbook *Electricity and Magnetism* that

at the present time there are two sciences of electricity – one that of the lecture-room and the popular treatise; the other that of the testing-office and the engineer's specification. The first deals with sparks and shocks which are seen and felt, the other with currents and resistances to be measured and calculated. The popularity of the one science depends on human curiosity; the diffusion of the other is a result of the demand for electricians as telegraph engineers.[3]

Moreover, as Jenkin himself had said in the book Maxwell was reviewing, it was "not a little curious that the science known to the practical men was, so to speak, more scientific that the science of the text-books."[4] An examination of the work of British telegraph engineers and electricians in the 1850s and 1860s will reveal how and where an effective demand for resistance standards first appeared, as well as how closely science and technology were intertwined in the making of the first ohms.

Previous discussions of the development of electrical measurement in Britain have often, and understandably, focused on Thomson, who was a central figure among both scientists and engineers. Here we will broaden our focus to take in not only Thomson and Maxwell on the scientific side but leading telegraph engineers and electricians as well, particularly Jenkin, Latimer Clark, and Cromwell Fleetwood Varley. As working engineers and as writers on electrical measurement (and as members, along with Thomson and Maxwell, of the influential British Association Committee on Electrical Standards), Jenkin, Clark, and Varley helped lay the groundwork for much that was to follow in the field of electrical metrology, and their writings provide a valuable window into the aims and practices of British telegraph engineers in the mid-nineteenth century. Through them, we will see how the demands and opportunities presented by the submarine telegraph industry helped shape the techniques, and literally set the standards, that were to become the ordinary working tools of both physicists and electrical engineers in the last third of the nineteenth century. It was through their work and that of their associates that the art of electrical measurement was reduced to

---

Institution of Civil Engineers, 1884), 149–74, on 150, repr. in Thomson, *Popular Lectures and Addresses*, 3 vols. (London: Macmillan, 1889–1894), *1*: 73–136, on 75–76.

[3] [James Clerk Maxwell], "Longman's Text-Books of Science" [rev. of Fleeming Jenkin, *Electricity and Magnetism*], *Nature* (May 15, 1873) *8*: 42–43. Although this review is unsigned, its style and content (including some characteristic puns and several passages that closely parallel ones in the preface Maxwell had just written for his own *Treatise on Electricity and Magnetism*) make it clear that it was written by Maxwell; see Harman in Maxwell, *SLP 2*: 842–44.

[4] Fleeming Jenkin, *Electricity and Magnetism* (London: Longmans, Green, 1873), vi.

a set of standard practices, as their own expertise and authority came to be embodied in the resistance boxes they calibrated and distributed.

Latimer Clark, Cromwell Varley, and Early Telegraphic Measurement

Latimer Clark and Cromwell Fleetwood Varley were both in their 30s when the failure of the Atlantic and Red Sea cables and the formation of the Joint Committee to Inquire into the Construction of Submarine Telegraph Cables combined to focus public and professional attention on problems of electrical measurement. As, respectively, the chief engineer and the chief electrician of the Electric Telegraph Company, the first and long the largest telegraph company in Britain, they were already among of the most experienced and highly regarded figures in the telegraph industry and would remain active in it for decades to come. Both first worked mainly on the Electric Telegraph Company's system of landlines, but from the mid-1850s they turned increasingly to submarine telegraphy and eventually made it their main work as consulting engineers[5] (Figure 4.1).

The network of telegraph lines that began to spread across Britain in the 1840s was a mix of overhead wires strung on poles and underground lines insulated with tarred cotton or (after about 1850) gutta-percha. The overhead lines were electrically quite simple; as long as there were no actual breaks in the wire and the leakage of current at the supporting poles was kept within reasonable limits, operators could signal along them quite satisfactorily without worrying about measuring much of anything. As Clark later noted, "no very exact measurements are required to be made of overhead lines," and their spread gave relatively little stimulus to the development of precision measurement techniques.[6] Simple vertical galvanometers (often just ordinary needle telegraph receivers) sufficed to show

---

[5] On Clark, see the entry by A. F. Pollard in the *ODNB*; on Varley, see Richard Noakes, *Physics and Psychics: The Occult and the Sciences in Modern Britain* (Cambridge: Cambridge University Press, 2019), 64–66 and 203–13. As Noakes notes, both Clark and Varley were convinced of the reality of spiritualist phenomena, and Varley used telegraphic instruments in some of his investigations of the medium Florence Cook. In 1855 the Electric Telegraph Company for which they both worked absorbed the International Telegraph Company and became the Electric and International Telegraph Company, though it continued to be commonly known simply as "the Electric." See Jeffrey Kieve, *The Electric Telegraph: A Social and Economic History* (Newton Abbot: David & Charles, 1973), 52–53.

[6] Latimer Clark and Robert Sabine, *Electrical Tables and Formulae for the Use of Telegraph Inspectors and Operators* (London: E. and F. N. Spon, 1871), 7; cf. Robert Rosenberg, "Academic Physics and the Origins of Electrical Engineering in America," PhD dissertation, Johns Hopkins University (1990), 6–8, 39–43.

Figure 4.1 Latimer Clark (left) and Cromwell Fleetwood Varley (right) were leading cable engineers and strong proponents of accurate electrical measurement.
(From Louis Figuier, *Merveilles de la Science*, Vol. 2: 220 and 277, 1868.)

that enough current was getting through to produce readable signals, and that was the only real concern of the operators of the early overhead lines.

Underground lines were more complicated electrically and gave a first hint of some of the problems that were later to bedevil submarine cables. Long underground lines were troubled by electric induction and retardation – obstacles to rapid signaling that, as we have seen, were to play a major role in British thinking about electrical propagation.[7] They also suffered from flaws in their insulation, which led to most long underground lines being replaced by overhead wires by the late 1850s, and such "faults" were much harder to find and repair on buried lines than were simple breaks on overhead wires. The main concern on the early underground and submarine lines was the electrical continuity of the conductor, and very simple tests were used to check that enough current was getting through – recall that Charlton Wollaston, the engineer on the first Channel cable of 1851, sometimes used his tongue to detect the current, a procedure that was fairly common on early landlines as well.[8]

---

[7] Bruce J. Hunt, "Michael Faraday, Cable Telegraphy and the Rise of Field Theory," *History of Technology* (1991) *13*: 1–19.
[8] Charles Bright, *Submarine Telegraphs: Their History, Construction, and Working* (London: Lockwood, 1898), 6n; see also G. B. Prescott, *History, Theory, and Practice of the Electric*

Clark and Varley did not use anything quite so crude, but in the early days they and their colleagues generally tested the continuity of an underground line or submarine cable by simply connecting it to a voltaic cell and seeing whether it could pass enough current to deflect the needle of an uncalibrated vertical galvanometer. They checked the leakage through the insulation by connecting the same galvanometer across the tarred cotton or gutta-percha in series with 100 or 200 cells; if the deflection of the needle remained fairly small and steady, they judged the insulation to be "good enough."[9]

More precise electrical measurement first acquired a commercial value in connection with methods for locating faults in insulated wires. When the first underground lines were laid in Britain in the late 1840s, such faults – breaks in the wire itself or gaps in its insulation – were typically found by pulling up the wire, cutting it, and testing which side the fault was on. This was repeated, halving the distance each time, until the fault was isolated. The procedure was slow and expensive, and the many cuts and splices inevitably damaged the wire and its insulation, making the line even more prone to future troubles.[10] If engineers could find a way to determine the location of a fault, even approximately, by electrical tests on the ends of a line, the saving in time and money would be enormous.

Several people worked out electrical methods for locating faults in the late 1840s and early 1850s, first on underground lines and then on submarine cables. Varley, as a very junior electrician at the Electric Telegraph Company, was among the first. In 1846–1847, while still in his teens, he was put in charge of a new underground line, insulated with tarred cotton and enclosed in lead pipes, that ran beneath the streets of London from the Nine Elms railway station to the Strand and the Admiralty. The line was plagued by bad joints, and Varley soon found that by connecting a battery and galvanometer to existing "good" wires, he could determine what length of wire gave the same deflection (i.e., the same current with the same battery power) as the faulty line. By simply

Telegraph (Boston: Ticknor and Fields, 1863), 268, 282. Werner Siemens, *Inventor and Entrepreneur: Recollections of Werner von Siemens* (London: Lund Humphries, 1966), 72, describes a method in which workmen located faults in cables during manufacture by putting their fingers into a tub of water and taking shocks as faulty sections were passed through the tub.

9 Willoughby Smith, *The Rise and Extension of Submarine Telegraphy* (London: J. S. Virtue, 1891), 23, describing the "recognised test for insulation" at the Gutta Percha Company in the early 1850s.

10 C. F. Varley, January 5, 1860, in *Joint Committee Report*, 149; see also Siemens, *Inventor and Entrepreneur*, 78, 89.

stepping off the corresponding distance along the course of the faulty line, he was able, he later said, to identify the bad joint "9 times out of 10."[11]

Varley's method did not require any deep understanding of currents and resistances; he simply set up a full-scale copy of the faulty circuit and walked off the corresponding length of wire. But while it was conceptually simple, Varley's method was cumbersome in practice (requiring as it did the use of enormous lengths of "good" wire) and not very precise. Werner Siemens worked out a somewhat better method for use on the Prussian underground lines in 1850, and in 1852 Charles Tilston Bright developed another, using a set of graduated resistance coils, for use on the network of underground lines the Magnetic Telegraph Company had built in northern England.[12] In the 1840s Charles Wheatstone, Moritz Jacobi, and other scientific researchers had introduced resistance coils calibrated in feet of copper wire or similar small units for laboratory use, but Bright, Varley, and other engineers required much larger units; as Jenkin later noted, "the first effect of the commercial use of resistance was to turn the 'feet' of the laboratory into 'miles' of telegraph wire," and Bright's coils were indeed calibrated in equivalents of a mile of sixteen-gauge copper wire, as were those Varley developed independently about the same time.[13] The replication and refinement of such resistance coils in the 1850s and early 1860s were crucial to the spread of precision electrical measurement among both engineers and physicists. It is hard to gauge how exact these early fault-location methods were in practice, but Varley later said that tests with his coils could not reliably bring him any closer than to within about 5 percent or 10 percent of the true position, partly because the actual resistance per mile of the conducting wire had not been accurately measured and recorded in terms of his standard, and partly because the resistance of the faults themselves could vary unpredictably.[14]

The earliest submarine cables were built and operated without much reference to precision measurement, as the story of Wollaston's tongue indicates. But several failures on cables laid in in the mid-1850s led those in charge to begin taking somewhat more care, both in locating faults on cables they had already laid and in testing the quality of the wire and insulation during manufacture. F. C. Webb of the Electric Telegraph

[11] C. F. Varley, January 5, 1860, in *Joint Committee Report*, 149.

[12] Siemens, *Inventor and Entrepreneur*, 80; Prescott, *Electric Telegraph*, 286–88.

[13] Jenkin, "Report on the New Unit," in Smith, *Reports*, 280.

[14] C. F. Varley, January 5, 1860, in *Joint Committee Report*, 169–70; see also Siemens, *Inventor and Entrepreneur*, 132–33, and William Thomson, "On the Forces Concerned in the Laying and Lifting of Deep-Sea Cables," *Proceedings of the Royal Society of Edinburgh* (December 1865) 5: 495–509, on 500, repr. in Thomson, *MPP 2:* 153–67, on 158.

Company and Willoughby Smith of the Gutta Percha Company, makers of virtually all of the insulated wire used in cables, were among the first to institute more careful electrical tests during manufacture – though by later standards these were still quite rudimentary, consisting mainly in using somewhat more sensitive horizontal galvanometers and recording the readings. Smith later said that some of the more rough and ready "practical men" at first derided these efforts as electrical "high farming" and a waste of time, but his tests proved of considerable use in detecting and eliminating faults and were taken up more widely in the later 1850s.[15]

Most of these early electrical tests involved gauging the resistance of the wire (and sometimes also the insulation), at first simply by noting the deflection that the current from a given battery produced on a given galvanometer. This worked well enough for comparisons using the same apparatus, but the results could not be readily related to those of tests performed with other batteries or instruments. Later, and more reliably, electricians began to use a differential galvanometer or Wheatstone bridge to compare the resistance of the line being tested to that of an arbitrarily chosen standard, typically a mile of ordinary sixteen-gauge copper wire.[16]

As better methods of applying gutta-percha came into use later in the 1850s, the leakage current on well-made cables became too small to deflect the needle of even a fairly sensitive galvanometer, and other ways had to be found to test the resistance of the insulation. In one widely used method, a length of cable was charged to a high tension (as indicated on an electrometer) and its ends kept insulated. The investigator then allowed the cable to discharge by leakage, and by noting the time it took to fall to half its original tension, could readily calculate the leakage rate and so the insulation resistance.[17] A related technique involved partially discharging the cable at intervals into short lengths of insulated cable or calibrated condensers. Electricians sometimes also estimated the inductive capacity of a cable (an important determinant of the maximum rate of signaling it could handle), either by discharging the cable through a galvanometer and noting the "throw," or maximum deflection, of the needle, or by partially discharging it into a calibrated condenser or a length of insulated cable and noting how much of the total charge remained.[18]

[15] Smith, *Rise and Extension*, 25, 331.
[16] Cromwell F. Varley, "On Some of the Methods Used for Ascertaining the Locality and Nature of Defects in Telegraphic Conductors," *BA Report* (1859), 252–55.
[17] Varley, January 5, 1860, *Joint Committee Report*, 164.
[18] See Latimer Clark, "Appendix 2," in *Joint Committee Report*, 310; Latimer Clark, *Elementary Treatise on Electrical Measurement* (London: E. and F. N. Spon, 1868), 113; and Siemens, *Inventor and Entrepreneur*, 132.

By the late 1850s, Clark, Varley, Bright, Smith, and other leading British cable engineers and electricians were using calibrated resistance coils on a regular basis and were beginning to use roughly calibrated condensers as well. To that extent, Thomson's 1883 remark was correct. The engineers had developed many of their techniques on their own and had borrowed others from the small number of laboratory researchers who had worked on electrical measurement in the 1820s, '30s, and '40s – particularly G. S. Ohm, Charles Wheatstone, and Wilhelm Weber – and then adapted them to fit their own needs, especially for the comparison of relatively large resistances. By the late 1850s cable engineers and electricians were performing reasonably precise electrical measurements far more often and on a far larger scale than laboratory physicists ever had. But neither engineers nor physicists had yet begun to state their results in "ohms" or "microfarads." The move to such standardized units did not come until the 1860s, in the wake of the string of dramatic reverses that were to beset the cable industry.

Specifications and Standards

The failure of the first Atlantic cable in September 1858 and the collapse of the Red Sea line two years later raised questions that went to the heart of the future development of submarine telegraphy, as the Joint Committee investigations clearly showed. Among the most important of these questions – one that would do much to shape the interaction between electrical physics and practical telegraphy in the 1860s, and indeed for the rest of the century – was that of engineering standards and specifications.

William Thomson's involvement with the Atlantic cable is well known: he was a director of the Atlantic Telegraph Company, he sailed on all of its laying expeditions, and his theory of signal propagation and his new instruments – particularly his mirror galvanometer – contributed greatly to the eventual success of the project in 1866 and the subsequent spread of the cable network around the globe. Less noticed but in some ways just as important was the key role he played in the adoption of electrical specifications for cable conductors and insulation. When in 1857 Thomson and a group of his Glasgow students used a Wheatstone bridge and a tangent galvanometer to compare various samples of wire intended for use in the Atlantic cable with one they had chosen as a standard, they had been surprised to find that the resistance of supposedly identical "pure" copper wires sometimes differed by nearly a factor of two.[19]

[19] William Thomson, "On the Electrical Conductivity of Commercial Copper of Various Kinds," *Proc. RS* (June 1857) 8: 550–55, repr. in Thomson, *MPP 2*: 112–17. Smith (*Rise*

Thomson's ensuing investigation of the varying conductivity of "commercial copper" had far-reaching consequences, particularly for issues of standardization and quality control. Subsequent analysis showed that small impurities, particularly of arsenic, could greatly impair the conductivity of copper, a point the London chemist Augustus Matthiessen would later follow up in detail.[20]

It is worth noting that Thomson's discovery did not require unusual instruments or especially precise measurements; the differences in conductivity were large enough to have been found by anyone who systematically compared the resistance of different samples of copper wire – that is, if they trusted their standards and instruments. But before reliable standards had been established, such trust was rare; it was easy, when an investigator found discordant results, to ascribe them to some unspecified flaw in one's procedures or instruments. Once Thomson and his students had established that the conductivity of different samples of copper in fact varied widely, their discovery gave an important stimulus to electrical measurement as a means of quality control. Before 1857, cable contracts had specified only the weight or gauge of the wire, with perhaps a vague reference to its chemical purity; they had said nothing about its electrical characteristics. The original contract for the first Atlantic cable, for example, had simply called for the conductor to consist of seven twenty-two-gauge copper wires stranded together; it said not a word about the electrical qualities of the copper, nor did it require its conductivity to be tested in any way.[21] By the time Thomson discovered the variation in the conductivity of commercial copper, the main length of the original Atlantic cable had already been manufactured, and it was only after a determined campaign that he managed to convince his fellow directors of the Atlantic Telegraph Company to require that only high-conductivity copper be used in any wire still to be delivered. After balking at first, the Gutta Percha Company and its copper suppliers instituted a program of testing and, for a small premium, began to supply wire of the required conductivity.[22] The last few hundred miles of cable to be manufactured, including the length ordered to replace that lost in the abortive 1857

and Extension, 116–17) later found that the conductivity of wire used in various early cables ranged from 17 percent to 75 percent of that of pure copper.

[20] On Matthiessen's work, see Graeme J. N. Gooday, *The Morals of Measurement: Accuracy, Irony, and Trust in Late Victorian Electrical Practice* (Cambridge: Cambridge University Press, 2004), 82–127.

[21] ATC Minute Book, entry for November 10, 1856, 7.

[22] William Thomson, "Analytical and Synthetical Attempts to Ascertain the Cause of the Differences of Electric Conductivity Discovered in Wires of Nearly Pure Copper," *Proc. RS* (February 1860) *10*: 300–9, repr. in Thomson, *MPP 2*: 118–28; see "Note of date June 27, 1884," 125n. See also Smith, *Rise and Extension*, 47, 66.

laying attempt, was thus of substantially higher conductivity than that delivered earlier. Nothing could be done about the relatively poor conductivity of the older length of cable; moreover, since the resistance per mile of its different portions had never been measured or recorded, there was no way, by measurements made on its ends, to determine with any real accuracy the location of faults that might appear in its insulation, as Varley and Thomson found when they arrived at Valentia to diagnose the ailing cable in August 1858.[23]

The Joint Committee traced the failure of the first Atlantic cable to a long list of causes, including hasty manufacture, rough handling, and a lack of proper testing of the completed cable. In its final report, the committee put particular emphasis on the failure of the Atlantic Telegraph Company to draw up proper specifications for the materials used in the cable and its related failure to enforce adequate control over the quality of the work during manufacture.[24] Such specifications required reference to agreed standards, which would be binding on all parties to the contract. The committee recommended that all future cable contracts should stipulate that "the conductivity of the wire shall be equal to that of a standard wire," which should be made from a metal or alloy whose resistance did not vary much with temperature.[25] Clark and Varley, themselves both members of the Joint Committee, also endorsed the use of standard resistances, while in his own testimony to the committee Jenkin went further, calling for the adoption of a "well recognized standard of resistance, or resistance coil," for universal use and reference.[26]

The need for a shared standard of resistance was widely felt by the time Jenkin testified in December 1859, but exactly what should that standard be (Figure 4.2)? A welter of resistance standards were then in circulation, but none were entirely satisfactory. Wheatstone had proposed a foot of copper wire weighing one hundred grains in 1843, and Jacobi had sent copies of his longer "etalon" to various physicists in 1848, but neither of these standards had come into wide use – in part because, at the time they were proposed, there was little effective demand for them. As we have seen, when telegraph engineers began to make their own resistance measurements, they preferred coils calibrated in larger units: miles of sixteen-

[23] C. F. Varley, January 5, 1860, *Joint Committee Report*, 162, reporting that the conductivity of different portions of the Atlantic cable differed by as much as 30 percent.

[24] *Joint Committee Report*, ix, xxxii; C. A. Hempstead, "The Early Years of Oceanic Telegraphy: Technology, Science, and Politics," *Proceedings of the Institution of Electrical Engineers* (1989) *136A*: 297–305.

[25] *Joint Committee Report*, xxxiii.

[26] Jenkin, December 22, 1859, *Joint Committee Report*, 139; see also the testimony of C. F. Varley, January 5, 1860, 162, and Clark, February 2, 1860, 206.

Figure 4.2 Fleeming Jenkin was 26 when he made this pencil sketch of himself in April 1859.
(From Fleeming Jenkin, *Papers*, Vol. 1: lxvii; courtesy University of Texas Libraries.)

gauge copper wire in Britain (standardized at the Electric Telegraph Company as "Varley units"), kilometers of iron wire in France, and German miles of iron wire in Germany. Just as these various standards were coming into wider use in the late 1850s, their reliability was called into question by Thomson's discovery of the variation in the resistance of "commercial copper," and comparisons soon showed that supposedly identical standard coils often differed by several percent.[27] In his report to the Joint Committee, Jenkin complained that the only resistance coils available to him at Newall's cable works were three "very imperfect" ones (made by Siemens and Halske) that were nominally equivalent to 30, 60, and 90 German miles of telegraph wire, but which, when he compared them on a Wheatstone bridge, turned out actually to be in the ratios 30 : 59.15 : 88.27.[28] When Jenkin served as a juror for the London

[27] See the comparative table of standards in Jenkin, "Report on the New Unit," in Smith, *Reports*, opp. 288.
[28] Fleeming Jenkin, "On the Insulating Properties of Gutta-Percha," *Joint Committee Report*, 464–81, on 469.

International Exhibition in 1862, he reported substantial discrepancies in the resistances of supposedly identical coils submitted by different makers, including the London and Berlin branches of the Siemens firm.[29]

An alternative to arbitrary material standards of resistance already existed, at least on paper, in the "absolute" system based on units of force and motion that Weber, building on C. F. Gauss's earlier magnetic work, had published in 1851.[30] Gauss had defined a unit magnetic pole as one which, at unit distance, exerts on an identical pole a unit of force – that is, enough force to accelerate a unit mass to unit velocity in a unit of time. Weber defined a unit of electric charge in a similar way, as that charge which, at unit distance, exerts a unit of force on an identical charge. He could then define a unit of electric current in either of two ways: as the current produced by the passage of a unit of charge in a unit of time, or as the current that (properly arranged) would, at unit distance, exert a unit of force on a unit magnetic pole. The first definition of a current yielded the electrostatic system of units; the second, the electromagnetic system. The ratio between the two systems of units came out as a velocity and was a universal physical constant whose magnitude – roughly equal to the speed of light, according to Weber's measurements – was independent of the units in which it was expressed.[31] It was an important quantity in Thomson's theory of telegraphic propagation and, as we shall see, would come to play a central role in Maxwell's theory of the electromagnetic field.

Weber's absolute system was cohesive and comprehensive and, as Thomson emphasized, it tied electrical quantities directly to those of work and energy: one unit of current driven by one unit of electromotive force delivered one unit of energy per second (in later terms, 1 amp × 1 volt = 1 watt).[32] It was, however, difficult to explain in an elementary way how Weber's units of electromotive force and resistance

---

[29] Fleeming Jenkin, "Report on Electrical Instruments," *Reports by the Juries, International Exhibition, 1862* (London: Society of Arts, 1863), 44–98, on 82.

[30] Wilhelm Weber, "Messungen galvanischer Leitungswiderstände nach einem absoluten Maasse," *Annalen der Physik* (March 1851) 82: 337–69. See also André Koch Torres Assis, *Weber's Electrodynamics* (Dordrecht: Kluwer Academic, 1994).

[31] For a recent discussion of this ratio and its interpretation, see Daniel Jon Mitchell, "What's Nu? A Re-Examination of Maxwell's 'Ratio-of-Units' Argument, from the Mechanical Theory of the Electromagnetic Field to 'On the Elementary Relations between Electrical Measurements,'" *Studies in History and Philosophy of Science* (2017) 65–66: 87–98.

[32] [William Thomson], "Report of the Committee appointed by the British Association on Standards of Electrical Resistance," *BA Report* (1863), 111–24, repr. in Smith, *Reports*, 58–78, on 58–59; on Thomson's authorship, see Thompson, *Kelvin*, 1: 419. Although Weber stated his unit in millimeters, most British accounts of his work converted this to meters.

were defined, and even more difficult to embody them accurately in material standards. Rather than simply choosing a certain piece of wire to represent the unit of resistance and simultaneously to serve as a material standard for comparison, Weber's absolute system required a delicate measurement with special apparatus to determine the resistance of a given conductor, which in the electromagnetic system of units would be expressed as a velocity. This conductor, typically a small coil of wire, could then be treated as a standard, but it would always be an open question whether its resistance in fact accurately matched the abstract unit it was meant to embody. Given the uncertainty of Weber's measurements – some of his own determinations differed among themselves by several percent – and the fact that his unit of resistance, the millimeter per second, was ludicrously small (less than 1/100 billionth the resistance of a mile of ordinary copper wire), it is perhaps not surprising that the few telegraph engineers who knew about the absolute system in the 1850s regarded it as having little practical value.

In 1860, Werner Siemens attempted to resolve the problem by introducing a new unit based on the resistance, at a temperature of 0°C, of a tube of pure mercury one meter long and one square millimeter in section. Siemens's unit was arbitrary and bore no particular relationship to units of length, mass, time, or energy. It was, however, of a convenient magnitude (equal to the resistance of a few hundred feet of ordinary copper telegraph wire), and standards representing it could be measured and reproduced quite precisely – to within a very small fraction of one percent, according to Siemens. Matthiessen later criticized Siemens's standard, claiming that impurities dissolved from the connecting wires would alter the conductivity of the mercury, but when Siemens began to issue wire coils calibrated in terms of his mercury unit in 1860, he had every reason to think they would soon come into general use, as they in fact did in Germany.[33]

---

[33] Werner Siemens, "Vorschlag eines reproducirbaren Widerstandsmaasses," *Annalen der Physik* (June 1860) *110*: 1–20, trans. as "Proposal for a New Reproducible Standard Measure of Resistance to Galvanic Currents" in *Phil. Mag.* (January 1861) *21*: 25–38 and as "Proposal for a Reproducible Unit of Electrical Resistance" in Siemens, *Scientific and Technical Papers*, *1*: 162–80; Werner Siemens and C. W. Siemens, "Outline of the Principles and Practice Involved in Dealing with the Electrical Conditions of Submarine Electric Telegraphs," *BA Report* (1860), 32–34; Werner Siemens and C. William Siemens, "Outline of the Principles and Practice Involved in Testing the Electrical Condition of Submarine Telegraph Cables," in Siemens, *Scientific and Technical Papers*, *1*: 137–59; Augustus Matthiessen and C. Vogt, "On the Influence of Traces of Foreign Metals on the Electrical Conducting Power of Mercury," *Phil. Mag.* (March 1862) *23*: 171–79; see also Siemens, *Inventor and Entrepreneur*, 133, 160. On the controversies over Siemens's mercury unit, see Gooday, *Morals of Measurement*, 82–127, and Kathryn M. Olesko, "Precision, Tolerance, and Consensus: Local Cultures in

By the time Siemens proposed his mercury unit in 1860, the idea of a single universal standard of resistance was very much in the air. Varley and Clark had both cited the advantages of such a standard in the evidence they presented to the Joint Committee, and in his own testimony Thomson had declared that all cables should be tested "entirely by comparison with absolute standards of resistance"; he would, he said, "never think any testing apparatus at all satisfactory or complete without a very well arranged set of coils for standards of resistance." Appearing a few days later, Jenkin called for a "standard coil with which any resistance might be compared" to be deposited in "some public institution," much as the standard meter was kept in Paris and the standard yard in London.[34] A standard of electrical resistance should, he said, have the same official status and authority.

A set of standard resistance coils could be of considerable use on its own for locating faults and testing lines, but the real advantages of standardization came into play only when the process was extended beyond the coils themselves and made part of a general policy and practice of calibration and quality control. Resistance standards could then be used to measure, control, and record the properties of the copper and gutta-percha in a cable while it was being made and laid, a procedure whose importance Clark, Varley, and Smith all emphasized in their testimony to the Joint Committee. Varley was especially forceful on the question, declaring that "During the manufacture of a cable, every mile of that cable ought to be tested carefully with resistance coils, and the results accurately noted down."[35] By securing strict control over the quality of materials, engineers equipped with reliable standard coils could ensure that only copper and gutta-percha that met a specified standard were used – an important consideration, since a single tiny flaw could put an entire cable out of operation. Moreover, standardization of materials would enable engineers to locate faults in their cables far more exactly, since they could then be sure that the resistance of each mile of the conductor was strictly comparable to that of the coils with which it had been tested, rather than subject to the large variations typical in untested materials. Indeed, Varley said that uncertainty about the resistance of different portions of the Atlantic cable was what had kept him from being able to pinpoint more precisely the location of the fault that had killed it. "Had the resistance of the wire been measured as

German and British Resistance Standards," in Jed Z. Buchwald, ed., *Scientific Credibility and Technical Standards* (Dordrecht: Kluwer Academic, 1996), 117–56.

[34] Thomson, December 17, 1859, *Joint Committee Report*, 118; Jenkin, 22 Dec. 1859, *Joint Committee Report*, 139.

[35] C. F. Varley, January 5, 1860, *Joint Committee Report*, 162.

manufactured into cable," he said, "the distance of the fault would have been known to within ten miles," instead of only within a range of about fifty miles.[36]

Such standardization – first of resistance coils, then of production materials – would enable scientists and engineers to extend their networks of calculation and control into the world around them by simply making and sending out what were, in effect, little pieces of their laboratories and testing rooms.[37] They would then be able to travel around the world without, in a sense, ever having to leave their laboratories – as long as they were able to put certified copies of their standards and instruments wherever they had to go. The engineer sought to make his entire cable as nearly as possible a chain of little standard resistances strung end to end, with the resistance of each mile and yard of it known, controlled, and recorded. Guesswork would be eliminated, and the engineer would be able to dazzle his lesser brethren by specifying the location of a distant and unseen fault more precisely than the repair-ship captain could navigate to it – something cable engineers in fact began to do in the 1860s and 1870s.[38]

As useful as the precision and control afforded by standardization was within the confines of a single company's system, it became even more important when an exchange of materials was involved – that is, when standardization became a part of contract specifications. By providing fixed and agreed reference points in which both parties could have confidence and to which they would be legally bound, standard resistances were crucial to heading off possible disputes or to settling them once they arose. By enabling engineers to secure the comparability and even uniformity of their copper and gutta-percha, to identify and police deviations, and to reproduce the properties of successful cables in a predictable way, reliable standards were crucial to the growth and success of the cable manufacturing industry and to the efficient extension, operation, and maintenance of the global cable system.

The prospective advantages of this kind of standardization became increasingly clear after the failure of the Atlantic and Red Sea cables. In its final summary report, the Joint Committee stated that "in the contract for a telegraph cable, a wire affording a standard of resistance" should always be provided, citing as examples the ones devised by Varley, Siemens, and Matthiessen.[39] Adoption not just of a separate standard

[36] C. F. Varley, January 5, 1860, *Joint Committee Report*, 159.

[37] Bruno Latour, *Science in Action: How to Follow Scientists and Engineers through Society* (Cambridge, MA: Harvard Univ. Press, 1987), 247–57, esp. 251.

[38] Bright, *Submarine Telegraphs*, 182n; Siemens, *Inventor and Entrepreneur*, 132–33.

[39] *Joint Committee Report*, xvi–xvii.

for each contract but of a single uniform standard for all would clearly offer even more advantages, comparable to those that had long been secured by state-sanctioned standard weights and measures. By the time the Joint Committee completed its report in April 1861, the demands of cable telegraphy were rapidly bringing to a head the movement toward a common standard of electrical resistance.

Manchester, 1861

The growing call for a shared set of electrical standards culminated in the formation of a Committee on Standards of Electrical Resistance at the Manchester meeting of the British Association in September 1861. Over the next few years, this committee (which initially included Thomson, Jenkin, Wheatstone, and several eminent chemists, and later added Maxwell and other leading physicists and cable engineers) produced essentially the system of ohms, amps, and volts that is still used today. It was the most important point of intersection between physicists and telegraph engineers in the 1860s, and its work, especially on the ohm, had far-reaching effects on virtually all later work in precision electrical measurement.

Many accounts of the formation of the British Association committee have focused on a paper, "On the Principles Which Should be Observed in the Formation of Standards of Measurement of Electrical Quantities and Resistance," that Latimer Clark and Sir Charles Tilston Bright presented at the Manchester meeting.[40] Although the paper bore the names of both men, who had recently become partners in a cable consulting firm, it was clearly rooted in the long report on "The Laws Which Govern the Propagation of the Electric Current in Long Submarine Telegraph Cables" that Clark had submitted a few months earlier to the Joint Committee (indeed, it first appeared in print as an appendix to his separately issued edition of that report), and he later said that the "original ideas" in it had all emanated from him.[41] Those ideas were certainly

---

[40] Latimer Clark and Sir Charles Bright, "On the Principles Which Should be Observed in the Formation of Standards of Measurement of Electrical Quantities and Resistance," in Clark, *Experimental Investigation of the Laws Which Govern the Propagation of the Electric Current in Long Submarine Telegraph Cables* (London: George E. Eyre and William Spottiswoode, 1861) 49–50, on 49. The paper appeared under the title "Measurement of Electrical Quantities and Resistance" in *Elec.* (November 9, 1861) *1*: 3–4. A brief abstract appeared in *BA Report* (1861), 37–38, under the title "On the Formation of Standards of Electrical Quantity and Resistance," while the *Athenæum* gave its title as "On Standards of Measurement of Electrical Quantities and Resistances," *Athenæum* (September 28, 1861) 412.

[41] Latimer Clark to William Thomson, May 3, 1883, quoted in Crosbie Smith and Norton Wise, *Energy and Empire: A Biographical Study of Lord Kelvin* (Cambridge:

remarkable and presaged in many ways the recommendations that would eventually be adopted by the British Association committee. After noting that "The science of electricity and the art of telegraphy have both now arrived at a stage of progress at which it is necessary that universally received standards of electrical quantities and resistances should be adopted, in order that precise language and measurement may take the place of the empirical rules and ideas now generally prevalent," Clark outlined a connected system of units of electrical tension, "quantity" (i.e., charge), current, and resistance. He proposed as his unit of tension the electromotive force produced by a single Daniell's cell; of static electricity, the charge stored when a unit of tension was applied across two metal plates one square meter in area and held one millimeter apart; of current, the passage of one unit of charge per second; and of resistance, a wire that, when subjected to one unit of tension, "will conduct one unit of electricity in one unit of time."[42] To fit the magnitudes of these units to the needs of cable engineers, Clark suggested employing prefixes ("kilo," "millio," and "billio") to indicate decimal multiples and submultiples – though he was soon criticized for not applying these in the proper way.[43] He also suggested that "for this temporary purpose," the units might be named after "some of our most eminent philosophers": "ohma" for tension, "farad" for charge, "galvat" for current, and "volt" for resistance. While the names given to the various quantities would soon be switched around, Clark's proposal was evidently the beginning of the now ubiquitous practice of naming units of measurement for distinguished figures in science and technology. The paper concluded by declaring that the adoption of this or a similar system of units and standards would be a "great a boon to science and to the art of telegraphy," and called on the British Association to form a committee, "with power to confer with English or foreign philosophers," that would promulgate such a system and prepare electrical standards "for public use and reference." If the Association were to launch such an effort, Clark said, he was confident it "would meet with the hearty co-operation and assistance of practical electricians."[44]

Cambridge University Press, 1990), 687. For earlier accounts of the formation of the committee, see A. C. Lynch, "History of the Electrical Units and Early Standards," *Proceedings of the Institution of Electrical Engineers* (1985) *132A*: 564–73; Bright, *Submarine Telegraphs*, 61; Thompson, *Kelvin*, *1*: 417–18; and Graeme Gooday, "Precision Measurement and the Genesis of Physics Teaching Laboratories in Victorian Britain," *British Journal for the History of Science* (1990) *23*: 25–51, on 34.

[42] Clark and Bright, "Principles," 49.
[43] "Standards of Electrical Measurement," *Elec.* (November 9, 1861) *1*: 3.
[44] Clark and Bright, "Principles," 50.

Given all this, it is not surprising that Clark and Bright's paper has often been cited as the direct stimulus for the formation at the same meeting of what later became the British Association Committee on Electrical Standards.[45] But several facts call this straightforward story into question. First, neither Bright nor Clark were named to the committee when it was formed in 1861, which would be odd if they were indeed its instigators. (Bright was added to it in 1862 and Clark not until 1866.) Secondly, the committee as originally constituted was asked to report only on "Standards of Electrical Resistance" – though one of Clark's main points had been his call for a connected system of standards of tension, quantity, and current as well. (The committee's terms of reference were broadened to include these other electrical standards in 1862.)[46] In addition, several of the best-informed accounts from the time explicitly state that the committee was formed at *Thomson's* suggestion – and Thomson was not at the Manchester meeting; having broken his leg in a fall on the ice while curling some months before, he remained laid up in Scotland and so could not have been responding to Clark and Bright's paper.[47] Finally, it turns out that Clark and Bright did not appear at the Manchester meeting until near its end, after the resolution establishing the committee had already been adopted.[48] What, then, was the relationship, if any, between Clark and Bright's paper and the formation of the British Association Committee on Standards of Electrical Resistance?

The answer, or most of it, can be found in two letters from Jenkin to Thomson, now held in Thomson's papers at the Cambridge University Library.[49] Although they are undated, their contents show that Jenkin wrote them during the Manchester meeting, at which he acted on Thomson's behalf. Jenkin's remarks reflect the fact that Thomson, motivated by concerns similar to Clark and Bright's, had launched his own effort to persuade the British Association to set up a committee on

---

[45] See, for example, Lynch, "Electrical Units," 564; Smith and Wise, *Energy and Empire*, 687; and "Mr. Latimer Clark" (obituary), *Science* (November 18, 1898), *8*: 704–5, all of which credit Clark and Bright's 1861 paper with prompting the formation of the British Association committee.

[46] The texts of the relevant resolutions appear in *BA Report* (1861), xxxix–xl, and (1862), xxxix.

[47] Jenkin, "Report on the New Unit," in Smith, *Reports*, 281; Thompson, *Kelvin*, *1*: 418; Smith and Wise, *Energy and Empire*, 687.

[48] Clark and Bright delivered their paper on September 10, 1861, while the General Committee meetings that established the Committee on Standards of Electrical Resistance were held on September 4 and September 9; see *BA Report* (1861), xxxi, xxxix–xl, and l.

[49] Fleeming Jenkin to William Thomson, J36 and J37 (undated, but between September 4 and September 11, 1861; J37 was written a little before J36), Kelvin Collection, CUL 7342.

electrical units, or at least resistance standards, well before the Manchester meeting began, and that he was already campaigning for the adoption of an "absolute" system based on Weber's. That June, Thomson had sent the Royal Society a paper, "On the Measurement of Electric Resistance," in which he described how to use a new form of Wheatstone bridge to make very precise resistance measurements. In a long footnote to the paper he praised Weber's system for the way it tied electrical units to those for work and energy, and urged the Royal Society and the British Association to help promote "a proper mutual understanding between electricians and national scientific academies, in all parts of the world," on the adoption of a common system of electrical units and standards.[50] Thomson had also written to Matthiessen and apparently also to Wheatstone, who drafted the resolution ("not altogether a bad one," according to Jenkin) calling for the Association to establish a committee on standards of electrical resistance and provide funds to support its work.[51] Clark and Bright had prepared their own proposal without knowing of Thomson's efforts, and by the time they arrived at the Manchester meeting and delivered their paper on September 10, the next to last day of the meeting, the committee had in fact already been appointed.

Bright's brother and son later said that Clark and Bright's paper "formed the sequel to a letter addressed by Bright to Prof. J. Clerk Maxwell, F.R.S., some months previously, on the whole subject of electrical standards and units," but they cited no evidence to support this claim, which on its face seems implausible. Maxwell was not then known for any work on electrical measurement and it would have been surprising for Bright to have written to him out of the blue for information about the subject. It was, however, at just this time – the summer of 1861 – that Maxwell first sought to use measurements of electric and magnetic constants to work out the speed at which electromagnetic disturbances would propagate in his vortex ether. He may at some point have written to Bright for information on such measurements, as he did to Faraday and Thomson later in 1861, and the Brights' reference may have been to Bright's reply, though this would be more likely to have occurred after Clark and Bright presented their paper at the Manchester meeting rather than some months before.[52] Maxwell later became one of the central

---

[50] William Thomson, "On the Measurement of Electric Resistance," *Proc. RS* (June 1861) *11*: 313–28, on 315n.

[51] Jenkin to Thomson, J37, Kelvin Collection, CUL 7342.

[52] Edward Brailsford Bright and Charles Bright, *The Life of the Late Sir Charles Tilston Bright*, 2 vols. (London: Archibald Constable, 1899), 2: 21; Maxwell to Michael Faraday,

figures in the British Association committee, but the Brights' claim not-withstanding, he does not appear to have been involved in its formation.

When Jenkin encountered Clark and Bright at the Manchester meet-ing, he immediately sought to enlist them behind Thomson's effort to launch a committee on resistance standards. He apparently had some success, telling Thomson that "Latimer Clark looked delighted" when Jenkin told him of the plan and was "eager to have it all explained."[53] But it was not all smooth sailing; Jenkin said later in his first letter to Thomson that he was "writing in a very great hurry after hot argument with Sir C. and Latimer," and in his next letter he said "They will no doubt be easily converted."[54] Jenkin did not say what Clark and Bright needed to be converted *from*, but he presumably meant they were skeptical about simply adopting Weber's system of absolute units. Jenkin told Thomson that it was only "by force of telling others about them" that he was "beginning really to understand" absolute units himself, and they cer-tainly were not then well understood or accepted by other telegraph engineers.[55] Indeed, Clark told Thomson more than twenty years later that in 1861 he "knew nothing of Weber's work." He had known a little about James Joule's work on the heating effects of electrical currents, he said, but had not yet grasped the importance of being able to relate electrical measurements directly to units of work and energy, a point that would become especially salient with the rise of the electrical power industry in the 1880s. Clark told Thomson that in 1861 "I was not mathematician enough to see the enormous value of an absolute system, founded on mass, time, and space"; only later would he recognize this as the feature that had "gained for the British System of Electrical Measurement its universal acceptance by mankind" – something that in 1861 still very much lay in the future.[56]

In the early 1860s, few telegraph engineers cared much about consid-erations of work and energy or the intricacies of Weber's absolute system; they simply wanted a concrete material standard – a coil of wire – whose resistance would be of a magnitude suited to the measurements they

October 19, 1861, and Maxwell to William Thomson, December 17, 1861, in Maxwell, *SLP 1*: 683–88 and 699–702.

[53] Jenkin to Thomson, J37, Kelvin Collection, CUL 7342.
[54] Jenkin to Thomson, J37 and J36, Kelvin Collection, CUL 7342.
[55] Jenkin to Thomson, J36, Kelvin Collection, CUL 7342.
[56] Latimer Clark to William Thomson, May 3, 1883, C91, Kelvin Collection, CUL 7342. Having seen that Thomson was to speak that evening at the Institution of Civil Engineers on "Electrical Units of Measurement," Clark sent him a copy of the paper he and Bright had delivered in 1861, noting that it was "so exceedingly rare that I have only one copy of it myself and it is more than probable that you have never seen it." Clark enclosed a stamped envelope so that Thomson could return the paper after he had read it.

made in their daily work and that would perhaps bear a simple relation to comparable units of tension and current. Siemens's mercury unit was about the right size and was defined in a simple and understandable way, but the fact that it lacked any direct relationship to units of tension or current left an opening for an alternative system. When the first issue of the *Electrician*, a new weekly "journal of telegraphy," appeared in November 1861, it carried both the full text of Clark and Bright's paper and a short piece on "Standards of Electrical Measurement" in which the editors welcomed the appointment of the British Association committee and urged its members to bear in mind the needs of practical telegraphists. The need for "standards, to which the tension and quantity of a current, and the resistance of a conductor, might be conveniently and intelligibly referred, had long been felt," the editors said,

and there is no doubt that their employment might greatly contribute to the degree of accuracy which is now beginning to be evidenced in the practice of electro-telegraphy. To be of any general utility, however, the proposed system of measurement much necessarily be sufficiently simple and easy of application, to meet the requirements of telegraphists. Glancing at what has already been published in reference to this important subject, we fear there is some danger that a system may be devised, which will be followed exclusively by the eminent gentlemen at whose recommendation it is put forward.[57]

The editors did not say what existing literature on electrical standards they had glanced at, but if it included the translation of Weber's "Messungen galvanischer Leitungswiderstände nach einem absoluten Maasse" that had recently appeared in the *Philosophical Magazine*, their fear that such a system would not appeal to telegraphists is understandable. Although Weber's paper had first been published in the *Annalen der Physik* as long ago as 1851, it began to draw attention in Britain only in the wake of the publication of the *Joint Committee Report* and Thomson's praise for the value of absolute electrical measurements. Noting "the great scientific and practical importance that the determination of electric resistance has of late acquired," the editors of the *Philosophical Magazine* commissioned a translation of Weber's paper, which appeared in the September and October 1861 issues under the title "On the Measurement of Electric Resistance According to an Absolute Standard."[58] But while it was now available in English, Weber's text remained notoriously forbidding. Few telegraphists would have been able to pick their way through his many equations, or known what to make of his definition of the absolute unit of

---

[57] "Standards of Electrical Measurement," *Elec.* (November 9, 1861) *1*: 3.
[58] Wilhelm Weber, "On the Measurement of Electric Resistance," *Phil. Mag.* (September 1861) *22*: 226–40 and (October 1861) *22*: 261–69; see the editorial note on 226.

electromotive force as "that electromotive force which the unit of measure of the earth's magnetism exerts upon a closed conductor, if the latter is so turned that the area of its projection on a plane normal to the direction of the earth's magnetism increases or decreases during the unit of time by the unit of surface."[59] Clark's proposal to take a single Daniell's cell as the unit of tension would no doubt have struck them as much simpler and more practical. Nor were telegraphists seeking further light on the subject likely to be helped much by the source to which Weber pointed them, Gauss's 1833 paper "Intensitas Vis Magneticæ Terestris ad mensuram absolutam revocata." As the editors of the *Electrician* observed, working telegraphists already had their own methods of measurement, "however imperfect and miscellaneous," and while they could no doubt be improved upon, "they will certainly not be relinquished for any plan, however perfect in theory, characterised by abstruseness, and difficulty in application."[60]

To they extent they knew of it at all, most telegraphists certainly regarded Weber's system as both abstruse and difficult to apply. Its definitions were abstract and hard to follow, the instruments and procedures required to put them into practice were complex and unfamiliar, and the magnitudes of the resulting units bore little relationship to the needs of telegraphists. Clark himself weighed in on the question in a letter to the *Electrician* in January 1862, writing that the men initially appointed to the British Association committee were "but little connected with practical telegraphy, and there is a fear that while bringing the highest electrical knowledge to the subject, and acting with the best motives, they may be induced simply to recommend the adoption of Weber's absolute units, or some other units of a magnitude ill adapted to the peculiar and various requirements of the electric telegraph."[61]

Choosing a Unit

The British Association Committee on Standards of Electrical Resistance was well aware of fears like those voiced by Clark and the editors of the *Electrician*, and it proved itself fully intent on devising a system that would appeal to telegraph engineers as well as laboratory researchers. This was in part because the members of the committee – especially Thomson and Jenkin, who took the lead in most of its early activities – sincerely wished to serve the needs of the cable industry, and in part because they knew that no new system they might propose would be able either to displace the existing patchwork of *ad hoc* resistance standards or to fend off rivals

---

[59] Weber, "Electric Resistance," 227.   [60] "Standards," *Elec.*, 3.
[61] Latimer Clark, (letter), *Elec.* (January 17, 1862) *1*: 129.

like Siemens's mercury unit unless it could appeal to working telegraph-ists. Telegraphy, especially cable telegraphy, was the chief arena for electrical measurement in the mid-nineteenth century and so provided the principal potential market for the work of the British Association committee. By devising a system that would meet the needs of both telegraphists and laboratory researchers, the committee sought to unify the practices of both communities on a single basis and so give their shared set of units and standards a broader and more comprehensive authority than either group could have commanded on its own.

The committee met several times in 1861–1862 and hammered out a set of guiding principles, which Jenkin laid out in the report he drew up for the 1862 Cambridge meeting of the British Association. He began by apologizing for the inability of the committee to complete its work in a single year, a delay he ascribed to "the inherent difficulty and import-ance of the subject."[62] Neither he nor the other members of the commit-tee could then foresee that it would in fact continue until 1870, much less that it would be reconstituted in 1881 and carry on for decades after that. Jenkin next carefully explained the difference between a *unit* of resistance and a *standard*, noting that if the committee were, for instance, to recom-mend "a unit based on Professor Weber's or Sir Charles Bright and Mr. Latimer Clark's system, this decision would not affect the question of construction," while on the other hand if it were to decide "in favour of any particular arrangement of mercury or gold wire as the best form of standard, this choice would not affect the question of what the absolute magnitude of the unit was to be." The committee had, Jenkin said, "arrived at a provisional conclusion" regarding the choice of a unit of resistance but had not yet decided on how to embody it in a concrete standard. In the event, choosing "the best form and material for the standard," and then actually constructing and certifying the resulting resistance coils, would prove to be a long, difficult, and contentious process.[63]

In choosing a unit of resistance that would be "the most convenient ... for all purposes, both practical and purely scientific," the committee was guided, Jenkin said, by five principles: the unit should be of a magnitude that "would lend itself to the more usual electrical measurements," par-ticularly those made by telegraph engineers, "without requiring the use of extravagantly high numbers of cyphers or of a long string of decimals"; it should, along with associated units of charge, current, and electromotive

[62] Fleeming Jenkin, "Provisional Report of the Committee Appointed by the British Association on Standards of Electrical Resistance," *BA Report* (1862), 125–35, on 125; repr. in Smith, *Reports*, 1–16, on 1.
[63] Jenkin, "Provisional Report," in Smith, *Reports*, 1.

force, "form part of a complete system of electrical measurements"; the resulting system should "bear a definite relation to the unit of work, the great connecting link between all physical measurements"; the unit of resistance should be "perfectly definite, and should not be liable to require correction or alteration from time to time"; and it should be "reproducible with exactitude," even if the original standard were some-how injured or lost.[64] As Jenkin noted, the first and especially the second of these principles reflected the influence of Clark and Bright's paper, with its emphasis on the merits of a connected system of electrical units with magnitudes suited to the needs of telegraph engineers. The third principle reflected Thomson's focus on work and energy as the founda-tion of all areas of physics, while the fourth and fifth principles, both of which concerned the relationship between units and standards, would in time prove to be the most problematic.

As the committee debated what system of electrical units to recom-mend, Thomson's support for an "absolute" system like Weber's carried the day. Thomson himself later admitted that "absolute" was perhaps not the best name for such a system, as it erroneously suggested that meas-urements made with it were absolutely correct or exact. In Weber's sense, however, an "absolute" measurement was simply one that was based on fundamental units of length, mass, and time rather than on simply com-paring a quantity with an arbitrary unit of its own type. Thus measuring the rate at which work is performed in foot-pounds per second was "absolute" in a way that measuring it in horsepower was not. It might have been better, Thomson said, to call Weber's electrical units "derived" or "mechanical" rather than "absolute," but since the latter term was already in use, he thought it best not to change it.[65]

In Thomson's eyes, the main advantage of absolute units was the way they formed a connected system that related the various electrical units not just to each other but to basic mechanical units of force and motion, and above all to those of work and energy. But what mechanical units should one use? And how should the committee adjust the magnitudes of the resulting electrical units to suit them to what Clark had called "the peculiar and various requirements of the electric telegraph"? One option, which Thomson had already adopted in his own work and in which Jenkin had followed him, was to base all measurements on the customary British units of feet, grains, and seconds. This yielded what Jenkin had dubbed the "Thomson's unit" of resistance, which when multiplied by 10^8 gave a "practical" unit equal to the resistance of about

---

[64] Jenkin, "Provisional Report," in Smith, *Reports*, 1–2.
[65] [Thomson], "Report of the Committee" (1863), in Smith, *Reports*, 60–61.

1200 feet of sixteen-gauge copper wire.[66] If, on the other hand, one started with basic metric units of meters, grams, and seconds, the resulting absolute unit of resistance, the meter per second, gave, when multiplied by 10^7, a "practical" unit equal to the resistance of about 320 feet of sixteen-gauge copper wire. Moreover, its value fell within a few percent of Siemens's mercury unit.[67] This closeness to Siemens's unit tipped the balance, and after careful deliberation the British Association committee voted unanimously to found its system of electrical units on what it called "the French metrical system" rather than on the customary British units. This was the first real incursion of the metric system into British measures, and it would prove crucial to the eventual wide success of the British Association system of electrical units. As Jenkin later noted, the committee believed that "while there is a possibility that we may accept foreign measures, there is no chance that the Continent will adopt ours."[68] Had the committee instead opted to base its system of absolute electrical units on feet, grains, and seconds, it is unlikely they would have won acceptance outside a few English-speaking countries or would ever have been widely used by scientific researchers.

Once the committee had chosen a metric basis for its system of units and set the magnitude of its practical unit of resistance, its units of charge, current, and electromotive force followed quite directly, adjusted as needed by appropriate decimal multiples to meet the needs of telegraphists. The committee was reassured that its proposed system would indeed meet those needs by a letter it received in September 1862 from Ernst Esselbach, a German-born cable engineer then working for the London branch of the Siemens firm. "The Committee attach high importance to this communication," Jenkin wrote, "showing as it does that a practical electrician had arrived at many of the very same conclusions as the Committee, quite independently and without consultation with any of its members."[69] Besides endorsing adoption of the metric form of Weber's absolute system, with decimal multiples to yield practical units of resistance and current, Esselbach noted that a decimal multiple of the absolute unit of electromotive force would be very close to that of a Daniell's cell, already in wide use as a practical standard of electrical tension; indeed, he suggested that by adjusting the concentration of the sulfuric acid used in such a cell, one could make it effectively equal to the required practical unit.[70]

---

[66] Jenkin, "Provisional Report," in Smith, *Reports*, 2–3, 9.

[67] Jenkin, "Provisional Report," in Smith, *Reports*, 9, 15.

[68] Jenkin, "Report on the New Unit," in Smith, *Reports*, 283.

[69] Jenkin, "Provisional Report," in Smith, *Reports*, 14.

[70] Ernst Esselbach to A. W. Williamson, September 18, 1862, appendix to Jenkin, "Provisional Report," in Smith, *Reports*, 44–46. Esselbach joined the British Association committee in 1862 but served on it for only a little more than a year; in February 1864, during a cable-laying expedition in the Gulf of Oman, he jumped

Esselbach also recommended, as the committee had, that "in order to avoid confusion" and circumlocution, the practical units should be given short distinctive names like those Clark had suggested in 1861.[71] It was not until 1865, however, that the unit of resistance began to be called an "Ohmad," soon shortened to "ohm"; in the meantime, it was generally referred to simply as the "B. A. unit."[72] By the early 1870s the British Association unit of electrostatic capacity had been dubbed the "farad" (though it was so large that most measurements were given in microfarads), and the unit of electromotive force the "volt."[73]

When the British Association committee issued its first report in the fall of 1862, it had settled on a *unit* of resistance but did not yet have a *standard*. As Jenkin noted, "Weber's unit has no material existence," and apart from Weber's own determinations, no concrete standards had yet been constructed to represent it.[74] While the members of the committee held Weber in high regard and thought it "probable that his determinations are very accurate," they believed that "in a matter of this importance, the results of no one man could be accepted without a check."[75] The committee therefore resolved to make its own independent determination of the absolute unit of resistance, using a new spinning coil method devised by Thomson. He and Jenkin, who was to assist him, had hoped to be able to report their results in time for the October 1862 meeting of the British Association, but their work as jurors for the London International Exhibition left them no time to make the required measurements. Jenkin optimistically declared that they still hoped to be able to complete the work "by Christmas," but it was not to be.[76] It would take nearly two and a half years before the committee was able to complete its measurements and issue certified resistance standards.

The British Association committee recognized from the first that uncertainties in its measurements would be inevitable, and however much care Thomson, Jenkin, and their colleagues might take, they

overboard and drowned after an episode of "delirium" reportedly brought on by fever; see *Telegraphic Journal* (March 19, 1862) *1*: 144 and (April 9, 1862) *1*: 178.

[71] Esselbach, in appendix to Jenkin, "Provisional Report," in Smith, *Reports*, 46.

[72] Jenkin, "Report on the New Unit," in Smith, *Reports*, 284.

[73] On the ohm, volt, and farad, see J. D. Everett, "First Report of the Committee for the Selection and Nomenclature of Dynamical and Electrical Units," *BA Report* (1873), 222–25, on 223. Clark called the British Association unit of inductive capacity the "Farad" as early as 1870; see his June 20, 1870, report on the laying of the British Indian cable from Bombay to Aden, DOC/CFC/3/27, C&W Archive, Porthcurno. On naming the microfarad and other British Association units, see William Thomson to James Clerk Maxwell, August 24, 1872, in Maxwell, *SLP 2*: 749n.

[74] Jenkin, "Provisional Report," in Smith, *Reports*, 6.

[75] Jenkin, "Provisional Report," in Smith, *Reports*, 9.

[76] Jenkin, "Provisional Report," in Smith, *Reports*, 10.

could never be entirely sure how close the standard they produced really came to its intended value of ten million meters per second. Did this mean that, once they had issued a standard, they would need to revise it (and all measurements made with it) every time someone managed to make a more precise determination of the absolute unit? Rendering their electrical standard impermanent in this way would raise a multitude of problems, and the committee resolved that once it had made a reasonably accurate determination and issued a standard, it would lock that in as permanently defining the unit of resistance. After that the unit of resistance would no more be altered in response to improved absolute electrical measurements than the standard meter in Paris was altered in response to new geodetic measurements. This principle was later abandoned, however, no doubt in part because the standard the committee issued in 1865, though billed as likely to be within 0.1 percent of its intended value of ten million meters per second, was in fact off by more than 1 percent.[77] In the event, the value of the British Association resistance unit, by then called the ohm, would be adjusted several times over the next few decades in response to improved measurements.[78]

Once Thomson's spinning coil apparatus was ready toward the end of 1862, he passed the task of making the actual measurements to a subcommittee consisting of Jenkin and two new members of the standards committee: James Clerk Maxwell and Balfour Stewart (Figure 4.3). The trio made numerous spins in the spring and summer of 1863 at King's College London, where Maxwell was the professor of natural philosophy, but the difficulty of performing absolute measurements of the required precision meant it was late in the year before the committee felt able to offer even tentative resistance standards, which it did through Elliott Brothers, the London instrument makers. The committee agreed to issue these preliminary coils in answer to what it said was the urgent demand for practical standards, and because "defective systems" (that is, Siemens's unit) were "daily taking firmer root" in the absence of concrete exemplars of the British Association unit.[79]

In 1864 Stewart was succeeded on the subcommittee by Charles Hockin, a recent Cambridge graduate who would go on to become

---

[77] Jenkin, "Report on the New Unit," in Smith, *Reports*, 284; on the response to the later discovery of the error in the standard issued in 1865, see "Interim Report of the Committee for Constructing and Issuing Practical Standards for Use in Electrical Measurements," *BA Report* (1881), 423–25, repr. in Smith, *Reports*, 293–96.

[78] Simon Schaffer, "Accurate Measurement Is an English Science," in M. Norton Wise, ed., *The Values of Precision* (Princeton: Princeton University Press, 1995), 135–72.

[79] [Fleeming Jenkin], "Report of the Committee on Standards of Electrical Resistance," *BA Report* (1864), 345–49, repr. in Smith, *Reports*, 159–66, on 159, 162.

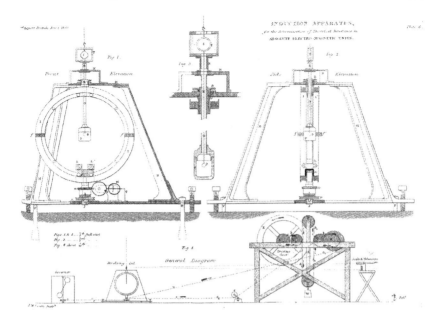

Figure 4.3 In 1863 and 1864 James Clerk Maxwell and Fleeming Jenkin used the spinning coil apparatus pictured here to determine the value of the British Association standard of electrical resistance, later dubbed the ohm. The coil was just over a foot in diameter.
(From *British Association Report*, 1863, p. 150; courtesy University of Texas Libraries.)

a leading cable engineer. After months more of making painstaking spins, tracking down sources of error, and reducing the resulting data, followed by Matthiessen's careful analysis of alloys suitable for constructing permanent standards, the committee was finally ready in February 1865 to issue its official resistance coils, certified to be accurate embodiments of the British Association unit. In a brief notice in the *Philosophical Magazine*, Jenkin, as secretary of the committee, announced that copies of the standard resistance were now available, and that "A unit coil and box will be sent on the remittance of £2 10s"[80] (Figure 4.4). As Maxwell wrote a few weeks later to his friend P. G. Tait, "the true origin of Electrical Resistance as expressed in BA units is Fleeming Jenkin Esqre 6 Duke Street Adelphi W.C. price £2.10s in

---

[80] Fleeming Jenkin, "Electrical Standard" (letter, dated February 7, 1865), *Phil. Mag.* (March 1865) 29: 248.

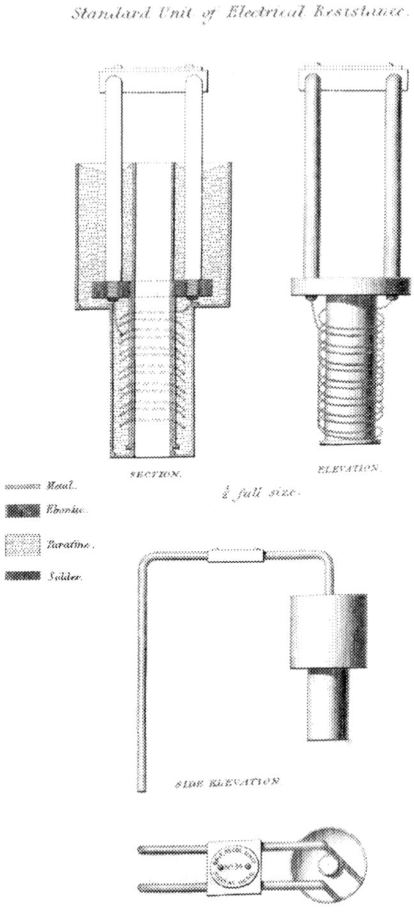

Figure 4.4 The British Association Committee on Electrical Standards issued the first certified copies of its resistance standard – soon dubbed the ohm – early in 1865. Each platinum-silver coil was embedded in paraffin wax and connected to stout copper rods of very low resistance. (From *British Association Report*, 1865, plate 10; courtesy University of Texas Libraries.)

a box."[81] It was an oddly quotidian source for such an ostensibly absolute standard.

What purchasers were really buying, of course, was not just bits of wire in a box but the concentrated expertise the coils embodied and the

---

[81] James Clerk Maxwell to P. G. Tait, March 7, 1865, in Maxwell, *SLP 2*: 214.

certified authority of the British Association committee that stood behind them. The committee sent a number of coils gratis to researchers in Britain and Germany, but it sold or gave most of them to telegraph companies and government telegraph departments around the world. "In distributing the coils," the committee said in its 1865 report, "it was thought best not to give them to institutions, where they would probably have laid on a shelf useless and unknown, but rather to distribute them widely, where they might become available to practical electricians."[82] The committee clearly recognized that the key to securing the spread and adoption of its units and standards would be to put its coils into the hands of working engineers. Indeed, the committee proudly noted that its new standard was now to provide the basis for resistance measurements at the Electric and International Telegraph Company, the British and Irish Magnetic Telegraph Company, the Atlantic Telegraph Company (then preparing another attempt to span the ocean), the Indian Telegraph Department, and several British colonial telegraph departments.[83] It had earlier expressed confidence that its standards would also be "accepted in America" and around the world, though it judged that progress on the Continent was likely to be slow.[84]

The committee was happy to report that several instrument makers had purchased certified coils in order to make and sell copies for general use. In principle anyone could make a British Association unit from scratch, based just on the meter and the second, but the task would be extraordinarily laborious and expensive, and it was obvious from the first that in practice ohms would be multiplied not by independent redeterminations but, as Thomson had said in 1861, by the "*transportation and comparison of actual standards* between different experimenters in different places."[85] Making and sending out copies was indeed how the British Association standard spread after 1865, and the dissemination of the ohm provides

---

[82] [Fleeming Jenkin], "Report of the Committee on Standards of Electrical Resistance," *BA Report* (1865), 308–11, on 310, repr. in Smith, *Reports,* 190–95, on 193–94.

[83] [Jenkin], "Report of the Committee" (1865), in Smith, *Reports,* 193–94. The statement in Gooday, *Morals of Measurement,* 115, that Jenkin said in 1866 that the British Association unit had not yet been adopted by any of the big British telegraph firms is based on a misreading. In his "Reply to Dr. Werner Siemens's Paper 'On the Question of the Unit of Electrical Resistance,'" *Phil. Mag.* (September 1866) *32*: 161–77, on 163, Jenkin in fact said that when the British Association committee was formed in 1861, the Siemens mercury unit was not being used by any large English telegraph company.

[84] [Jenkin], "Report of the Committee" (1864), in Smith, *Reports,* 166; [Jenkin], "Report of the Committee" (1865), in Smith, *Reports,* 194.

[85] William Thomson, "On the Measurement of Electric Resistance," *Proc. RS* (June 1861) *11*: 313–28, on 315n.

a clear example of what has been aptly called "the creation of universality by the circulation of particulars."[86]

The success of the ohm and its associated system of electrical units was not immediate or uncontested, however. From the start, it faced stiff competition from Siemens's mercury unit, which had the powerful backing of Siemens and Halske, then the leading electrical firm in the world. Werner Siemens had made a strong pitch to the British Association committee in 1862 for it to adopt his unit rather than one based on Weber's absolute system, and he would keep up a steady campaign for years to come.[87] Coils calibrated in terms of his mercury unit came into wide use in the 1860s, in part simply because they were what Siemens and Halske had on offer; as Jenkin remarked in 1866, "People ordered coils from the most celebrated firm in Europe and took what was given them – the miles of copper wire before 1860, and the mercury units afterwards."[88] German telegraph systems adopted Siemens's unit, and it was also used for a time in tests on some British-made cables, notably the 1865 Atlantic cable. Certified British Association standards became available as work on that cable progressed, and tests were made on it using both kinds of units. W. H. Russell, who sailed on the 1865 laying expedition, noted in his book *The Atlantic Telegraph* that the electrical room on the *Great Eastern* was equipped with both "Siemens's and B. A. unit cases," while the song "The Lay of the Electricians," published in the shipboard newspaper, mentioned both "units of Siemens" (rhymed with "cunning of demons") and "units B. A." (numbered in "millions and trillions").[89]

The battle for supremacy grew sharper after the British Association standards were issued in 1865, fought out partly over which unit should be used and partly over the material standard in which it should be embodied. Siemens and Matthiessen engaged in some particularly nasty exchanges over the relative merits of mercury and solid wires that did not peter out until after Matthiessen's death in 1870. In the 1880s

[86] Joseph O'Connell, "Metrology: The Creation of Universality by the Circulation of Particulars," *Social Studies of Science* (1993) 23: 129–73, esp. 136–47; see also Latour, *Science in Action*, 247–57.
[87] Werner Siemens, "To the Committee appointed by the British Association to report on Standards of Electrical Resistance," *BA Report* (1862), 152–55, repr. in Smith, *Reports*, 39–44; see also Gooday, *Morals of Measurement*, 82–127, and Olesko, "Precision, Tolerance, and Consensus."
[88] Jenkin, "Reply to Siemens," 163.
[89] On the use of Siemens units to measure the insulation resistance of the 1865 Atlantic cable, see Smith, *Rise and Extension*, 145; on the use of both Siemens and B. A. resistance coils during the laying expedition, see W. H. Russell, *The Atlantic Telegraph* (London: Day and Son, 1865), 45. For the lyrics of "The Lay of the Electricians," see Smith, *Rise and Extension*, 340–41.

a compromise was hammered out in which the ohm was defined as the resistance of a thread of mercury one square millimeter in section and 106.3 centimeters long – that is, essentially a Siemens unit stretched to make its resistance match as closely as practicable the ten million meters per second of the British Association unit.[90]

The American reception of the ohm is revealing. With their relatively simple network of overhead wires, American telegraphists had little real need for the kind of exact electrical measurements that preoccupied British cable engineers and so were late to take up the whole issue of units and standards. In 1868 Cromwell Fleetwood Varley visited New York and made a series of tests on wires running to the north end of Manhattan, including measuring their resistance in ohms. The editors of the *Journal of the Telegraph*, an organ of the Western Union Telegraph Company, published excerpts from Varley's report and added a note explaining that "The term ohm is the name of the British unit used in measuring the power of electric currents, an ohm being the minimum degree."[91] Someone – perhaps Varley himself – pointed out how wrong this was, and the editors hastened to correct themselves, devoting a column in their next issue to the question "What is an 'Ohm'?" The ohm, they now explained, is "a *measure of resistance* (not of power) in the same way as an inch or yard is a measure of length"; it had been "adopted by a committee of the British Association of Electricians [*sic*] and is now the acknowledged standard throughout Europe."[92] The editors also reported that Western Union had "ordered several of these Ohm standards" and would soon begin using them to improve the "care and nicety of inspection" on its lines.

When the editors of the *Journal of the Telegraph*, no doubt relying on Varley, said in 1868 that the ohm was "now the acknowledged standard throughout Europe," they were exaggerating, as was Clark when he said that same year that "The measures now universally adopted are those of the British Association."[93] But despite continued competition from Siemens's mercury unit, the ohm and the other British Association units were clearly winning wider acceptance by the later 1860s. A key to this success was the adoption of the ohm by the big British cable companies, particularly amid the global boom in cable laying that followed the

[90] On the later history of the ohm and other units, see Larry Lagerstrom, "Putting the Electrical World in Order: The Construction of an International System of Electromagnetic Measures," PhD dissertation, University of California–Berkeley, 1992.

[91] "Effect of Cold on Insulation," *Journal of the Telegraph* (April 1, 1868) *1*: (9) 4.

[92] "What Is an 'Ohm'?," *Journal of the Telegraph* (April 15, 1868) *1*: (10) 4.

[93] Clark, *Elementary Treatise*, 43.

successful completion of the Atlantic cable in 1866.[94] Between 1868 and 1875, new cables were laid to India, Australia, China, Japan, and along the African and South American coasts, while the existing cables across the North Atlantic and through the Mediterranean were joined by supplementary lines.[95] These cables – almost all of them made and laid by British companies – provided a much-expanded arena for the practice of precision electrical measurement, and they were standardized and tested from the first in terms of British Association ohms.

Contract specifications provide perhaps the most concrete and consequential instance of electrical standards being put to use, and the specifications for cables laid during the boom of the late 1860s and early 1870s show very clearly the widening adoption of the ohm and other British Association units. Consider, for example, the specifications for the French Atlantic cable that was laid in 1869 (which, despite its name, was a thoroughly British project). Given that its engineers were Clark, Jenkin, Varley, and Thomson, it is perhaps not surprising that the electrical requirements for the cable were all stated in British Association units: the resistance of the copper conductor of its main length from Brest to St. Pierre was specified to be "not greater than 3.25 B.A. units" per nautical mile, and that of its gutta-percha insulation "not less than 250 millions of B.A. units" per nautical mile.[96] The specifications for the many cables built for what would become John Pender's vast Eastern and Associated group were stated in similar terms, and the trove of such specifications now held by the PK Porthcurno Museum of Global Communications in Cornwall for cables laid between 1868 and 1912 reveals not just the evolving adoption of British Association units but also some of the confusion that initially surrounded them.

The earliest specification in the Porthcurno collection, dated May 11, 1868, is for the cable laid later that year from Malta to Alexandria for the Anglo-Mediterranean Telegraph Company; it replaced the government-owned cable laid in 1861. Sir Charles Tilston Bright served as engineer and electrician for the Anglo-Mediterranean company, and the specification called for the resistance of its conductor to be "9 B.A. units per nautical mile when tested at 75° Fahrenheit," and its insulation resistance to be "not less than 200 millions of B.A. units per nautical mile, when tested at 75°

---

[94] On the growth of the cable network after 1866, see Daniel R. Headrick, *The Invisible Weapon: Telecommunications and International Politics, 1851–1945* (Oxford: Oxford University Press, 1991).

[95] Bright, *Submarine Telegraphs*, 106–45.

[96] The full specification appears in Smith, *Rise and Extension*, 212–26; quotations from 214. On the French Atlantic cable, see Bright, *Submarine Telegraphs*, 107–8.

Fahrenheit fourteen days after manufacture."[97] The specification for the Red Sea section of the British-Indian Submarine Telegraph, dated January 27, 1869, is more precise, calling for the "mean resistance" of the conductor to be "not more than 11.03 B.A. units per nautical mile when tested at 75° Fahrenheit," and the insulation resistance to be "not less than 200 millions of B.A. units per nautical mile" when tested at "75° Fahrenheit after twenty-four hours immersion fourteen days after manufacture."[98] The February 1870 specification for a cable from Marseille to Algeria is the first in the collection to speak of "ohms" and "megohms" rather than "B.A. units"; it called for the conductor to have a mean resistance of "not more than 12.15 ohms per nautical mile," and the core to have an insulation resistance of "not less than 150 megohms per nautical mile." Some confusion set in, however, about the leg of the cable that would run from Algeria to Malta; for it, the specification called for a copper conductor with a mean resistance of "not more than 12.75 *megohms* per nautical mile."[99] One hopes this error in a legally binding contract was caught before the cable was actually manufactured, or at least that the maker kept the resistance *well* under – say, one millionth of – the specified value. Evidently in 1870 "ohms" and "megohms" were not yet familiar terms to those drawing up specifications, or at least not to the typesetters and proofreaders.

Another kind of confusion cropped up in the August 1870 specification for a yet another cable to be laid from Malta to Alexandria. This was to be of somewhat lighter and cheaper construction than the one laid in 1868, and the specification evidently meant to call for a resistance of 12.15 ohms per nautical mile versus the 9 "B.A. units" of the old cable. But in the printed version, this came out as "not more than 12.15 shms. per nautical mile." Nor was this a random typographical error; the same specification called for the insulation resistance to be "not less than 150 megshms. per nautical mile."[100] Such confusion soon faded, however, as ohms and megohms were incorporated into the daily work of telegraph engineers. After the early 1870s, specifications clearly and consistently stated the maximum resistance for conductors in ohms and the minimum

[97] Specification for Anglo-Mediterranean Telegraph Co. cable (May 5, 1868), in "Eastern Telegraph Co. agreements with Telegraph Construction & Maintenance Co.," DOC/ETC/114, 1.6, C&W Archive, Porthcurno, hereinafter cited as "ETC Agreements."

[98] Specification, British-Indian Submarine Telegraph Co. cable (January 27, 1869), ETC Agreements, 5.36–37.

[99] Specification, Anglo-Mediterranean Telegraph Co., Marseille-La Calle cable (February 5, 1870), ETC Agreements, 9.7–8, emphasis added.

[100] Specification, Anglo-Mediterranean Telegraph Co. New Cable (August 15, 1870), ETC Agreements, 10.6–7.

resistance for insulation in megohms; some also specified the maximum allowable inductive capacity in microfarads per nautical mile.[101]

The provision of reliable standards put much of electrical measurement on a substantially new basis. Electricians sensed that they had entered a new era, and Jenkin noted in 1865 that "we have now reached a point where we look back with surprise at the rough and ready means by which the great discoveries were made on which all our work is founded."[102] An enormous amount of expertise was now built into the ordinary electrician's instruments, where it remained largely invisible – nowhere more so than in that humble but crucial "instrument," the resistance box.

When Jenkin declared that the resistance coil had become "as necessary to the electrician as the balance to the chemist," he was expressing a fundamental shift in the working world of both physicists and telegraph engineers.[103] A box of ohms became a central part of the practice of precision electrical measurement as it began to spread more widely and be performed more routinely from the mid-1860s. It was no coincidence that the first physics teaching laboratories appeared in Britain in this period (see Chapter 6), or that almost all of them strongly emphasized electrical measurement.[104] Telegraphy had provided the initial market for such measurement and it continued to stimulate the development of improved measurement techniques and tools. By 1886, the London physics professor Frederick Guthrie could declare that "Electricity, especially voltaic, lends itself perhaps more abundantly to exact measurement in the elementary laboratory than the other branches" of physics.[105] It was a statement that would have seemed absurd thirty years before, and that reflected the way in which electrical measurement, and with it the position of electricity relative to the rest of physics, had been transformed in the intervening years in response to the demands and opportunities presented by telegraphy.

Conclusion

The quotations from Thomson and Maxwell with which this chapter began reflect the sentiments of two men – really three, since Maxwell

---

[101] See, for example, the specification for the Suez-Perim-Aden cable (July 29, 1890), ETC Agreements, 27.13.

[102] Jenkin, "Report on the New Unit," in Smith, *Reports,* 280.

[103] Jenkin, "Report on the New Unit," in Smith, *Reports,* 279.

[104] See Gooday, "Precision Measurement," and Graeme J. N. Gooday, "Precision Measurement and the Genesis of Physics Teaching Laboratories in Victorian Britain," PhD dissertation, University of Kent at Canterbury, 1989.

[105] Frederick Guthrie, "Teaching Physics," *Journal of the Society of Arts* (May 1886) *34*: 659–63, on 662.

was mainly paraphrasing and endorsing remarks Jenkin had made in his book[106] – who had been in the middle of a major transition in electrical practice in the early 1860s, a transition associated first with the cable industry and then with the British Association Committee on Electrical Standards. All three sought, though in somewhat different ways, to encourage the alliance that had grown up between physics and telegraph engineering and to reinforce the increasing emphasis in both on precision electrical measurement. This is especially clear in Thomson's case – he was, after all, addressing an audience of engineers about "Electrical Units of Measurement" – and the enormous role practical measurement played in Thomson's thinking has now been well established.[107] Jenkin, too, was keenly aware of the importance of electrical standards to both scientists and engineers and wrote extensively on the subject.[108]

The immediate context of Maxwell's remarks is especially revealing. In 1863–1864, he had served with Jenkin on the British Association subcommittee that had determined the value of the ohm and had devoted great effort to working out, both experimentally and conceptually, the relationships between electrical measurements. As we shall see, this experience shaped his thinking in important ways at a crucial stage in the development of his electromagnetic theory.[109] By the time Maxwell wrote his review of Jenkin's book in the spring of 1873, he had just completed his own *Treatise on Electricity and Magnetism* and was busy supervising the preparation of the new Cavendish Laboratory at Cambridge. He was intent on installing *measurement*, particularly electrical measurement, as the chief activity of the laboratory, and even before the Cavendish opened early in 1874, he wrote to Jenkin about securing the transfer to it of the apparatus their British Association subcommittee had used in its work on determining the value of the ohm – apparatus Maxwell wished to install for research and training at

---

[106] Jenkin, *Electricity and Magnetism*, esp. v–vi.

[107] Smith and Wise, *Energy and Empire*; see also M. Norton Wise and Crosbie Smith, "Measurement, Work and Industry in Lord Kelvin's Britain," *Historical Studies in the Physical and Biological Sciences* (1986) *17*: 147–73.

[108] C. A. Hempstead, "An Appraisal of Fleeming Jenkin (1833–1885), Electrical Engineer," *History of Technology* (1991) *13*: 119–44.

[109] See the following chapter. Salvo d'Agostino explored a possible connection between Maxwell's work on the ohm and the development of his electromagnetic theory in "Esperimento e teoria nell'opera di Maxwell: Le misure per le unita assolute elettromagnetiche e la velocita della luce" ("Experiment and Theory in Maxwell's Work: The Measurements for Absolute Electromagnetic Units and the Velocity of Light"), *Scientia* (1978) *113*: 453–80. His argument is vitiated, however, by his mistaken attribution to Maxwell of the 1863 British Association committee report, which was written by Thomson; see Thompson, *Kelvin*, *1*: 419.

his new laboratory.[110] Maxwell clearly intended the electricity studied at the Cavendish not to be that of "sparks and shocks which are seen and felt," but instead that of "currents and resistances to be measured and calculated" – not that of "the lecture-room and the popular treatise," but ultimately that of "the testing-room and the engineer's specification." Thomson, Jenkin, Maxwell, and their collaborators had helped extend the physics laboratory into the cable industry, but they had also brought an important part of the cable industry, encapsulated in the resistance box, into the physics laboratory.

[110] James Clerk Maxwell to Fleeming Jenkin, [March 17], July 22, and November 18, 1874, in Maxwell, *SLP 3*: 51–52, 89–90, and 138–39. Maxwell listed the equipment in his "Report on the Cavendish Laboratory for 1874," *Cambridge University Reporter* (April 27, 1875), 352–54, repr. in Maxwell, *SLP 3*: 208–15, on 213. On the later history of electrical standards work at the Cavendish, see Simon Schaffer, "Late Victorian Metrology and Its Instrumentation: A Manufactory of Ohms," in Robert Bud and Susan Cozzens, eds., *Invisible Connections: Instruments, Institutions, and Science* (Bellingham, WA: SPIE Optical Engineering Press, 1992), 23–56.

5   The Ohm, the Speed of Light, and Maxwell's Theory of the Electromagnetic Field

One of the classic questions in the history of nineteenth-century physics centers on James Clerk Maxwell's attitude toward his mechanical models of the ether. Did he look on them as no more than heuristic tools, to be tossed aside once they had helped him find the proper field equations? Or did he instead regard them as steps toward a realistic representation of the actual structure of the electromagnetic medium? In his long paper "On Physical Lines of Force," published in installments in 1861–1862, Maxwell laid out an elaborate model of the ether, picturing it as an array of tiny spinning vortices interspersed with layers of even smaller "idle wheel" particles.[1] He showed that such a vortex medium could reproduce the main phenomena of electricity and magnetism, including the production of magnetic fields and the induction of electric currents, and could also convey transverse waves very much like – perhaps identical to – those of light. "Physical Lines" marked a major step in the development of field theory and the unification of optics with electromagnetism; with it, Maxwell appeared to be well on his way toward delineating the real mechanical structure of the electromagnetic ether. Yet just over two and a half years later, he sent the Royal Society of London his "Dynamical Theory of the Electromagnetic Field," in which he seemingly abandoned his vortex model and instead derived the equations of the electromagnetic field from general dynamical principles, without invoking any hypothetical mechanical microstructure.[2] Why the shift? Had Maxwell really renounced his vortex model? Had he ever really believed in it, or had he always regarded it as mere scaffolding, to be cast aside when no longer needed? Why had he taken one approach to electromagnetic theory in 1862, and such a seemingly different one in 1864? What does this sequence of moves by Maxwell tell us about the roots and development of his theory of the electromagnetic field, as well as about the deeper

---

[1] James Clerk Maxwell, "On Physical Lines of Force," *Phil. Mag.* (March 1861) *21*: 161–75, (April 1861) *21*: 281–91, (May 1861) *21*: 338–48, (January 1862) *23*: 12–24, and (February 1862) *23*: 85–95, repr. in Maxwell, *SP 1*: 451–513.

[2] James Clerk Maxwell, "A Dynamical Theory of the Electromagnetic Field," *Phil. Trans.* (1865) *155*: 459–512, repr. in Maxwell, *SP 1*: 526–97.

attitudes of Victorian physicists toward the nature of physical reality and the means by which they might best seek to grasp and describe it?

To understand the apparent shift Maxwell's thinking underwent between 1862 and 1864, an obvious first step is to look closely at what he was doing in 1863. It turns out he spent much of that year, and some months both before and after, working hard for the British Association Committee on Electrical Standards, establishing the value of the ohm, clarifying the relationships among electrical measurements, and laying the groundwork for a careful experimental determination of the ratio of electrostatic to electromagnetic units. Maxwell's close collaboration in this period with Fleeming Jenkin and other telegraph engineers led him to adopt, at least for a time and for the purposes at hand, an "engineering approach" to electrical questions in which he focused not on devising hypothetical mechanisms but on formulating demonstrable relations between quantities he could measure and manipulate.

As his work in this period shows, Maxwell was not wedded to just one way of doing physics. Sometimes he found it useful to devise hypothetical microscopic mechanisms and trace out their consequences, with hopes of penetrating to the real mechanical substructure of the physical world; other times, he sought instead to formulate macroscopic laws that would be independent of such hypotheses.[3] These shifts did not represent his abandonment of one approach or the other but rather his attempts to advance scientific understanding at different times along different fronts. Maxwell's work on the British Association Committee on Electrical Standards shaped his thinking not just about the ohm and electrical measurement but about the range of analytical approaches and expository strategies that could be useful in physics, and it became a significant thread not just in his "Dynamical Theory" but also in his *Treatise on Electricity and Magnetism* (1873) and other works.

Models and Standards

Over the years, many historians, philosophers, and physicists have discussed a sequence of three papers on electromagnetism that Maxwell published between the mid-1850s and the mid-1860s:

---

[3] Note that while the mechanism of the electromagnetic medium that Maxwell discussed in "Physical Lines" was characterized by equations of continuum mechanics that had the same form as the equations used for macroscopic bodies, it was differentiated microscopically by the motion of its parts, as was the version of the kinetic theory of gases Maxwell was developing at the same time; cf. M. Norton Wise, "The Maxwell Literature and British Dynamical Theory," *Historical Studies in the Physical Sciences* (1982) *13*: 175–205, esp. 188–89 and 200–1.

1856: "On Faraday's Lines of Force," in which he laid out a fluid-flow analogy to the distribution and interaction of lines of electric and magnetic force;

1861–62: "On Physical Lines of Force," in which he presented his vortex and idle wheel model of the electromagnetic ether and introduced the first "electromechanical" version of his electromagnetic theory of light;

1864: "A Dynamical Theory of the Electromagnetic Field," in which he formulated a set of electromagnetic field equations, including a fully electromagnetic theory of light, based on the general dynamics of a connected system, independently of any detailed model of the medium.

The move from "Physical Lines" to "Dynamical Theory" has drawn particular attention, and it will be our main focus here.[4] The key question has been whether Maxwell believed his vortex model represented the real structure of the ether, at least in part, or instead regarded it as little more than a convenient fiction he could use to help him find the field equations that, on this account, were always his real and final goal (Figure 5.1).

The evidence pulls in two directions. In "Physical Lines," Maxwell certainly spoke of the vortices very much as if he regarded them as real. Citing Michael Faraday's 1845 discovery of magneto-optic rotation, he declared that William Thomson's 1856 analysis of it proved that "the cause of the magnetic action on light must be a real rotation going on in the magnetic field," and he thought it sufficiently likely that the rotation was performed by tiny "molecular vortices" that in 1861 he had special apparatus built with which he tried to measure their expected gyroscopic

---

[4] P. G. Tait first drew attention to the sequence from "Faraday's Lines" to "Physical Lines" to "Dynamical Theory" in his *Sketch of Thermodynamics* (Edinburgh: Edmonston and Douglas, 1868), 74; W. D. Niven later emphasized it in his preface to Maxwell, *SP 1*: xix–xxii. Among subsequent discussions, see Joseph Turner, "Maxwell on the Logic of Dynamical Explanation," *Philosophy of Science* (1956) *23*: 36–47; Robert Kargon, "Model and Analogy in Victorian Science: Maxwell's Critique of the French Physicists," *Journal of the History of Ideas* (1969) *30*: 423–36; C. W. F. Everitt, *James Clerk Maxwell: Physicist and Natural Philosopher* (New York: Scribner's, 1975), 87–102; Ole Knudsen, "The Faraday Effect and Physical Theory, 1845–1873," *Archive for History of Exact Sciences* (1976) *15*: 235–81, on 248–55; Daniel Siegel, "The Origin of the Displacement Current," *Historical Studies in the Physical and Biological Sciences* (1986) *17*: 99–146; Daniel Siegel, *Innovation in Maxwell's Electromagnetic Theory: Molecular Vortices, Displacement Current, and Light* (Cambridge: Cambridge University Press, 1991); P. M. Harman, *The Natural Philosophy of James Clerk Maxwell* (Cambridge: Cambridge University Press, 1998), 71–90, 98–124; Crosbie Smith, *The Science of Energy: A Cultural History of Energy Physics in Victorian Britain* (Chicago: University of Chicago Press, 1998), 218–38; Bruce J. Hunt, *Pursuing Power and Light: Technology and Physics from James Watt to Albert Einstein* (Baltimore: Johns Hopkins University Press, 2010), 94–110; Malcolm Longair, "'... a paper ... I hold to be great guns': A Commentary on Maxwell (1865) 'A Dynamical Theory of the Electromagnetic Field,'" *Phil. Trans. A* (2015) *373*: 20140473; and Giora Hon and Bernard R. Goldstein, *Reflections on the Practice of Physics: James Clerk Maxwell's Methodological Odyssey in Electromagnetism* (London: Routledge, 2020).

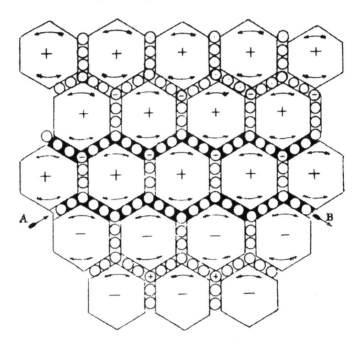

Figure 5.1 James Clerk Maxwell's vortex model of the ether. The spinning vortices, shown here as hexagons, represent a magnetic field; the displacements and motions of the smaller "idle wheel" particles, which serve to pass rotational motion from one vortex to the next, represent electrical fields and currents.
(From *Philosophical Magazine*, Vol. 21, plate, 1861; courtesy University of Texas Libraries.)

effect.[5] The experiment was inconclusive; terrestrial magnetism interfered with the expected effect and prevented Maxwell from establishing more than that the vortices, if they existed at all, must be extremely small. He still believed the magneto-optic evidence to be very strong, however, and he continued to speak of the existence of the vortices as highly probable. On the other hand, he was much more tentative about the idle wheel particles, presenting them as no more than a concrete and readily investigated way to connect the rotations of adjacent layers of vortices. As he acknowledged near the end of the second part of "Physical Lines,"

[5] Maxwell, "Physical Lines," *23*: 88, repr. in Maxwell, *SP 2*: 505. On Maxwell's gyroscopic vortex experiment, see Maxwell, "Physical Lines," *21*: 345n, repr. in Maxwell, *SP 1*: 485–86n; Maxwell to Michael Faraday, October 19, 1861, in Maxwell, *SLP 1*: 688 and Plate X; and Maxwell, *Treatise 2*: § 575.

The conception of a particle having its motion connected with that of a vortex by perfect rolling contact may appear somewhat awkward. I do not bring it forward as a mode of connexion existing in nature, or even as that which I would willingly assent to as an electrical hypothesis. It is, however, a mode of connexion which is mechanically conceivable, and easily investigated, and it serves to bring out the actual mechanical connexions between the known electro-magnetic phenomena; so that I venture to say that any one who understands the provisional and temporary character of this hypothesis, will find himself rather helped than hindered by it in his search after the true interpretation of the phenomena.[6]

This passage has often been cited as evidence that Maxwell regarded his entire vortex model as "awkward," "provisional and temporary," and frankly unrealistic.[7] But in fact Maxwell applied such terms only to the idle wheel particles and the supposition that they were in perfect rolling contact with the vortices; it was only the "mode of connexion" of the vortices that he presented as awkward and unrealistic, not the vortices themselves. From the time he first introduced the idle wheel particles in Part II of "Physical Lines," Maxwell explicitly distinguished their status, which he characterized as merely "provisional," from that of the vortices, whose existence he regarded as "probable."[8] Moreover, when he took up the Faraday effect in his *Treatise* in 1873, he returned to the vortices, saying he could find no other way to account for the rotational action of magnetism on polarized light.[9]

Such is the case for concluding that Maxwell believed his vortices really existed. On the other side, we have the fact that in "Dynamical Theory," he only briefly mentioned his vortex model and formulated his entire theory of the electromagnetic field simply in terms of the dynamics of a connected mechanical system, without reference to any hypothetical structure of the ether. Citing "Physical Lines," he said he had "on a former occasion attempted to describe a particular kind of motion and a particular kind of strain, so arranged as to account for the phenomena" but declared that "in the present paper I avoid any hypothesis of this kind"; his use in "Dynamical Theory" of such terms as "electric momentum" and "electric elasticity" was, he said, merely "illustrative, not ... explanatory."[10] However confidently Maxwell may have spoken of his

---

[6] Maxwell, "Physical Lines," *21*: 346, repr. in Maxwell, *SP 1*: 486.

[7] See, for example, John Hendry, *James Clerk Maxwell and the Theory of the Electromagnetic Field* (Bristol: Adam Hilger, 1986), 174, and Nancy J. Nersessian, *Faraday to Einstein: Constructing Meaning in Scientific Theories* (Dordrecht: Kluwer, 1984), 84.

[8] Maxwell, "Physical Lines," *21*: 282, repr. in Maxwell, *SP 1*: 468. Knudsen, "Faraday Effect," 551–53, and Siegel, *Innovation*, 47, emphasize that Maxwell continued to believe in the reality of the vortices.

[9] Maxwell, *Treatise*, 2: § 806–31, esp. § 831.

[10] Maxwell, "Dynamical Theory," 487, repr. in Maxwell, *SP 1*: 563–64.

vortices in "Physical Lines," it certainly appeared that by 1864 he had left them behind. Writing a few years later, the Edinburgh physicist P. G. Tait, who was presumably in a position to know his close friend's mind, declared that Maxwell had "discarded" his "particular hypotheses as to the molecular vortices" in favor of a theory of the electromagnetic field founded solely on general dynamical principles.[11] One can easily understand how later observers could conclude that Maxwell had "quietly abandoned" his vortex model.[12]

But why would Maxwell discard the vortices if he continued to believe that the Faraday effect gave strong evidence of their existence? Why, after 1862, did he not turn his efforts to digging out the true microscopic structure of the ether, rather than formulating field equations that ignored any such structure?

There are many reasons for the move Maxwell made in 1864, but an especially important though often overlooked one grew out of his work at just that time for the British Association Committee on Electrical Standards. As we saw in Chapter 4, the committee had been formed at the Manchester meeting of the British Association for the Advancement of Science in September 1861, largely in response to the perceived needs of the British submarine telegraph industry.[13] When British firms began laying telegraph cables in the early 1850s, they initially paid little attention to the electrical condition of their conductors and insulation; they simply covered a length of wire with gutta-percha, laid it beneath the sea, and hoped for the best. Several early successes were followed by a series of catastrophic failures, culminating in the breakdown of the first Atlantic cable in 1858 and the costly collapse of the Red Sea cables two years later. The humiliating failure of the Atlantic cable, along with concerns surrounding a British government plan in 1859 to lay a cable from Cornwall to Gibraltar, had prompted the Board of Trade and the Atlantic Telegraph Company to establish their Joint Committee to Inquire into the Construction of Submarine Telegraph Cables. The resulting *Joint Committee Report* had called, amongst much else, for the adoption of accurate and agreed standards of electrical resistance, noting that cable

[11] P. G. Tait, *Sketch of Thermodynamics* (Edinburgh: Edmonston and Douglas, 1868), 74; Tait repeated this passage in the 2nd ed. (1877), 90.

[12] L. Pearce Williams, *The Origins of Field Theory* (New York: Random House, 1966), 133.

[13] See also Bruce J. Hunt, "Michael Faraday, Cable Telegraphy, and the Rise of British Field Theory," *History of Technology* (1991) *13*: 1–19; Simon Schaffer, "Late Victorian Metrology and its Instrumentation: A Manufactory of Ohms," in Susan Cozzens and Robert Bud, eds., *Invisible Connections: Instruments, Institutions, and Science* (Bellingham, WA: SPIE Optical Engineering Press, 1992), 23–56; Bruce J. Hunt, "The Ohm Is Where the Art Is: British Telegraph Engineers and the Development of Electrical Standards," *Osiris* (1994) *9*: 48–63.

manufacturers and operators needed to be able to cite such standards in their contract specifications, while engineers needed reliable resistance coils they could use to help them monitor the condition of their cables and locate faults for repair.[14]

In 1860 Werner Siemens had sought to meet the demand for a convenient standard of electrical resistance by introducing his unit based on the resistance of a thread of mercury one meter long and one square millimeter in section.[15] Coils calibrated in terms of the new unit began to catch on, especially on the Continent, but Siemens's unit soon drew criticism in Britain. In the paper "On the Principles Which Should be Observed in the Formation of Standards of Measurement of Electrical Quantities and Resistance" that they presented at the September 1861 meeting of the British Association in Manchester, Latimer Clark and Sir Charles Tilston Bright said that their experience as cable engineers had convinced them that what was really needed was not just a stand-alone unit of resistance, like Siemens's, but a connected system of units of charge, current, and electromotive force as well.[16] William Thomson, too, had long favored adoption of a connected system of electrical units and standards, a view that had been reinforced by his experience on the Atlantic cable project and in his Glasgow laboratory, and that he had made clear in his testimony to the Joint Committee.[17] Like Clark and Bright, he hoped to enlist the British Association behind an effort to

[14] *Joint Committee Report*, xvi; see also the testimony of William Thomson, December 17, 1859, 118, and Fleeming Jenkin, December 22, 1859, 139.

[15] Werner Siemens, "Vorschlag eines reproducibaren Widerstandsmaasses," *Annalen der Physik* (June 1860) *110*: 1–20, trans. as Siemens, "Proposal for a New Reproducible Standard Measure of Resistance to Galvanic Circuits," *Phil. Mag.* (January 1861) *21*: 25–38. On the ensuing controversy between Siemens and the British, see Kathryn Olesko, "Precision, Tolerance, and Consensus: Local Cultures in German and British Resistance Standards," in Jed Z. Buchwald, ed., *Scientific Credibility and Technical Standards in 19th and Early 20th Century Germany and Britain* (Dordrecht: Kluwer, 1996), 117–56, and Graeme J. N. Gooday, *The Morals of Measurement: Accuracy, Irony, and Trust in Late Victorian Electrical Practice* (Cambridge: Cambridge University Press, 2004), 82–127.

[16] Latimer Clark and Sir Charles Bright, "On the Principles Which Should be Observed in the Formation of Standards of Measurement of Electrical Quantities and Resistance," in Clark, *Experimental Investigation of the Laws Which Govern the Propagation of the Electric Current in Long Submarine Telegraph Cables* (London: George E. Eyre and William Spottiswoode, 1861) 49–50, on 49; also published as "Measurement of Electrical Quantities and Resistance," *Elec.* (November 9, 1861) *1*: 3–4.

[17] On Thomson's early advocacy of a connected system of "absolute" electrical units, which he initially based on British units of feet, grains, and seconds, see Thomson, "Applications of the Principle of Mechanical Effect to the Measurement of Electro-motive Forces, and of Galvanic Resistances, in Absolute Units," *Phil. Mag.* (December 1851) *2*: 551–62, repr. with additional notes in Thomson, *MPP 1*: 490–502; see also Thomson's testimony, December 17, 1859, *Joint Committee Report*, 118.

develop and promulgate such a system of units, and he had taken steps earlier, independently of Clark and Bright, to organize what became the British Association Committee on Electrical Standards.[18] Crucially, Thomson believed that such a system should not be based on the properties of an arbitrary material standard (whether the resistance of Siemens's thread of mercury or the electromotive force of a Daniell's cell that provided the starting point for Clark and Bright's system) but instead on Wilhelm Weber's "absolute" system, in which electric and magnetic quantities were derived from the fundamental units for time, length, and mass. Thus, in Weber's system of electromagnetic units, electrical resistance came out as a velocity; as the committee later expressed it, "the resistance of a circuit is the velocity with which a conductor of unit length must move across a magnetic field of unit intensity in order to generate a unit current in the circuit."[19] Although some engineers balked at the idea that the resistance of a wire was in some sense really a velocity, Thomson eventually won the point by emphasizing that Weber's system had the great merit of tying all electric and magnetic units to the unit of work or energy, "the great connecting link between all physical measurements."[20] Thus, for instance, a unit current passing through a unit potential difference delivered one unit of power. Where the values of Weber's absolute units were inconveniently large or small, the committee adopted appropriate decimal multiples to adjust them to the needs of telegraphers; for example, it set the value of the "British Association unit of resistance" (later dubbed the ohm) at 10^7 meters per second, equal to the resistance of about a tenth of a mile of ordinary telegraph cable (and, as the committee noted, very close to the value of Siemens's mercury unit).[21] The traffic between the scientists and engineers on the committee thus ran both ways, as the scientists brought their expertise to bear on the needs of the telegraph industry, while also absorbing and transmitting to the broader scientific community some of the techniques and approaches to electrical measurement that prevailed among engineers.

Maxwell, then a young professor of natural philosophy at King's College London, was not initially a member of the British Association committee, but he was added to it in 1862 and soon took an active part in

[18] *BA Report* (1861), xxxix–xl; Hunt, "Ohm," 57–59.

[19] [William Thomson], "Report of the Committee Appointed by the British Association on Standards of Electrical Resistance," *BA Report* (1863), 111–24, on 118, repr. in Smith, *Reports*, 58–78, on 69.

[20] Fleeming Jenkin, "Provisional Report of the Committee Appointed by the British Association on Standards of Electrical Resistance," *BA Report* (1862), 125–35 on 126, repr. in Smith, *Reports*, 1–16, on 2.

[21] Jenkin, "Provisional Report," 126–27, 129, repr. in Smith, *Reports*, 2–3, 7.

its work. Why did he join up and what did he draw from the experience? Maxwell had long taken an interest in new technologies, including telegraphy, but mostly as an observer rather than, like Thomson, a direct participant.[22] Thomson, based in industrial Glasgow, would go on to develop a substantial and lucrative business as a consulting telegraph engineer and instrument manufacturer; Maxwell, a Scottish country gentleman with family roots in Edinburgh, always stood somewhat aside from trade and industry, and there is little evidence that he joined the British Association committee out of a desire to serve the needs of the cable industry, much less to find a place for himself within it.[23] Rather, he appears to have turned to the committee in 1862 in hopes of obtaining measurements to test and if possible confirm his new electromagnetic theory of light, particularly by verifying that the ratio of electrostatic to electromagnetic units was equal, as his theory required, to the speed of light. In this he was only partially successful, but the effort had other important consequences, as Maxwell's exposure to the work of the committee and to engineers' characteristic attitudes toward the quantities they measured led him to reframe his electromagnetic theory in fundamental ways.

Making Measurements

The most famous and important result in "Physical Lines" was Maxwell's identification of light with waves in his electromagnetic medium. This, however, was not at all part of the paper as he first conceived it. In its first installments, published in the *Philosophical Magazine* in March, April, and May 1861, he showed how his vortices and idle wheels could account mechanically for the existence and operation of magnetic fields, electric currents, and electromagnetic induction, and he closed the May installment with some general remarks about the use of models and analogies in physical theorizing. The tone and structure of this section make it clear that Maxwell considered the paper completed at this point; he gave no hint that there might be more parts to come. But while spending that summer at Glenlair, his estate in southwestern Scotland, he returned to his ether model and asked how he might make it account for electrostatic phenomena. His answer – by making the vortex cells elastic – carried with it an unexpected bonus: it turned out his newly elastic medium could

[22] On Thomson's deep involvement in technology and commerce, see Crosbie Smith and M. Norton Wise, *Energy and Empire: A Biographical Study of Lord Kelvin* (Cambridge: Cambridge University Press, 1989), 687–98.

[23] On Maxwell's family background, see John W. Arthur, *Brilliant Lives: The Clerk Maxwells and the Scottish Enlightenment* (Edinburgh: John Donald, 2016).

carry transverse waves very much like those of light. Moreover, after making some simplifying (and rather questionable) assumptions about the moduli of elasticity of his vortex cells, he found that the speed of the waves came out equal to the ratio of electrostatic to electromagnetic units, a quantity he came to designate as v.[24]

Maxwell later told Faraday that he had "worked out the formulae in the country," with no access to published measurements of v and no way to put an actual number on the speed with which waves would travel in his vortex medium; he knew only that it would equal the ratio of units.[25] Only after he returned to London, probably in late September 1861, did he find that in 1857 Wilhelm Weber and Rudolf Kohlrausch had measured the ratio to be $310,740 \times 10^6$ millimeters per second, or 193,088 miles per second. To his evident delight, he also found in Joseph A. Galbraith and Samuel Haughton's *Manual of Astronomy* that Hippolyte Fizeau had measured the speed of light to be 193,118 miles per second.[26] Weber and Kohlrausch had noted the closeness of the two quantities (though they cited a slightly different value for the speed of light), but said it had no real physical significance; in Weber's theory, the ratio of units was related to the speed at which the electrostatic attraction between two oppositely charged particles of electricity would balance their electromagnetic repulsion and had nothing to do with waves of light.[27] Maxwell saw the matter

---

[24] Maxwell, "Physical Lines," *23:* 15–22, repr. in Maxwell, *SP 1*: 492–500. Siegel, *Innovation*, 136–41, discusses the questionable steps in Maxwell's derivation of the wave speed; see also Simon Schaffer, "Accurate Measurement Is an English Science," in M. Norton Wise, ed., *The Values of Precision* (Princeton: Princeton University Press, 1995), 135–72, esp. 137 and 145. Siegel rightly notes that Maxwell had settled on identifying v with the speed of the waves *before* he saw Weber and Kohlrausch's measurement; see Daniel M. Siegel, "Author's Response," *Metascience* (1993) *4*: 27–33, on 30–31.

[25] Maxwell to Michael Faraday, October 19, 1861, in Maxwell, *SLP 1*: 683–89, on 685–86.

[26] Maxwell to Faraday, October 19, 1861, in Maxwell, *SLP 1*: 685; also Maxwell to C. J. Monro, c. October 20, 1861, in Maxwell, *SLP 1*: 690–91, on 690. Joseph A. Galbraith and Samuel Haughton, *Manual of Astronomy* (London: Longmans, 1855), 36, gave Fizeau's value of the speed of light as 169,944 "geographical miles" (of 6000 feet) per second; converting to statute miles, Maxwell arrived at 193,118 miles per second. Fizeau's own published value, reported in H. Fizeau, "Sur une expérience relative à la vitesse de propagation de la lumière," *Comptes Rendus* (1849) *29*: 90–92, was in fact 70,948 "lieues de 25 au degré"; this converts to 195,937 miles per second, but in "Physical Lines," *23*: 22, repr. in Maxwell, *SP 1*: 500, Maxwell mistakenly transcribed Fizeau's number as 70,843, and so gave the speed of light as 195,647 miles per second; see Siegel, *Innovation*, 211–12, n. 21, and Schaffer, "English Science," 145–47.

[27] On Weber and Kohlrausch's measurement of the ratio of units, see Oliver Darrigol, *Electrodynamics from Ampère to Einstein* (Oxford: Oxford University Press, 2000), 66. Darrigol also discusses Gustav Kirchhoff's 1857 derivation, based on Weber's theory, of the speed with which electricity would propagate in a thin wire, and Weber's reasons for concluding that the closeness of this value to the speed of light was not physically significant; see 72–73. Note that Weber's conception of electric currents as

very differently; as he wrote to Faraday on October 19, 1861, he was convinced that "this coincidence is not merely numerical" but reflected a deep connection between light and electromagnetism. "I think we have now strong reason to believe," he declared, "whether my theory is a fact or not, that the luminiferous and the electromagnetic medium are one."[28]

That v should equal the speed of light was not the only empirically testable consequence of Maxwell's vortex theory, and his immediate motive in writing to Faraday was to obtain other experimental data. As he told Faraday, his vortex model had led him to "some very interesting results, capable of testing my theory, and exhibiting numerical relations between optical, electric and electromagnetic phenomena, which I hope soon to verify more completely."[29] In particular, Maxwell's model predicted that the specific inductive capacity (or, as it is now called, the relative permittivity) of a transparent dielectric would be equal to the square of its index of refraction, and he asked Faraday if he knew of any good measurements of those quantities for different substances. Maxwell's theory of the Faraday effect also implied that the rotation of the plane of polarization of a beam of polarized light would be proportional to the strength of the magnetic field through which it passed and so to the size, density, and rotational speed of the underlying vortices, and he asked if Faraday could point him toward any measurements bearing on the question. Faraday pencilled "Verdet" in the margin of Maxwell's letter, a reference to the experiments on magneto-optical phenomena that Émile Verdet had been publishing since 1854.[30] By the time Maxwell sent an account of his vortex model to William Thomson on December 10, 1861, he had worked Verdet's results into his developing theory, suggesting, for example, that he could explain the seemingly anomalous results Verdet had found for light passing through ferro-magnetic solutions by assuming that in them the magnetic vortices set the iron molecules themselves spinning in the opposite direction.[31]

Maxwell wrote up two new installments of "Physical Lines," published in January and February 1862, in which he showed how making the vortex cells elastic would enable his model to account for light waves

counterflowing streams of oppositely charged particles led him to focus on the ratio of electrostatic to electrodynamic units, which was larger than Maxwell's ratio of electrostatic to electromagnetic units by a factor of $\sqrt{2}$.

[28] Maxwell to Faraday, October 19, 1861, in Maxwell, *SLP 1*: 685–86.

[29] Maxwell to Faraday, October 19, 1861, in Maxwell, *SLP 1*: 683.

[30] Maxwell to Faraday, October 19, 1861, in Maxwell, *SLP 1*: 684–85 and n. 10.

[31] Maxwell to William Thomson, December 10, 1861, in Maxwell, *SLP 1*: 692–98, esp. 697, n. 28 and n. 29; Maxwell, "Physical Lines," *23*: 89–90, repr. in Maxwell, *SP 1*: 507–8.

and magneto-optic rotation.[32] In a much quoted passage, he cited the closeness between the ratio of units and the speed of light and declared that "we can scarcely avoid the inference that *light consists in the transverse undulations of the same medium which is the cause of electric and magnetic phenomena.*"[33] He clearly thought he was onto something big and eagerly sought more and better experimental evidence. Even when he found that other published measurements of the speed of light did not match the ratio of units nearly as closely as did Galbraith and Haughton's report of Fizeau's, Maxwell emphasized that the various values lay on either side of Weber and Kohlrausch's number, and he strongly implied that he expected the values to converge as measurements of both quantities improved. He also continued to seek better measurements of specific inductive capacity, writing to Thomson that "I think Fleeming Jenkin has found that of gutta percha caoutchouc &c." and asking "where can one find his method, and what method do you recommend."[34] This was Maxwell's first known mention of Jenkin; it would be far from the last.

Very few letters to Maxwell from this period have survived, and we have no record of any responses he may have received from Faraday or Thomson. Nor do we have evidence of any direct contact between Maxwell and Jenkin at this time. Both men were then in London, however, and Jenkin was just beginning his work as secretary of what was then known as the British Association Committee on Standards of Electrical Resistance. Given Maxwell's growing interest in electrical measurements and his question to Thomson, it seems likely that he got in touch with Jenkin in the early months of 1862. In any case, we know that Maxwell formally joined the British Association committee in October 1862 and thereafter worked closely with Jenkin, first assembling the necessary apparatus and then performing the experiments with which they first established the value of the ohm.[35]

---

[32] Maxwell, "Physical Lines," *23:* 12–24, 85–95, repr. in Maxwell, *SP 1*: 489–513.

[33] Maxwell, "Physical Lines," *23:* 22, repr. in Maxwell, *SP 1*: 500.

[34] Maxwell to William Thomson, December 10, 1861, in Maxwell, *SLP 1*: 697. Although Maxwell and Jenkin both attended the Edinburgh Academy in the 1840s, there is no evidence they were acquainted then; see Gillian Cookson and Colin Hempstead, *A Victorian Scientist and Engineer: Fleeming Jenkin and the Birth of Electrical Engineering* (Aldershot: Ashgate, 2000), 16–18. On Jenkin's work on the inductive capacities of gutta-percha and rubber, see Bruce J. Hunt, "Insulation for an Empire: Gutta-Percha and the Development of Electrical Measurement in Victorian Britain," in Frank A. J. L. James, ed., *Semaphores to Short Waves* (London: Royal Society of Arts, 1998), 85–104, on 98–99.

[35] On the addition of Maxwell to the Committee on Standards of Electrical Resistance, see "Recommendations Adopted by the General Committee," *BA Report* (1862), xxxix–xl, on xxxix.

In its 1862 report, the British Association committee had adopted 10^7 meters per second as its unit of electrical resistance, but that was only the prelude to the difficult task of constructing a physical standard (essentially a coil of wire) that would accurately embody the new unit.[36] In the wake of the 1862 meeting, the committee assigned that task to a London-based subcommittee consisting of Jenkin, Maxwell, and Balfour Stewart, the director of the Kew observatory and an expert on terrestrial magnetism. They planned to use a method devised by Thomson in which a small magnet attached to a needle was suspended at the center of a large coil of wire. Initially the needle simply pointed northward, but as the coil was spun around a vertical diameter, its motion through the earth's magnetic field induced a current within it that deflected the needle. The neat trick here was that since both the original force directing the needle to the north and the new one arising from the induced current were proportional to the strength of the earth's magnetic field, the net deflection was independent of that field and depended solely on the size of the coil, its speed of rotation, and, crucially, its electrical resistance. The spinning coil method was conceptually simple but carrying it out to high precision involved many subtleties. For instance, the need for a very steady rate of spin prompted both Maxwell and Jenkin to take up the theory of governors, while Maxwell's observations of the decaying oscillations of the deflected needle led him to experiments on the viscosity of air that later provided a crucial test of his kinetic theory of gases.[37]

Once they had set up and tested the apparatus, Maxwell, Jenkin, and Stewart spent May and June 1863 spinning their coil and taking readings in a room at King's College London. As Maxwell later described the procedure, "the Secretary" (Jenkin) cranked the driving wheel, "the Astronomer" (Stewart) timed the rotation of the coil, and Maxwell himself tracked the deflection of the needle through a telescope.[38]

---

[36] The British Association committee initially intended to construct a material standard whose resistance was as close to 10^7 meters per second as they could manage and then treat that as the unit, in much the same way that the meter, though in principle set at 10^{-7} of the distance from the equator to the pole, came to be defined as a length marked on a metal bar kept near Paris. See Jenkin, "Provisional Report, 1862," 129–30, repr. in Smith, *Reports*, 7. Later the committee simply defined the "B. A. unit," or ohm, as 10^7 meters per second, though it had continuing difficulties constructing a material standard that matched its defined value.

[37] Maxwell to William Thomson, September 11, 1863, in Maxwell *SLP 2*: 112–16, includes discussion of "Jenkin's governor" and Maxwell's own; see also James Clerk Maxwell, "On Governors," *Proc. RS* (1868) *16*: 270–83, repr. in Maxwell, *SP 2*: 105–20. Concerning Maxwell's work on the viscosity of air, see Maxwell to G. G. Stokes, June 9, 1863, in Maxwell, *SLP 2*: 96, n. 10.

[38] Maxwell to Robert Dundas Cay, August 21, 1863, in Maxwell, *SLP 2*: 103; James Clerk Maxwell, Balfour Stewart, and Fleeming Jenkin, "Description of an Experimental Measurement of Electrical Resistance, Made at King's College," Appendix D to

Unexpected complications soon turned up, arising from such things as small shifts in the direction of the earth's magnetic field and difficulties in regulating the temperature of the coil; even the passing of steamers on the Thames could affect the sensitive magnetic needle. Maxwell's biographer, Lewis Campbell, wrote in 1882 that Jenkin had preserved "a mass of correspondence, containing numerous suggestions made by Maxwell from day to day in 1863–64," but apart from one quoted by Campbell, those letters have evidently all been lost.[39] Maxwell also consulted with Thomson, sending him accounts of the "spins" and inviting him to observe a few and give advice when he was in London in early June.[40]

As Maxwell and Jenkin refined their procedures, they became confident they could bring their results to a high level of precision. Aided now by Charles Hockin, a young Cambridge graduate who would later become a leading cable engineer, they made more spins over the winter and spring of 1863–1864 and set their sights on being able to issue certified coils at the British Association meeting in September 1864.[41] Delays in duplicating the standards kept them from meeting that target, however, and that fall Elliott Brothers, the London instrument makers, began to sell unofficial coils to "persons who were unwilling to wait for the final experiments by the Committee."[42] Certified standards were finally ready in February 1865, when Jenkin sent the *Philosophical Magazine* a brief note announcing that, for a price of £2 10s each, "copies of the Standard of Electrical Resistance chosen by the Committee on Electrical Standards appointed by the British Association in 1861, can now be procured by application to me as Secretary to the Committee."[43] By

[Thomson], "Report, 1863," 163–76, on 174, repr. in Smith, *Reports*, 156. See also I. B. Hopley, "Maxwell's Work on Electrical Resistance: The Redetermination of the Absolute Unit of Resistance," *Annals of Science* (1957) 13: 265–72. On the King's College laboratory, see Jordi Cat, *Maxwell, Sutton, and the Birth of Color Photography: A Binocular Study* (New York: Palgrave Macmillan, 2013).

39 Lewis Campbell and William Garnett, *The Life of James Clerk Maxwell* (London: Macmillan, 1882), 317 and 336–37.

40 Maxwell to William Thomson, May 29, 1863, June 1863, and July 31, 1863, in Maxwell, *SLP 2*: 88–92, 93–94, and 98–101.

41 See the obituary of Hockin in *Elec.* (May 6, 1882) 8: 409–10.

42 On the coils made by Elliott Brothers in 1864, as well as a set made by Siemens and Halske for the government telegraphs in India, see [Fleeming Jenkin], "Report of the Committee on Standards of Electrical Resistance," *BA Report* (1864), 345–49, on 345, repr. in Smith, *Reports*, 159–66, on 159. See Maxwell to Hockin, September 7, 1864, in Maxwell, *SLP 2*: 164, expressing "hope there will be resistance coils at the British Association," and Gooday, *Morals*, 107–10, on Augustus Matthiessen's efforts with Hockin to make reliable duplicates of the resistance standards.

43 Fleeming Jenkin, "Electrical Standard" (letter, dated February 7, 1865), *Phil. Mag.* (March 1865) 29: 248.

that fall the committee had sold sixteen coils; it also sent standard coils gratis to government telegraph departments around the world and to several eminent physicists, including Weber, Gustav Kirchhoff, and James Joule.[44]

The new standard was soon taken up by instrument makers, scientific laboratories, and perhaps most importantly by the companies that made and laid telegraph cables. In his 1873 review of Jenkin's *Electricity and Magnetism*, a textbook aimed at "practical men," Maxwell had made it clear that he believed the real call for precision measurement and accurate electrical standards had come from engineers and businessmen rather than from academic scientists. "Sparks and shocks which are seen and felt" might be enough for "the lecture-room and the popular treatise," but "the testing-office and the engineer's specification" demanded accurate measurements of well-defined quantities.[45] Agreed standards and reliable measurements were vital to any commercial exchange, and those drawing up contracts for expensive telegraph systems and submarine cables wanted to be sure they were getting exactly what they paid for. Siemens's mercury unit was cited in some telegraph specifications in the early 1860s, but it largely gave way to the new British Association unit later in the decade, particularly in cable telegraphy. The cable industry was dominated by British firms and at the time constituted by far the largest market for precision electrical measurement; the British Association committee knew that if it could persuade the cable industry to adopt its standards, the battle would practically be won. In 1865 the committee boasted that "The new unit has been actually employed to express the tests of the Atlantic Telegraph Cable," but certified standards were not yet available when its specifications (both for the cable that snapped during laying in August 1865 and the one that was successfully completed the following year) were drawn up.[46] There was a brief lull in cable laying after 1866, owing mainly to a financial panic that had dried up capital markets, but in 1868 companies led by "cable king" John Pender launched a wave of new projects that would set the pattern for the entire industry. Beginning with the contract for the Malta–Alexandria cable, drawn up in May 1868, these called for the resistance of both the copper conductor and the gutta-percha

[44] [Fleeming Jenkin], "Report of the Committee on Standards of Electrical Resistance," *BA Report* (1865), 308–11, on 310–11, repr. in Smith, *Reports*, 190–95, on 193–94.

[45] [James Clerk Maxwell], "Longman's Text-Books of Science" [review of Fleeming Jenkin, *Electricity and Magnetism*], *Nature* (May 15, 1873) 8: 42–43, repr. in Maxwell, *SLP 2:* 842–44, on 842. Maxwell's remarks echoed ones Jenkin had made in Fleeming Jenkin, *Electricity and Magnetism* (London: Longmans, Green, and Co., 1873), vii.

[46] [Jenkin], "Report" (1865), 311, repr. in Smith, *Reports*, 194.

insulation to be measured in "B. A. units" or, as they soon came to be called, "ohms" and "megohms."[47]

Problems with the British Association resistance standards began to turn up in the 1870s, as we shall see, and their values later had to be adjusted. But in the late 1860s the new system appeared to be a clear success; its units and standards had taken firm root in the cable industry and were rapidly becoming central to the daily practice of telegraph engineers. Its work seemingly done, in 1870 the British Association committee dissolved itself.

"Elementary Relations"

Maxwell's work for the British Association committee had two sides, experimental and conceptual. His experimental work led to the establishment of the ohm and the associated system of units and standards; his conceptual work resulted in an important but often overlooked paper, "On the Elementary Relations Between Electrical Measurements." Written with Jenkin and first published as an appendix to the committee's 1863 report, "Elementary Relations" was devoted to formulating the basic principles of electrical measurement as clearly and simply as possible, with a minimum of speculation or hypothesis. It fits right between "Physical Lines" and "Dynamical Theory," and adding it to our previous list of three papers helps clear up some otherwise puzzling aspects of Maxwell's apparent shift in approach between 1862 and 1864, and illuminates themes that would run through his later work.[48]

In 1901 Richard Glazebrook, a leading figure for many years at the Cavendish Laboratory in Cambridge, the first head of the National Physical Laboratory (effectively the British standards bureau), and a great expert on precision electrical measurement, declared with perhaps only slight exaggeration that Maxwell and Jenkin's 1863 paper laid "the foundation of everything that has been done in the way of absolute

---

[47] See the cable contracts in DOC/ETC/1/114, C&W Archive, Porthcurno, and the discussion of them in Chapter 4.

[48] James Clerk Maxwell and Fleming Jenkin, "On the Elementary Relations between Electrical Measurements," *BA Report* (1863), 130–63, and in the separately issued *Report of the Committee Appointed by the British Association on Standards of Electrical Resistance* (London: Taylor and Francis, 1864), which retained the same pagination. "Elementary Relations" appeared with minor revisions in *Phil. Mag.* (June 1865) *29*: 436–60 and (supp. 1865) *29*: 507–25, and with more extensive changes in Fleeming Jenkin, ed., *Reports of the Committee on Electrical Standards Appointed by the British Association for the Advancement of Science* (London: E. and F. N. Spon, 1873), 59–96; the latter version was reprinted in Smith, *Reports*, 86–140. Citations here will be to the version in the 1863 *BA Report*.

electrical measurement since that date."[49] Surveying the field in 1907, E. B. Rosa and N. E. Dorsey of the National Bureau of Standards in Washington called the paper "masterly" and quoted long passages from it; it long remained a touchstone for those engaged in precision electrical measurement.[50] "Elementary Relations" has attracted relatively little attention from historians, however, no doubt in part because it was omitted from W. D. Niven's 1890 edition of Maxwell's *Scientific Papers* and also from Peter Harman's 1990–2002 edition of Maxwell's *Scientific Letters and Papers*. Simply getting its title straight has proven tricky; in his excellent *Dictionary of Scientific Biography* article on Maxwell, later published separately as *James Clerk Maxwell: Physicist and Natural Philosopher*, Francis Everitt inexplicably cited it as "On the Elementary Relations of Electrical Quantities," an error John Hendry, Daniel Siegel, and Harman all later repeated.[51]

Set next to the path-breaking "Physical Lines" and the magisterial "Dynamical Theory," it is easy to see how "Elementary Relations" could be overlooked. It is not a flashy paper, and at least on the surface, it contains little that would have seemed obviously new; long stretches are devoted just to carefully defining such things as magnetic induction and electromotive force and to clarifying how such quantities are related to each other. Instead of advancing dazzling new experimental results or bold theoretical ideas, "Elementary Relations" sought simply to make readers' ideas about electrical and magnetic phenomena clearer and more definite and to relate those ideas as closely as possible to quantities that could actually be measured. Its ultimate aim was to facilitate fruitful collaboration and exchange. "Whenever many persons are to act together," Maxwell and Jenkin observed, "it is necessary that they should have a common understanding of the measures to be employed," and their aim in "Elementary Relations," they said, was "to assist in attaining this

---

[49] R. T. Glazebrook, *James Clerk Maxwell and Modern Physics* (London: Cassell, 1901), 55–56.

[50] E. B. Rosa and N. E. Dorsey, "A Comparison of the Various Methods of Determining the Ratio of the Electromagnetic to the Electrostatic Unit of Electricity," *Bulletin of the Bureau of Standards* (1907) *3*: 605–22, on 605.

[51] Everitt, *Maxwell*, 100; Hendry, *Maxwell*, 200, 294; Siegel, *Innovation*, 215 n. 17; Harman, in Maxwell, *SLP 2*: 8. The paper is briefly discussed (under the correct title) in Harman, *Natural Philosophy*, 64–65, mainly in connection with dimensional analysis, and in Thomas K. Simpson, *Maxwell on the Electromagnetic Field: A Guided Study* (New Brunswick, NJ: Rutgers University Press, 1997), 409 n. 9, and more fully in Salvo d'Agostino, "Esperimento e teoria nell'opera di Maxwell: Le misure per le unita assolute elettromagnetiche e la velocita della luce" ("Experiment and Theory in Maxwell's Work: The Measurements for Absolute Electromagnetic Units and the Velocity of Light"), *Scientia* (1978) *113*: 453–80, esp. on 472.

common understanding as to electrical measurements."[52] They were in effect seeking to carve out what Peter Galison has called a "trading zone," a realm in which scientists and engineers who might have very different practical aims and theoretical commitments could nonetheless act in concert when making and comparing electrical measurements.[53]

Aside from a note toward the end that is credited to Maxwell, it is hard say which of the coauthors wrote any particular part of "Elementary Relations," and it is perhaps best to treat all of it as reflecting both men's views.[54] Maxwell and Jenkin evidently saw eye to eye on most subjects, and they certainly became close friends. Maxwell later took to calling Jenkin "the Fleemingo," and Jenkin asked Maxwell to serve as the godfather of his youngest son, Bernard Maxwell Jenkin, born in 1867.[55]

In the opening lines of the paper, Maxwell and Jenkin placed their effort squarely within a technological context. "The progress and extension of the electric telegraph," they declared, "has made a practical knowledge of electric and magnetic phenomena necessary to a large number of persons who are more or less occupied in the construction and working of the lines, and interesting to many others who are unwilling to be ignorant of the use of the network of wires which surrounds them." The situation called for a careful analysis of foundations, they said, for "between the student's mere knowledge of the history of discovery and the workman's practical familiarity with particular operations which can only be communicated to others by direct imitation, we are in want to a set of rules, or rather principles," to guide us in applying abstract laws to achieve specific ends.[56]

The thrust of "Elementary Relations" is strongly anti-hypothetical, in places what we might, a little anachronistically, call "operationalist." Thus Maxwell and Jenkin defined "electric quantity" or "charge" not by invoking an imponderable fluid, or even strains in a surrounding field, but by noting that "When two light conducting bodies are connected with the same pole of a voltaic battery, while the other pole is connected to the earth, they may be observed to repel one another.... Bodies, when in a condition to exert this peculiar force one on the other, are said to be electrified, or charged with electricity. These words are mere names given

---

[52] Maxwell and Jenkin, "Elementary Relations," 130.

[53] On "trading zones," see Peter Galison, *Image and Logic: A Material Culture of Microphysics* (Chicago: University of Chicago Press, 1997), 803–5.

[54] Maxwell and Jenkin, "Elementary Relations," 160–61; this "Note," including a discussion of how magnetic measurements would differ if made in "a sea of melted bismuth" rather than in air, was omitted when the paper was reprinted in 1873 and 1913.

[55] Maxwell to Tait, November 7, 1874, in Maxwell, *SLP 3*: 135; on Maxwell as godfather to Bernard Maxwell Jenkin, see Cookson and Hempstead, *Jenkin*, 17.

[56] Maxwell and Jenkin, "Elementary Relations," 130.

to a peculiar condition of matter."[57] For the purposes of their paper, Maxwell and Jenkin did not wish to commit themselves to anything beyond what could be seen and measured; an electric current, they said, did not necessarily represent the flow of anything material and should be regarded as simply a state into which certain bodies are thrown under certain circumstances, as when a wire is connected across the poles of a battery. Similarly, they said, "in speaking of a quantity of electricity, we need not conceive it as a separate thing, or entity distinct from ponderable matter, any more than in speaking of sound we conceive it as having a distinct existence." We nonetheless often find it convenient to speak of the "velocity of sound"; similarly, they said, "we may speak of electricity, without for a moment imagining that any real electric fluid exists."[58] Jenkin would take the same tack ten years later in his *Electricity and Magnetism*, saying that while for convenience he sometimes spoke of electricity as flowing like a fluid, "it is quite unnecessary to assume that the phenomena are due to one fluid, two fluids, or any fluid whatever."[59]

The members of the British Association committee knew that to win the widest possible acceptance for their proposed system of units and standards, they needed to take care that it not be seen as tied to any one national tradition or any particular theory of electromagnetism, whether Faraday and Maxwell's field theory on the one hand or Weber's action-at-a-distance theory on the other. The committee sought, Maxwell and Jenkin said, to form a system that "bears the stamp of the authority, not of this or that legislator or man of science, but of nature."[60] Toward that end, Maxwell and Jenkin formulated their definitions and accompanying equations in ways that simply related together macroscopic objects and forces that one could measure and manipulate, while founding the entire system on universal units of mass, length, and time. It was an approach that George Chrystal and others would later describe as "businesslike"; moreover, it reflected a characteristic engineering approach to such problems.[61]

As Edwin Layton and others have observed, engineers typically focus on macroscopic phenomena and seek to formulate empirically verifiable relations between quantities they can measure and control. They are concerned with practical outcomes and generally see little value in speculating about unseen entities that will not affect the result. Physicists, on

---

[57] Maxwell and Jenkin, "Elementary Relations," 136.
[58] Maxwell and Jenkin, "Elementary Relations," 136.
[59] Jenkin, *Electricity and Magnetism*, 1.
[60] Maxwell and Jenkin, "Elementary Relations," 131.
[61] George Chrystal, "Clerk Maxwell's "Electricity and Magnetism,"" *Nature* (January 12, 1882) 25: 237–40, on 238.

this account, are more concerned with ferreting out the physical micro-structure underlying such macroscopic phenomena and are not averse to introducing hypotheses about molecules and the like to help them do so; indeed, Layton pointed to the contrast between a macroscopic and a microscopic focus as one of the "mirror-image" differences between engineers and scientists.[62] Jenkin himself embodied both sides of this divide. Well known for the great breadth of his interests, he was on occasion happy to play the natural philosopher and speculate about such things as the structure of atoms and the inner workings of the ether, notably in an influential 1868 essay on Lucretian atomism.[63] In his engineering writings, however, Jenkin left such microphysical hypotheses firmly aside; his concern there was solely with what he could measure and manipulate. His *Electricity and Magnetism* contains scarcely a word about the ultimate nature or microphysical foundations of its ostensible subjects; he opened it not by asking "What Is Electricity?" but by describing how to detect and measure "Electric Quantity."[64]

In an important 1987 paper on the ways science and technology interacted in the development of the induction motor, Ronald Kline argued that in this case Layton's distinction did not hold, since everyone involved, both scientists and engineers, used equations drawn from Maxwell's 1864 "Dynamical Theory" that were already cast in macroscopic form and did not involve any hypothetical microstructure. Kline wrote that "Maxwell [had] himself made the fundamental translation of knowledge between electrophysics and electrotechnology," driven by a desire to understand the workings of certain electrical instruments, particularly ones involving spinning coils like those he had used on the British Association committee to establish the

---

[62] Edwin Layton, "Mirror-Image Twins: The Communities of Science and Technology in 19th-Century America," *Technology and Culture* (1971) *12*: 562–89, on 569. See also David Channell, "The Harmony of Theory and Practice: The Engineering Science of W. J. M. Rankine," *Technology and Culture* (1982) *23*: 39–52, on 51–52, and Walter G. Vincenti, *What Engineers Know and How They Know It: Analytical Studies from Aeronautical History* (Baltimore: Johns Hopkins University Press, 1990), 129–36. Of course, some branches of physics, notably thermodynamics, focus on macroscopic phenomena and leave aside microphysical considerations, but note that thermodynamics had strong engineering roots in the study of steam engines; see D. S. L. Cardwell, *From Watt to Clausius: The Rise of Thermodynamics in the Early Industrial Age* (Ithaca: Cornell University Press, 1971).

[63] Fleeming Jenkin, "The Atomic Theory of Lucretius," *North British Review* (March 1868) *48*: 211–42, repr. as "Lucretius and the Atomic Theory" in Sidney Colvin and J. A. Ewing, eds., *Papers Literary, Scientific, &c. by the late Fleeming Jenkin, F. R. S.*, 2 vols. (London: Longmans, Green, and Co., 1887), *1*: 177–214; see also Crosbie Smith, "Engineering the Universe: William Thomson and Fleeming Jenkin on the Nature of Matter," *Annals of Science* (1980) *37*: 387–412.

[64] Jenkin, *Electricity and Magnetism*, 1.

value of the ohm.[65] We can now see that Maxwell's focus on such instruments, and his characteristic way of dealing with them, were themselves driven by a larger technological context, in particular that associated with the formation of the British Association committee itself. Maxwell's 1861–1862 "Physical Lines of Force" is about the hypothetical microstructure of the ether; his 1864 "Dynamical Theory of the Electromagnetic Field" is not. His 1863 paper with Jenkin on "Elementary Relations" was in effect a transition piece that served to carry him, at least for a time and for specific purposes, to a less hypothetical, more macroscopic – and more engineering-oriented – approach to electrical phenomena.

In "Elementary Relations," Maxwell and Jenkin sought to place electrical and magnetic measurements on a concrete, unhypothetical foundation that all scientists and engineers could endorse. They saw this as a first step toward winning universal acceptance for the British Association system of units and standards, particularly in the face of Siemens's competing mercury unit. They also needed to explain in a simple and accessible way the principles behind "absolute" electric and magnetic measurement – a subject that, as originally presented by Gauss and Weber, struck most telegraph engineers as abstruse, forbidding, and of little practical value. An arbitrary material standard of resistance (such as Siemens's column of mercury) combined with a similarly arbitrary standard of electromotive force (such as that given by a Daniell's cell or other stable voltaic battery) would have provided a coherent system of electrical units, as the British Association committee freely admitted, and would have met the main needs of telegraphers.[66] Few engineers in the 1860s thus saw much value in relating their electrical measurements to mechanical units of length, mass, and time, or to units of work and energy, as was done in Weber's system; telling them the resistance of a length of wire could be expressed as so many meters per second had little appeal to telegraphers who simply wanted to compare the resistance of one piece of wire to that of another.

Latimer Clark was among those who initially dismissed Weber's units as impractical, saying in 1862 that he hoped the newly formed British Association committee would not "recommend the adoption of Weber's absolute units, or some other units of a magnitude ill adapted to the peculiar and various requirements of the electric telegraph."[67] In "Elementary Relations," however, Maxwell and Jenkin reiterated that

[65] Ronald Kline, "Science and Engineering Theory in the Invention and Development of the Induction Motor, 1880–1900," *Technology and Culture* (1987) *28*: 283–313, on 313.
[66] [Thomson], "Report, 1863," 114, repr. in Smith, *Reports*, 62–63.
[67] Latimer Clark (letter), *Elec.* (January 17, 1862) *1*: 129.

simple decimal multiples could be used to define practical units of convenient size, while also emphasizing, as had Thomson, that the absolute system had the great merit of linking all physical measurements to units of energy. For Thomson, energy (or work) was the ultimate measure of value, grounded in an overarching view of what really counts in the physical world.[68] Moreover, in urging that the system of electrical units be tied to the concept of energy, Thomson and the committee showed what proved to be valuable foresight, for although energy considerations rarely came up in telegraphy, the rise of the electric power industry in the years after 1880 made them fundamental to electrical engineering practice. Indeed, Clark himself later acknowledged what he called "the enormous value of an absolute system" of electrical units.[69]

Along with energy considerations, Maxwell and Jenkin made analysis of the dimensions of physical quantities a central feature of "Elementary Relations." Each time they introduced a quantity, they took care to express its dimensions in terms of products and ratios of length, time, and mass, or L, T, and M, and showed how attention to dimensions clarified the relationships among quantities. Thus energy, with the dimensions L^2M/T^2, could be regarded as the product of force, LM/T^2, and distance, L, or of momentum, LM/T, and velocity, L/T. Joseph Fourier had drawn up a table of dimensions as early as 1822, but Maxwell and Jenkin appear to have worked out the underlying principles independently, and their treatment of dimensional analysis in "Elementary Relations" played a major part in spreading knowledge of the subject more widely. They introduced the now familiar bracket notation, $[LM/T]$, in the 1873 reprint of the British Association committee reports, edited by Jenkin.[70]

Maxwell and Jenkin made a special point of establishing the dimensions of v, the ratio of electrostatic to electromagnetic units. By analyzing the force between two charges, they had already shown that in the electrostatic system, charge has the dimensions $L^{3/2}M^{1/2}/T$; similarly, by

[68] Smith, *Science of Energy*, 121–25.

[69] Latimer Clark to William Thomson, May 3, 1883, in Smith and Wise, *Energy and Empire*, 687.

[70] On the history of dimensional analysis, including Fourier's contributions, see John J. Roche, *The Mathematics of Measurement: A Critical History* (London: Athlone Press, 1998), 188–207, and the 1877 article by J. C. M[axwell], "On Dimensions," *Encyclopedia Britannica*, 9th ed. (Edinburgh: Adam and Charles Black, 1875–89), 7: 240–41, repr. in Maxwell, *SLP 3*: 517–20. Although Maxwell and Jenkin did not mention Fourier's earlier work in the original 1863 version of "Elementary Relations," they added a note about it to the 1873 revision, 89 n., repr. in Smith, *Reports*, 132 n. For an illuminating discussion of issues in the history of dimensional analysis, see Sybil G. de Clark, "The Dimensions of the Magnetic Pole: A Controversy at the Heart of Early Dimensional Analysis," *Archive for History of Exact Sciences* (2016) *70*: 293–324.

analyzing the force between two current-carrying wires, they had shown that in the electromagnetic system, charge has the dimensions $L^{1/2}M^{1/2}$. The ratio of the two systems' units thus came out as L/T, a velocity – one whose magnitude, they emphasized, was independent of any particular theory of electrical action or choice of basic units.[71] Citing Weber and Kohlrausch's measurement of v as 310,740,000 meters per second, they noted that this was "a velocity not differing from the estimated velocity of light more than the different determinations of the latter quantity differ from each other."[72] Although Maxwell and Jenkin made no overt reference in "Elementary Relations" to the electromagnetic theory of light, they were clearly dropping hints.

Weber and Kohlrausch had found v by measuring a quantity of electricity first in electrostatic units, by charging a condenser of known capacity to a potential that they measured with an electrometer, and then in electromagnetic units, by discharging the condenser through an electrodynamometer and gauging the degree and duration of its deflection.[73] Over the summer of 1863, Maxwell and Jenkin discussed two other methods, based on finding common measures of electromotive force and of resistance, and in "Elementary Relations" they added two more, based on finding common measures of current and of capacity; the capacity method, they said, "would probably yield very accurate results," as indeed later proved to be the case.[74]

Like Thomson before him, Maxwell emphasized that the ratio of units was an important quantity quite apart from the electromagnetic theory of light. It came into play whenever energy passed between its electromagnetic and electrostatic forms, as in a submarine cable carrying a pulse of current and its concomitant wave of voltage, and so figured in calculations

---

[71] Maxwell and Jenkin, "Elementary Relations," 149; see also Maxwell, *Treatise*, 2: § 768, and James Clerk Maxwell, "On a Method of Making a Direct Comparison of Electrostatic and Electromagnetic Force; with a Note on the Electromagnetic Theory of Light," *Phil. Trans.* (1868) *158*: 643–57, on 643–44, repr. in Maxwell, *SP 2*: 125–43, on 125–26. See also Daniel Jon Mitchell, "What's Nu? A Re-Examination of Maxwell's 'Ratio-of-Units' Argument, from the Mechanical Theory of the Electromagnetic Field to 'On the Elementary Relations between Electrical Measurements,'" *Studies in History and Philosophy of Science* (2017) *65–66*: 87–98; note, however, that although in some fonts the symbols look very similar, Maxwell always used v (vee), not ν (nu), to represent the ratio of units.

[72] Maxwell and Jenkin, "Elementary Relations," 149.

[73] Darrigol, *Electrodynamics*, 66.

[74] Maxwell to Fleeming Jenkin, August 27, 1863, in Campbell and Garnett, *Maxwell*, 336–37; Maxwell and Jenkin, "Elementary Relations," 153–54; Rosa and Dorsey, "Comparison," 616–17. In a letter to G. G. Stokes, October 15, 1864, in Maxwell, *SLP 2*: 188, Maxwell said he and Jenkin planned to measure v by the capacity method, but they never did so.

of the limits on the speed of signaling.[75] Noting the practical and scientific importance of an accurate knowledge of the ratio of units, Maxwell and Jenkin announced in "Elementary Relations" that "a redetermination of v will form part of the present Committee's business in 1863–64."[76] However, the most promising methods for finding v depended on comparing it to a resistance whose value was already known in absolute units (so that v would actually be measured in "B. A. units," or ohms), and the redetermination of v thus had to wait until the committee had completed its work on the new resistance standard.

In the meantime, Maxwell set about refashioning his electromagnetic theory to take maximum advantage of the new determination of v once it became available. His experience on the British Association committee and especially in writing "Elementary Relations" had brought home to him the persuasive value of basing his theoretical claims as directly as possible on measurable quantities and the demonstrable relations among them. A speculative hypothesis like his vortex and idle wheel model might be a wonderfully fruitful source of new ideas, but it was unlikely to carry much weight with skeptical critics. Once he had a reliable measurement of the ratio of units in hand, Maxwell wanted to be able to cite it as evidence that light was indeed waves in the electromagnetic medium without facing objections that he had derived the supposed connection between v and the speed of light from a fanciful mechanical model. As he and Jenkin had done in their examination of "The Elementary Relations between Electrical Measurements," Maxwell sought to strip his theory of its hypothetical elements and reduce it to what could, he argued, be derived from measurable phenomena. The result was his "Dynamical Theory of the Electromagnetic Field."

"Dynamical Theory" and the Ratio of Units

Maxwell and Jenkin finished writing up "Elementary Relations" in the fall of 1863 and, with Hockin, completed their spinning coil experiments the following spring.[77] While laying plans to measure the ratio of units, Maxwell turned to composing his "Dynamical Theory of the Electromagnetic Field." He began by justifying its name: it was a theory of the field, he said, because it concerned the space surrounding electric

---

[75] Maxwell, "Direct Comparison," 644, repr. in Maxwell, *SP 2*: 126; see also Schaffer, "English Science," 149.

[76] Maxwell and Jenkin, "Elementary Relations," 149.

[77] James Clerk Maxwell, Fleeming Jenkin, and Charles Hockin, "Description of a Further Experimental Measurement of Electrical Resistance Made at King's College," Appendix A to [Jenkin], "Report, 1864," 350–51, repr. in Smith, *Reports*, 166–67.

and magnetic bodies, and a dynamical theory "because it assumes that in that space there is matter in motion, by which the observed electromagnetic phenomena are produced."[78] Unlike in "Physical Lines," however, he did not proceed to lay out a detailed mechanical model of the medium and then set about explicating its workings. Instead, he simply started with the laws of energy that governed any connected dynamical system and, drawing on what had become the standard tools of Cambridge mathematical physics, used a Lagrangian analysis to derive what he argued were the necessary relations among electric and magnetic quantities. Maxwell's methods were those of a Cambridge wrangler, but his motivation for using them as he did was rooted in the measurement-based approach of "Elementary Relations."

Maxwell wrote up most of "Dynamical Theory" at Glenlair over the summer and added the finishing touches after returning to London in the fall; on October 27, 1864, he submitted it to the Royal Society for publication in the *Philosophical Transactions*, traditionally the favored repository for the least speculative contributions to science.[79] The way he described his paper in letters to friends is revealing. Writing to Hockin from Glenlair on September 7, Maxwell said he had now "cleared the electromagnetic theory of light from all unwarrantable assumption, so that we may safely determine the velocity of light by measuring the attraction between bodies kept at a given difference of potential, the value of which is known in electromagnetic measure" – that is, by measuring the ratio of units.[80] Back in London on October 15, he sent Thomson detailed plans for such a measurement, adding: "I can find the velocity of transmission of electromagnetic disturbances indep[enden]t of any hypothesis now & it is = v ...."[81] Writing the same day to G. G. Stokes, Maxwell said he now had "materials for calculating the velocity of transmission of a magnetic disturbance through air founded on experimental evidence without any hypothesis about the structure of the medium or any mechanical explanation of electricity or magnetism."[82]

In all of these letters, Maxwell framed his "Dynamical Theory" not as his definitive formulation of field theory, or even of the electromagnetic

---

[78] Maxwell, "Dynamical Theory," 460, repr. in Maxwell, *SP 1: 527*.
[79] Maxwell moved from Glenlair to London between September 27, 1864, and October 15, 1864; see his letters to William Thomson of those dates, in Maxwell, *SLP 2*: 172 and 175. The Royal Society of London received "Dynamical Theory" on October 27, 1864; it was "read" at the meeting of December 8, 1864, and finally printed in the *Philosophical Transactions* in June 1865; see Maxwell, *SLP 2*: 189 n. 3.
[80] Maxwell to Charles Hockin, September 7, 1864, in Campbell and Garnett, *Maxwell*, 340, repr. in Maxwell, *SLP 2*: 164.
[81] Maxwell to William Thomson, October 15, 1864, in Maxwell, *SLP 2*: 180.
[82] Maxwell to G. G. Stokes, October 15, 1864, in Maxwell, *SLP 2*: 187–88.

theory of light, but primarily as a way to link the ratio of units to the speed of light without relying on a mechanical hypothesis like that in "Physical Lines." After citing the wave theory of light to establish that space is filled with a medium capable of storing and conveying energy by the motion and elasticity of its parts, he used a Lagrangian analysis to show that any medium capable of exerting the known electric and magnetic forces would also carry waves at a speed given by the ratio of units – which Weber and Kohlrausch's measurements had shown to be the speed of light, or something very close to it. Maxwell emphasized that in "Dynamical Theory" he was able to do all of this without invoking a detailed model like the vortices and idle wheels of "Physical Lines." In a sense he was extending to the electromagnetic medium the macroscopic, measurement-based approach that he and Jenkin had followed in "Elementary Relations," though now assisted by the full apparatus of analytical dynamics.

Maxwell did not leave the vortices out of "Dynamical Theory" altogether, however. Having drawn on the wave theory of light to establish that space must be filled with a medium possessing both density and elasticity, he cited Thomson's analysis of the Faraday effect as proof that, in a magnetic field, portions of this medium must be rotating around the lines of force – that is, must form vortices. This led us, Maxwell said, "to the conception of a complicated mechanism," filling all space and "capable of a vast variety of motion," yet with its parts all connected in definite but as yet unknown ways.[83] But instead of proceeding to imagine a specific mode of connection, as in "Physical Lines," Maxwell now asked how, on general dynamical principles, the observable electric and magnetic phenomena must be related to one another, whatever the details of the underlying machinery.

Tait was right when he later wrote that Maxwell had based his "Dynamical Theory" on energy principles rather than a detailed mechanical model, but he went too far when he said that Maxwell had "discarded" his hypotheses about molecular vortices.[84] After seeing a draft of Tait's chapters, Maxwell wrote to him in December 1867. "There is a difference," he said,

between a vortex theory ascribed to Maxwell at p. 57, and a dynamical theory of Electromagnetics by the same author in Phil Trans 1865. The former is built up to show that the phenomena are such as can be explained by mechanism. The nature of this mechanism is to the true mechanism what an orrery is to the Solar System.

[83] Maxwell, "Dynamical Theory," 464, repr. in Maxwell, *SP 1*: 533.
[84] Tait, *Sketch*, 74.

The latter is built on Lagranges Dynamical Equation and is not wise about vortices.[85]

The point of the last remark, of course, was that Maxwell *was* wise about vortices. While he acknowledged that the connecting mechanism he had described in "Physical Lines" was as awkward and artificial as that of an orrery, he had scarcely any more doubt that the vortices really existed and were spinning on their axes than he did that the planets really circled the sun. Even as he shifted his mode of formulating electromagnetic theory, Maxwell remained convinced that the Faraday effect proved that molecular vortices (or something very much like them) must exist in a magnetic field. His aim in "Dynamical Theory" was not to do away with the vortices but to see how far he could get in explaining optical and electromagnetic phenomena without them.

Using Lagrangian methods, Maxwell proceeded to develop the equations of the electromagnetic field, taking the "electromagnetic momentum" or vector potential as his starting point. Early on he introduced the "displacement" that an electromotive force produces in a dielectric, comparing it to the elastic yielding that machinery undergoes when subjected to a force. Though not an ongoing electric flow, such displacement was "the commencement of a current," he said, and its variation constituted a transient current that, when added to the conduction and convection currents, served to close otherwise open circuits.[86] After deriving a long list of relations among various electric and magnetic quantities, Maxwell combined several of them to form a wave equation, from which he extracted what he regarded as his most important result: that transverse waves of magnetic force would propagate through the electromagnetic medium at a speed of $\sqrt{k/4\pi\mu}$ – in free space, simply the ratio of units.[87]

Maxwell drew several experimentally testable consequences from this result, including that the square of the index of refraction of a transparent medium would be equal to the product of its specific dielectric capacity and its specific magnetic capacity.[88] He noted with evident satisfaction that his theory explained why most transparent solids are good insulators while most good conductors are opaque and suggested that apparent exceptions, such as the transparency of many electrolytes and the

[85] Maxwell to P. G. Tait, December 23, 1867, in Maxwell, *SLP 2*: 337.

[86] Maxwell, "Dynamical Theory," 462, 480, repr. in Maxwell, *SP 2*: 531, 554.

[87] Maxwell, "Dynamical Theory," 497–99, repr. in Maxwell, *SP 1*: 577–80.

[88] Maxwell, "Dynamical Theory," 466, repr. in Maxwell, *SP 1*: 535. When Ludwig Boltzmann measured these quantities for various materials in 1873, he found they fit Maxwell's relationship reasonably well; see Darrigol, *Electrodynamics*, 259 n. 129, and Jed Z. Buchwald, *The Creation of Scientific Effects: Heinrich Hertz and Electric Waves* (Chicago: University of Chicago Press, 1994), 208–14.

anomalously low opacity of gold leaf, might be traced to lower resistive losses at very high frequencies.[89] He also discussed propagation in crystalline media, but said nothing about reflection or refraction, having failed to satisfy himself concerning the proper boundary conditions. Recalling that Stokes, who was far more expert in the intricacies of the wave theory of reflection, had once told him that "the subject was a stiff one to the best skilled in undulations," Maxwell decided it would be best, given the uncertainties, simply to leave it aside.[90] Nor did he address magneto-optic rotation in "Dynamical Theory" – an ironic omission, since it had been Thomson's analysis of the Faraday effect that first put Maxwell on the path that led him to "Physical Lines of Force" and so to the electromagnetic theory of light. Maxwell found, however, that the general Lagrangian methods to which he had restricted himself in "Dynamical Theory" were not competent on their own to account for magneto-optic rotation; as he wrote to Thomson in January 1873, while drafting the corresponding section of his *Treatise*, "It is very remarkable that in spite of the *curl* in the electromagnetic equations of all kinds Faradays twist of polarized light will not come out without what the schoolmen call local motion."[91] In "Dynamical Theory," Maxwell had tried to see how far he could get toward explaining optical and electromagnetic phenomena without invoking his molecular vortices. Here was his answer: pretty far, but not all the way to magneto-optics.

Having shown theoretically in "Dynamical Theory" that the electromagnetic medium would carry waves at a speed given by v, Maxwell next began marshalling the experimental evidence that the measured value of v in fact matched the measured speed of light. The best available value of the latter was 298,000 kilometers per second, measured just two years before by Léon Foucault using a rotating mirror apparatus. In "Dynamical Theory," however, Maxwell also cited Fizeau's older figure of 314,858 kilometers per second, as well as the value of 308,000 kilometers per second derived from the aberration of starlight, perhaps because including the two latter numbers yielded an average that, as he said, "agrees sufficiently well" with the only measurement of v then

[89] Maxwell, "Dynamical Theory," 504–5, repr. in Maxwell, *SP 1*: 587.

[90] Maxwell to G. G. Stokes, October 15, 1864, in Maxwell, *SLP 2*: 186; see P. M. Harman, "Through the Looking-Glass, and What Maxwell Found There," in A. J. Kox and Daniel M. Siegel, eds., *No Truth Except in the Details: Essays in Honor of Martin J. Klein* (Dordrecht: Kluwer, 1995), 78–93. See also Maxwell, *SLP 2*: 182–85, a draft section, not included in the published "Dynamical Theory," in which Maxwell wrestled with the boundary conditions for reflection. Later in the 1870s H. A. Lorentz and G. F. FitzGerald independently found ways to include reflection and refraction in Maxwell's theory; see Darrigol, *Electrodynamics*, 190–92, 323–24.

[91] Maxwell to William Thomson, January 22, 1873, in Maxwell, *SLP 2*: 784.

available, Weber and Kohlrausch's figure of 310,740 kilometers per second.[92] Maxwell, however, was clearly looking toward new and better measurements of v, as he and Jenkin had declared in "Elementary Relations" and as he had outlined in his letters to Hockin and Thomson. Indeed, the inclusion of the last section of "Dynamical Theory," on calculating the self-induction of a coil of wire, made little sense except in this context. The problem had no obvious connection to the rest of the paper, but it was important for the experimental determination of the British Association ohm and so for Maxwell's planned new determination of v.[93]

Maxwell hoped and expected that the measured values of v and the speed of light would converge as new and better experiments were performed, but for a long time, they stubbornly refused to do so. Once they had certified "B. A." resistance standards in hand, Thomson and a team of his Glasgow students, several of whom later became cable engineers, set about carefully measuring the electromotive force of a battery with both an electrometer and an electrodynamometer. By 1868 they had arrived at a value for v of 28.25 ohms or (taking the ohm at its intended value of 10^7 meters per second) 282,500 kilometers per second.[94] That same year Maxwell and Hockin, using a method that balanced the electrostatic attraction of two disks against the electromagnetic repulsion of two coils, found v to be 28.8 ohms, or 288,000 kilometers per second (Figure 5.2). Given the great difficulty of the measurements involved, this was arguably not a bad match with Foucault's 298,000 kilometers per second for the speed of light, and it was certainly a striking fact that the two quantities should be even approximately equal.[95] But the

---

[92] Maxwell, "Dynamical Theory," 499, repr. in Maxwell, *SP 1*: 580.

[93] Maxwell, "Dynamical Theory," 506–12, repr. in Maxwell, *SP 1*: 589–97. See also Kline, "Induction Motor."

[94] W. F. King, "Description of Sir Wm. Thomson's Experiments Made for the Determination of v, the Number of Electrostatic Units in the Electromagnetic Unit," *BA Report* (1869), 434–36; see also Schaffer, "English Science," 150. King worked on several of Thomson's cable projects and later became chief engineer of the Western and Brazilian Telegraph Company before taking up the manufacture of dynamos and other electrical power equipment; see his obituary in *Proceedings of the Royal Society of Edinburgh* (1929) *49*: 386–87.

[95] Maxwell, "Direct Comparison," 644, 651, repr. in Maxwell, *SP 2*: 126, 135. See also I. B. Hopley, "Maxwell's Determination of the Number of Electrostatic Units in One Electromagnetic Unit of Electricity," *Annals of Science* (1959) *15*: 91–108. Schaffer, "English Science," 153, suggests that Maxwell took steps to "force" the measured value of v to match the known speed of light more closely, but Maxwell insisted that the measurements he rejected "were condemned on account of errors observed while they were being made," before he had calculated the values of v they implied; Maxwell, "Direct Comparison," 650, repr. in Maxwell, *SP 2*: 135. Jenkin's 1873 reprint of *Reports of the Committee on Electrical Standards*, 92, included Foucault's value for the speed of light, given as 29.8 ohms, in the list of values of v.

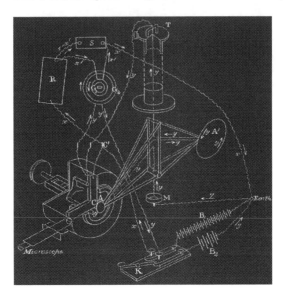

Figure 5.2 A schematic of the apparatus James Clerk Maxwell and Charles Hockin used in 1868 to measure the value of v, the ratio of electrostatic to electromagnetic units.
(From Maxwell, *Scientific Papers*, Vol. 2: 129.)

3 percent gap between Foucault's value for the speed of light and Maxwell and Hockin's for v was a far cry from the margin of just 0.02 percent that had fired Maxwell's enthusiasm in 1861 when he first compared Fizeau's reported value for the speed of light with Weber and Kohlrausch's for v. Throughout the 1860s and 1870s, the measured values of the two quantities remained too discrepant and uncertain to win over those like Thomson who doubted that light was really waves in the electromagnetic medium.

A further complication arose from the fact that the values of v found by Maxwell, Thomson, and several later experimenters were based on the putative value of the ohm, and as time went on it became increasingly clear that all was not well with the original British Association standards. Although the committee had expressed confidence that the resistance of its coils lay within 0.1 percent of their intended value, tests in the 1870s by German, British, and American physicists indicated the certified standards were off by ten or even twenty times that amount.[96] In 1880 the

---

[96] For the claim that the resistance of the British Association coils "does not probably differ from true absolute measurement by 0.08 per cent," see [Jenkin], "Report, 1864," 346,

British Association Committee on Electrical Standards was reconstituted and Lord Rayleigh, Maxwell's successor at the Cavendish Laboratory in Cambridge, set out to repeat the determination of the ohm using the original spinning coil apparatus, which Maxwell had secured for the laboratory in 1874. He also carefully reviewed Maxwell and Jenkin's procedures and calculations, and though he was unable to pinpoint the precise source of the discrepancy, he found they had made several significant errors, including transposing some numbers when measuring their spinning coil and miscalculating its self-induction. Rayleigh concluded that the resistance of the 1865 British Association standards was 1.3 percent too low – about 9870 kilometers per second instead of the intended 10,000.[97] As scientists in Britain and around the world developed improved methods of absolute electrical measurement, they gradually hammered out a corrected set of standards in their laboratories and in a series of international congresses.[98]

By the time measurements of the ratio of units and the speed of light definitively began to converge around 1890, Maxwell's theory had already largely won out on other grounds, notably those provided by Heinrich Hertz's dramatic experimental detection of electromagnetic waves in 1888. Had the gap between measurements of v and the speed of light persisted, it would no doubt have raised serious problems for Maxwell's theory, but in the event the eventual convergence in the values, though welcome, played relatively little role in drawing adherents to the theory.[99]

The strategy Maxwell pursued in the mid-1860s of trying to clinch the case for his electromagnetic theory of light by showing that the ratio of units equalled the speed of light met with only equivocal success. Discrepancies in measurements and uncertainties over the precise value of the ohm proved too great to overcome the skepticism of Thomson and

repr. in Smith, *Reports*, 161; for later summaries of the evidence that the discrepancy was in fact much larger, see Schaffer, "English Science," 161–62, and Olesko, "Precision," 137–43.

[97] Lord Rayleigh and Arthur Schuster, "On the Determination of the Ohm in Absolute Measure," *Proc. RS* (May 1881) *32*: 104–41.

[98] Larry Lagerstrom, "Constructing Uniformity: The Standardization of International Electromagnetic Measures, 1860–1912" (PhD dissertation, University of California–Berkeley, 1992).

[99] On improvements in the measurement of v between the 1870s and 1907 and their convergence with measurements of the speed of light, see Schaffer, "English Science," 163, and Rosa and Dorsey, "Comparison." On the growing acceptance of Maxwell's theory in the 1880s and 1890s, see Jed Z. Buchwald, *From Maxwell to Microphysics: Aspects of Electromagnetic Theory in the Last Quarter of the Nineteenth Century* (Chicago: University of Chicago Press, 1985), and Bruce J. Hunt, *The Maxwellians* (Ithaca: Cornell University Press, 1991).

others, and when he addressed the issue in his *Treatise* in 1873, Maxwell felt he could say no more than that his theory was "not contradicted" by the available measurements.[100] But what he had seen as a preliminary step in that strategy, of showing on general dynamical grounds that the electromagnetic medium would carry waves at a speed given by the ratio of units, proved remarkably fruitful, for it played a large part in leading him to formulate his general equations of the electromagnetic field.

Conclusion: "A Treatise on Electrical Measurement"

George Chrystal was a Scottish-born Cambridge wrangler who worked closely with Maxwell at the Cavendish Laboratory from 1874 until 1878; he later became professor of mathematics at the University of Edinburgh. He knew Maxwell and his work well, and when the second edition of Maxwell's *Treatise* appeared in 1881, two years after Maxwell's death, Chrystal was asked to review it for *Nature*. The resulting review is valuable and interesting in many ways, not least for a striking remark bearing on Maxwell's work for the British Association committee. Maxwell's *Treatise on Electricity and Magnetism*, Chrystal declared, is "in the strictest sense a Treatise on Electrical *Measurement*," for it "looks at electrical actions almost exclusively as measurable." Indeed, he said, much of the *Treatise* represented "a continuation of the labours of its author in conjunction with the rest of the distinguished band of electricians who formed the Committee of the British Association on Electrical Measurements [*sic*]."[101]

Chrystal's remark reinforces the point that Maxwell's work for the British Association Committee on Electrical Standards was not a mere side project for him, but lay firmly in the main line of his scientific development, and had a lasting effect on the way he approached electrical questions – including in his *Treatise*. The *Treatise* has customarily been seen as a work of electromagnetic *theory*, indeed, as the prime expression of Maxwell's own field theory and the foundation of its later development.[102] It was certainly that. But anyone who sits down and actually reads through its two thick volumes is likely to be struck by how little of the *Treatise* focuses on field theory at all, much less on Maxwell's own distinctive contributions to the subject. Instead one finds its pages

[100] Maxwell, *Treatise*, 2: § 787.
[101] Chrystal, "Clerk Maxwell's 'Electricity and Magnetism,'" 238.
[102] This is reflected, for instance, in Thomas K. Simpson, *Figures of Thought: A Literary Appreciation of Maxwell's "Treatise on Electricity and Magnetism"* (Santa Fe, NM: Green Lion Press, 2005), which focuses entirely on ten chapters in Part IV of Maxwell's *Treatise* and thus leaves aside 85 percent of the book.

filled with long accounts of one and two fluid electrical theories, extended discussions of the mathematical intricacies of such things as spherical harmonics and confocal surfaces, and detailed descriptions of the workings of guard ring electrometers and other electrical and magnetic instruments. As Andrew Warwick has shown, different groups read the *Treatise* in very different ways and for very different purposes, even just within Cambridge. Those attending W. D. Niven's intercollegiate lectures might delve into the physics of Maxwell's theory (as did George Francis FitzGerald, Oliver Heaviside, Oliver Lodge, and a scattered band of "Maxwellians" outside Cambridge), but most Tripos coaches turned to the *Treatise* solely for its treatment of mathematical techniques and told their students to skip almost everything we would now regard as "Maxwell's theory." Experimenters at the Cavendish, on the other hand, generally focused on Maxwell's accounts of electrical instruments and how to use them.[103]

Besides the college lecture hall, the Tripos coaching room, and the nascent Cavendish Laboratory, Maxwell's *Treatise* should also be seen in relation to a fourth context: the testing rooms of the great cable telegraph companies. The British Association Committee on Electrical Standards had been set up in 1861 largely to meet the needs of the growing British cable industry, and in his review of the *Treatise*, Chrystal emphasized how profoundly the development of electrical units and standards, and the ideas associated with them, had affected technological practice. "Instead of the old vague, unscientific, and still more, unbusinesslike statements of quantity and intensity," he noted, "we have the precise ideas of electromotive force, resistance, current, and so on, measured in their respective units, the volt, the ohm, the ampère." Now, he declared with evident satisfaction, "electrical commodities can be bought and sold by rule and measure, as heretofore cloth, coals, or horse-power."[104] Here Chrystal was echoing remarks Maxwell had made in his 1873 review of Jenkin's *Electricity and Magnetism* and that Thomson had often made about the commercial measurement of physical quantities.[105] Crosbie Smith, Norton Wise, and Simon Schaffer have all drawn attention to the ways commercial values and concerns informed the work of the British Association Committee on Electrical

---

[103] Andrew Warwick, *Masters of Theory: Cambridge and the Rise of Mathematical Physics* (Chicago: University of Chicago Press, 2003), 286–356; on the reception and development of Maxwell's theory outside Cambridge, see Hunt, *Maxwellians*.

[104] Chrystal, "Clerk Maxwell's 'Electricity and Magnetism,'" 238.

[105] [Maxwell], "Review of Jenkin, *Electricity and Magnetism*." Thomson argued that the farad should be defined in terms of a "real purchaseable tangible object" (i.e., a condenser of specified capacity) rather than a quantity of charge; see William Thomson to Maxwell, August 24, 1872, in Maxwell, *SLP 2*: 749 n. 8.

Figure 5.3 R. H. Campbell's 1929 portrait of James Clerk Maxwell depicts him seated beside the spinning coil resistance apparatus he and Fleeming Jenkin used in 1863–1864 to determine the value of the ohm. This painting hung for many years in the London headquarters of the Institution of Electrical Engineers, now the Institution of Engineering and Technology.
(Courtesy Institution of Engineering and Technology.)

Standards.[106] Chrystal's point (and mine) is that these values and concerns shaped Maxwell's wider work as well and are reflected both in his *Treatise* and in the crucial shifts his thinking underwent in the 1860s and 1870s.

The most striking of these shifts came, of course, in 1863–64, when Maxwell was fresh off his work on "Physical Lines" and had just joined the British Association Committee on Electrical Standards. At this crucial juncture, as he sought to capitalize on his discovery that light was evidently a disturbance in the electromagnetic medium and that its speed

---

[106] Smith and Wise, *Energy and Empire*, 687–98; Smith, *Science of Energy*, 277–87; Schaffer, "Late Victorian Metrology," 24–26; and Schaffer, "English Science," 135–36.

was given by the ratio of electrostatic to electromagnetic units, he considered how he could present his results in a way that would win them the widest possible support. He never gave up his belief that magnetic fields are filled with molecular vortices; indeed, when he returned to the Faraday effect toward the end of his *Treatise,* he said he still thought "we have good evidence for the opinion that some phenomenon of rotation is going on in the magnetic field" and that "this rotation is performed by a great number of very small portions of matter, each rotating on its own axis."[107] But Maxwell's collaboration with Jenkin on "Elementary Relations" and on the determination of the ohm, and his exposure to engineers' characteristic ways of thinking, had brought home to him the value of framing his results not just in terms of unseen microstructures but as far as possible also in terms of relations among measurable quantities. He was thus led to step back from the vortex model of "Physical Lines" and instead formulate the general equations of his "Dynamical Theory."

Maxwell was a man of many sides. As the titles of at least two books about him attest, he is often – and rightly – viewed as a "natural philosopher," perhaps one of the last in a long British line.[108] But he had another side, too, as seen in his work for the British Association Committee on Electrical Standards, and while he was no William Thomson, knee-deep in Glasgow commerce and industry, Maxwell developed closer links to the practicalities of electrical engineering than has often been recognized. He was, for instance, an early member of the Society of Telegraph Engineers and took an active part in its work promoting accurate electrical measurement. It is fitting that the portrait of Maxwell that was presented in 1929 to the Institution of Electrical Engineers, as the Society of Telegraph Engineers had by then been renamed, and that for many years adorned the "Maxwell Room" at the Institution's London headquarters, depicts him sitting with the spinning coil apparatus that he and Jenkin had used to determine the ohm, and that had done so much to shape his thinking on electrical questions[109] (Figure 5.3).

---

[107] Maxwell, *Treatise,* 2: § 831.   [108] Everitt, *Maxwell*; Harman, *Natural Philosophy.*
[109] On Maxwell's early membership in the Society of Telegraph Engineers, see Rollo Appleyard, *The History of the Institution of Electrical Engineers, 1871–1931* (London: Institution of Electrical Engineers, 1939), 43, and Maxwell to P. G. Tait, March 25, 1875, in Maxwell, *SLP 3*: 201 n. 8. The portrait of Maxwell by R. H. Campbell was presented to the IEE in 1929 by L. B. Atkinson; as of 2020, it is in storage after renovations to the headquarters of the Institution of Engineering and Technology (successor to the IEE). I thank Jonathan Cable and Aisling O'Malley of the IET Archives for this information.

6   To Rule the Waves
Britain's Cable Empire and the Making of "Maxwell's Equations"

The fortunes of oceanic submarine telegraphy were at a low ebb in the early 1860s. Battered by the failure of the first Atlantic cable and the collapse of the Red Sea line, the public had lost confidence in the once promising technology, and investors showed little appetite for tossing more of their money into the sea. "Englishmen are accustomed to pride themselves upon their perseverance and resolution," the *Times* observed in October 1861, yet when it came to submarine telegraphy, "after some abortive trials, we gave in."[1] It seemed Britain had abandoned the dream of spanning the world with undersea cables.

Amid this gloom, Cyrus Field and a few others fought to keep the Atlantic cable project alive. After years of strenuous effort, and leaning heavily on the encouraging findings of the *Joint Committee Report*, they managed to raise enough money to make another try, and in July 1865 the *Great Eastern* set out to lay a new and more carefully made cable across the North Atlantic – only to have it snap two-thirds of the way across.[2] Even this latest reverse left industry insiders remarkably unshaken; believing they now saw exactly what errors to avoid, they dug deep and raised the money for yet another attempt. This time all went smoothly and on July 27, 1866, the *Great Eastern* landed its cable at the aptly named Heart's Content, Newfoundland. Even more impressively, the great ship then returned to mid-ocean, grappled up the broken end of the 1865 cable, spliced it, and completed it to Newfoundland as well, thus establishing that cables could be recovered even from the depths of the

---

[1] *Times* (October 11, 1861), 6.
[2] On the 1865–1866 Atlantic cables, see especially Gillian Cookson, *The Cable: The Wire That Changed the World* (Stroud: Tempus, 2003); Chester G. Hearn, *Circuits in the Sea: The Men, the Ships, and the Atlantic Cable* (Westport, CT: Praeger, 2004); Donard de Cogan, *They Talk Along the Deep: A Global History of the Valentia Island Telegraph Cables* (Norwich: Dosanda Publications, 2016); H. M. Field, *The Story of the Atlantic Telegraph* (New York: Scribner's, 1893); Charles Bright, *The Story of the Atlantic Cable* (London: George Newnes, 1903); and W. H. Russell, *The Atlantic Telegraph* (London: Day and Son, 1866).

ocean. Europe and North America have been linked by cables of one kind or another ever since.

The success of the Atlantic cables soon set off a boom in cable laying around the world. British firms laid new cables in the Mediterranean and across the North Atlantic, quickly followed by lines to India, Singapore, China, Australia, and Brazil. By the mid-1870s nearly every major seaport in the world was in telegraphic contact with London, linked in a vast network that many compared to "the nervous system of the British Empire."[3] The web of wires continued to grow and ramify through the 1880s and 1890s, binding the global economy more and more closely together and ushering in an age in which world events and even daily life would be increasingly shaped by the rapid flow of information along wires and cables.

Electrical science, too, was deeply affected by the growth of the global cable system. James Clerk Maxwell completed his *Treatise on Electricity and Magnetism* in the early 1870s, at the height of the cable boom, and in its opening pages he noted that the rise of the telegraph industry had produced both a new "demand for electrical knowledge" and new "experimental opportunities for acquiring it."[4] British colleges and universities responded to that demand by establishing their first physics teaching laboratories, including Cambridge's Cavendish Laboratory, headed by Maxwell himself. These laboratories emphasized mastery of the techniques of precision electrical measurement, and they turned out substantial numbers of students equipped to work in cable company testing rooms or to teach those who would.[5]

In his *Treatise*, Maxwell also observed that the telegraph industry had given electrical researchers access to "apparatus on a scale which greatly transcends that of any ordinary laboratory," and so had exposed them to phenomena, notably those associated with the propagation of pulses of current along cables, that they would not have encountered on their own.[6] The seeds that had been planted by Latimer Clark's discovery of retardation in the early 1850s now grew to fruition as Maxwell and his successors developed new ways to treat propagation phenomena in terms

---

[3] On the cables laid between 1868 and 1875, see Charles Bright, *Submarine Telegraphs, Their History, Construction, and Working* (London: Lockwood, 1898), 106–30; on "the nervous system of the British Empire," see Percy A. Hurd, "Our Telegraphic Isolation," *Contemporary Review* (June 1896) 69: 899–908, on 899, and George Peel, "Nerves of Empire," in *The Empire and the Century* (London: John Murray, 1905), 249–87.

[4] Maxwell, *Treatise*, *1*: viii.

[5] Graeme Gooday, "Precision Measurement and the Genesis of Physics Teaching Laboratories in Victorian Britain," *British Journal for the History of Science* (1990) 23: 25–51.

[6] Maxwell, *Treatise*, *1*: viii.

of the action of the surrounding electromagnetic field. Fittingly, the final steps toward formulating what came to be called "Maxwell's equations" of the electromagnetic field, and of working out the theory of how electromagnetic waves propagate along wires and through space, were taken in the mid-1880s by Oliver Heaviside, a former telegrapher. The version of Maxwell's field theory that passed into the textbooks in the last years of the nineteenth century had been deeply shaped by Britain's global network of telegraph cables.

Try, Try Again

The Atlantic Telegraph Company fell on hard times after the failure of its 1858 cable. Amid sniping and recriminations, the company did little more over the next year or so than poke around the shore ends of its dead cable in fruitless attempts to revive it. By early 1860 it was virtually broke; having sunk its capital of £460,000 in the sea, the company had just £150 left to its name. The directors were forced to meet ongoing office expenses out of their own pockets, and that March the new chairman, James Stuart-Wortley, warned shareholders that "we are now … almost in the crisis of our existence."[7] He held out a faint hope that they might be saved by an unconditional government guarantee like the one the Red Sea company had secured, but the collapse of that line soon soured the government on offering similar support to any other cable projects. Shares that had gone for £1000 in 1856 were now finding few buyers at £30.[8] By the fall of 1860, the Atlantic Telegraph Company looked dead.

On the other hand, those who had studied the question most closely were convinced that with proper care, long oceanic cables – including ones across the Atlantic – could be made, laid, and operated with complete success. The Joint Committee to Inquire into the Construction of Submarine Telegraph Cables had said as much in its official *Report*, published in the summer of 1861, and had backed up its conclusions with voluminous experimental evidence. Some doubted, however, that the Atlantic Telegraph Company would be the best ones to do the job. Drawing heavily on the *Joint Committee Report*, an article on "Long-Sea Telegraphs" that appeared in *All the Year Round* in March 1862 reviewed the recent history of "wretched failures" in the cable industry and the

---

[7] "Atlantic Telegraph Company. Verbatim Minutes of Proceedings at an Extraordinary General Meeting," March 28, 1860, WC–NYPL; George Saward, *The Trans-Atlantic Submarine Telegraph* (London: privately printed, 1878), 46.

[8] In 1864 Stuart-Wortley said he had never lost faith in the company and had bought £1000 shares for as little as £30; see "Atlantic Telegraph Company," *Daily News* (April 1, 1864).

resulting loss of public confidence but then expressed cautious optimism for the future. "Unreasoning confidence," the article said,

has been succeeded by unreasoning distrust; private enterprise and the public purse have both been so severely taxed by the failures of rash ignorance and unscrupulous jobbery, that submarine-cable communications have fallen into undeserved disrepute. Yet it will not be difficult to show that, with existing materials and existing experience, properly employed, the most distant civilised regions may be brought into telegraphic communication with this country.[9]

The article traced the failure of the both the Atlantic and Red Sea cables not to anything intrinsic to submarine telegraphy, but to the incompetence and cupidity of the projects' promoters, who were distinguished, it said, more by their skill in securing exclusive landing rights from the authorities in Newfoundland and Constantinople than by any mastery of the art and science of cable telegraphy. The key to the future success of submarine telegraphy, the article concluded, lay simply in putting competent people in charge and avoiding the "jobbing concessionaires" who had repeatedly led the industry into disaster.[10]

A few cable-laying efforts continued through the early 1860s, but their record was mixed. The string of cables the British government laid from Malta to Alexandria in 1861 worked well enough, but the route had deliberately avoided deep waters. In 1860 and 1861 the French government contracted with a series of London-based firms to attempt to span the deeper waters from Toulon to Algiers, but the cables repeatedly snapped during laying and attempts to recover them failed. Working on behalf of the Spanish government, Sir Charles Tilston Bright managed in 1860 to lay four short cables to the Balearic Islands in waters as deep at 1400 fathoms, and Glass, Elliot laid one the next year from Toulon to Corsica in similar depths, but most other attempts to span deep parts of the Mediterranean failed.[11] Although the public remained understandably skeptical of deep-sea telegraphy, cable engineers contended that they had learned much from their repeated failures and would soon be able to solve the problem of oceanic submarine telegraphy.[12] It was far from clear, however, exactly when that might be.

[9] "Long-Sea Telegraphs," *All the Year Round* (March 15, 1862) 7: 9–12 and (March 22, 1862) 7: 39–45, on 9.
[10] "Long-Sea Telegraphs," 45.
[11] Bright, *Submarine Telegraphs*, 62–72; for lists of working and failed cables, see G. B. Prescott, *History, Theory, and Practice of the Electric Telegraph*, 3rd ed. (Boston: Ticknor and Fields, 1866), 473–74.
[12] George Saward to Cyrus Field, July 5, 1861, in Isabella Judson Field, *Cyrus W. Field, His Life and Work, 1819–1892* (New York: Harper & Brothers, 1896), 131.

In the meantime, the demand for rapid communications continued to grow. As the "Long-Sea Telegraphs" article noted, with a revealing mix of commercial and imperial concerns, "It is self-evident that telegraphic communication with our colonists and customers in Asia, Africa, and America is one of the most pressing wants of the age."[13] The outbreak of the American Civil War sharpened that demand, especially in the wake of the *Trent* affair. In November 1861, the captain of a US naval vessel stopped the *Trent*, a British mail packet, and seized two Confederate diplomats en route to Europe. This violation of maritime law sparked calls for war between the United States and Britain and set off a serious financial panic. Tensions mounted for several weeks as official protests passed slowly back and forth across the Atlantic by ship before the United States resolved the affair by disavowing the captain's actions and releasing the Confederate diplomats. Many observers at the time said it could all have been settled very quickly had an Atlantic cable been available; "what tears and anxieties – what stagnation of trade and depreciation of public securities – and what national animosities arising from the intemperance of English and American writers, would then have been prevented!," the *Daily News* wrote.[14] Other analysts, both then and later, argued that the slowness of shipborne communications actually helped cool tempers, and that quick exchanges via a cable might have exacerbated the situation.[15] Whatever effect a cable would really have had, the *Trent* affair clearly stoked the public demand for a new cable across the Atlantic, and Field and other proponents of the project lost no opportunity to cite it in their efforts to attract support.[16]

Field, Stuart-Wortley, George Saward, and other leaders of the Atlantic Telegraph Company tried hard to revive the project throughout 1862, winning official expressions of support from the US government, including a mention in President Lincoln's state of the union message, and securing endorsements from merchants' associations in both America and Britain.[17] Money, however, proved harder to come by.

---

[13] "Long-Sea Telegraphs," 9.

[14] "Ocean Telegraphy," *Daily News* (February 11, 1862). Similar sentiments appeared in "Prospects for the Atlantic Telegraph," *Morning Chronicle* (January 6, 1862); "The Atlantic Telegraph Revived," *Times* (February 4, 1862), 10, repr. from *The Observer*; and "Prospects of a New Atlantic Telegraph," *New York Times* (March 31, 1862), 4.

[15] David Paull Nickles, *Under the Wire: How the Telegraph Changed Diplomacy* (Cambridge, MA: Harvard University Press, 2003), 65–78.

[16] Cyrus W. Field, *Prospects of the Atlantic Telegraph. A Paper Read Before the American Geographical and Statistical Society, at Clinton Hall, New York, May 1, 1862* (New York: privately printed, 1862), 3.

[17] "Prospects of a New Atlantic Telegraph," *New York Times* (March 31, 1862), 4; "The President's Message," *New York Times* (December 2, 1862), 8; "The Atlantic Telegraph," *Liverpool Mercury* (December 20, 1862).

Field later recounted how the merchants of Boston listened attentively to his pitch, praised his tireless efforts on behalf of "one of the grandest enterprises ever undertaken by man," and passed a resolution warmly supporting the project – *"But not a man subscribed a dollar!"*[18] Thomson, Varley, and other leading electricians and engineers expressed their confidence in the practicability of a new cable, and Glass, Elliot offered to make and lay one on highly favorable terms, but what Saward called "the monetary public" still held back.[19] Only after cable industry insiders, including the leaders of Glass, Elliot and the Gutta Percha Company, stepped forward with substantial sums was the Atlantic company able to announce in June 1863 that it had secured the £300,000 it needed to start work and seek bids from potential contractors.[20] The board of directors referred the resulting bids to a committee of "the most eminent electricians and mechanicians in the kingdom," including Thomson, Joseph Whitworth, and Douglas Galton, who proceeded to test the mechanical strength and electrical properties of the submitted cable samples. Not surprisingly, given both its accumulated technical expertise and the substantial investment it had made in the project, Glass, Elliot was awarded the contract in March 1864, with the Gutta Percha Company to supply the core.[21]

Glass, Elliot's bid to make and lay the cable came in at £700,000, however, well above the amount the Atlantic Telegraph Company had earlier estimated, and while Stuart-Wortley told his shareholders that the extra expense would be "amply compensated" by the high quality of the resulting cable, the company was once again forced to seek additional capital.[22] The difficulty was soon met by John Pender, a Manchester cotton merchant who had been an early investor in the Atlantic company, and Daniel Gooch of the Great Western Railway Company, who took steps that spring to assemble a company with the financial muscle to make and lay the new Atlantic cable and, they hoped, many more. Armed with Pender's personal guarantee of £250,000, they were able to attract

[18] Field, *Story of the Atlantic Telegraph*, 238.
[19] Saward, *Trans-Atlantic Submarine Telegraph*, 47; "Facts for Shareholders," Atlantic Telegraph Company, November 26, 1862, WC–NYPL. In March 1862 Stuart-Wortley and Field led a large delegation to appeal to Prime Minister Palmerston to offer the project an unconditional guarantee, but they met with a decidedly chilly reception; see "Verbatim Minutes of Proceedings between the Right Hon. Viscount Palmerston, K.G., and a Deputation from the Atlantic Telegraph Company, at Cambridge House, Piccadilly, March 21st, 1862," IET.
[20] "Money-Market and City Intelligence," *Times* (June 2, 1863), 12; "Atlantic Telegraph Company," *Bristol Mercury* (June 6, 1863); "Atlantic Telegraph," *New York Times* (May 23, 1863).
[21] "Atlantic Telegraph Company," *Daily News* (March 17, 1864) and (April 1, 1864).
[22] "Atlantic Telegraph Company," *Daily News* (April 1, 1864).

enough capital to swing the deal, and in April 1864 Glass, Elliot absorbed the Gutta Percha Company to form the Telegraph Construction and Maintenance Company, a giant firm that would go on to dominate the cable-making industry for decades to come.[23] Taking over the contract Glass, Elliot had just signed, TC&M agreed to take most of its payment from the Atlantic company in the form of shares and bonds, with a bonus of additional shares if it successfully completed the cable.[24] In effect, the Atlantic Telegraph Company agreed to sell much of itself to TC&M in return for making and laying the cable that had so long been its goal.

Another key part of the plan was the ship that would lay the cable. Gooch controlled the company that had recently bought the *Great Eastern* out of bankruptcy, and finding remunerative employment for the grand but star-crossed ship provided much of his motivation for getting involved in the cable project.[25] TC&M quickly arranged to charter the *Great Eastern*, and over the summer and fall of 1864 Gooch had the ship refitted with cable tanks and laying gear while TC&M began manufacturing the cable at its Greenwich works.

The copper conductor of the new cable weighed 300 pounds per nautical mile – nearly three times as much as that of the 1858 cable – and was covered with gutta-percha weighing 400 pounds per nautical mile, half again as much as the insulation of the old cable. With its much lower resistance and capacitance, the new cable should, according to Thomson's theory, be able to carry signals at a substantially higher rate than the old one, as indeed proved to be the case, while also being much sturdier and more reliable. Willoughby Smith's electrical staff at TC&M initially measured the resistance of both the conductor and insulation in Siemens units, but when British Association resistance coils became available early in 1865, they began using those as well.[26] Manufacture of the cable started slowly at first, but by the spring of 1865 hundreds of

---

[23] On the formation of TC&M, see Daniel Gooch, *Diaries of Sir Daniel Gooch* (London: Kegan Paul, Trench, Trübner, & Co., 1892), 82, and Pender to the Earl of Wilton, October 10, 1866, in Stewart Ash, *The Cable King: The Life of John Pender* (London: Stewart Ash, 2018), 104–8, on 105. There is no adequate history of the company, which was later generally known as "Telcon"; G. L. Lawford and L. R. Nicholson, *The Telcon Story, 1850–1950* (London: Telegraph Construction & Maintenance Co., 1950) is thin and unreliable.

[24] On the financial arrangements between the Atlantic Telegraph Company and Glass, Elliot, which TC&M soon took over, see "Atlantic Telegraph Company," *Daily News* (April 1, 1864). Although TC&M was not officially formed until April 6, 1864, its principals had already been involved in the negotiations between the Atlantic Telegraph Company and Glass, Elliot for some time.

[25] Gooch, *Diaries*, 82–83.

[26] Willoughby Smith, *The Rise and Extension of Submarine Telegraphy* (London: J. S. Virtue, 1891), 124, 135; in his "Diary of the Atlantic Telegraph Expedition, 1866," *Daily News* (August 18, 1866), John C. Deane referred to the resistance unit as an "ohmad."

miles of it were being ferried out to the *Great Eastern* at its moorings off Sheerness.

The ship was fitted with three huge iron tanks so that the 2500 miles of coiled cable could be kept under water to prevent oxidation of the gutta-percha and facilitate testing for faults. As the *Times* noted, "Electricians are constantly employed on board in a portion of the *Great Eastern* appropriated for their accommodation, and by means of the most sensitive and delicate instruments every portion of the cable is subjected to the most careful and rigid tests . . . in order that the most trifling defect may be discovered."[27] No significant flaws were found and both TC&M and the Atlantic Telegraph Company touted the cable as the finest ever made.

The great ship and its cargo of cable offered an intriguing spectacle, and Pender hosted gala visits to the *Great Eastern* by dukes and dignitaries and, on May 24, by the Prince of Wales. Thomson, Varley, and other "scientific gentlemen" were called upon to explain the various operations to the prince, and while it is not clear how much he really took in, he was reportedly gratified to witness the transmission of a short message – "God save the Queen" – through 1200 miles of stowed cable, and he wished the project well.[28]

Loaded with coal and cable, the *Great Eastern* set off for Ireland on July 15 under the command of James Anderson, an experienced captain from the Cunard line. "The most entire confidence is expressed by the naval and scientific men," the *Times* reported; "success," they said, "is almost certain on this occasion."[29] A smaller vessel, the *Caroline*, landed the heavy shore end at Valentia on July 19, an event that W. H. Russell, the celebrated correspondent TC&M had commissioned to chronicle the expedition, said was marked with speeches and "hearty cheers."[30] Four days later the *Great Eastern* made its splice to the shore end and began steaming westward, accompanied by HMS *Sphinx* and HMS *Terrible*, while keeping in constant touch with Valentia through the cable.

The laying started smoothly, but about eighty miles out, the electricians detected a fault in the insulation of the cable, forcing the *Great Eastern* to begin the laborious task of hauling it back in. In an omen of troubles to come, the hauling-in machinery proved barely adequate to the task, but

[27] "The Atlantic Telegraph Cable," *Times* (April 25, 1865), 14.
[28] "The Atlantic Telegraph Cable," *Times* (May 25, 1865), 14; "Visit of the Prince of Wales to the Great Eastern," *Glasgow Herald* (May 27, 1865). Russell, *Atlantic Telegraph*, 41, said the Prince sent the message "I wish success to the Atlantic cable" through 1395 nautical miles of cable.
[29] "The Atlantic Telegraph Expedition," *Times* (July 17, 1865), 9.
[30] Russell, *Atlantic Telegraph*, 49. On the commissioning of Russell by TC&M, see the documents on the "Atlantic Cable" website at atlantic-cable.com/Books/Russell/index.html.

the crew finally pulled the faulty section of cable aboard, cut it out, and replaced it.[31] On examination, the engineers found that a short piece of iron wire had somehow penetrated the gutta-percha. They took it to be an isolated accident, however, and continued laying the cable.

The *Great Eastern* steamed placidly on for four more days, until on July 29 the electricians detected another and more serious fault – a "dead earth" that cut off all communication with Valentia. The *New York Times*, relying on accounts arriving from England by ship, blared a headline on August 10 reporting "Probable Failure of the Enterprise" and remarked that no one in America was much surprised by such an outcome.[32] Aboard the *Great Eastern*, however, the crew simply hauled the faulty section back in and replaced it with a fresh length; within a day, they had resumed paying out cable and were soon more than half way to Newfoundland. The mood on the ship darkened, however, after Samuel Canning, the chief engineer, examined the damaged section of cable. He found that it, too, had been pierced by an iron wire, and in a manner that suggested sabotage. Suspicion fell on the workmen in the cable tank and a rotation of "gentlemen," including Field himself, was assembled to keep watch against further mischief.[33] The laying continued, but everyone was on edge.

The crisis came on August 2 – "a sad and memorable day," Russell said, "in the annals of Atlantic telegraphy."[34] A poorly tempered armoring wire had snapped and, in passing through the paying-out machinery, was driven into the insulation of the cable – this, rather than sabotage, had probably caused the earlier faults as well. The damaged section passed overboard before the ship could be stopped, and after electrical tests established the seriousness of the fault, the crew began slowly hauling in the cable. This time the hauling-in gear proved balky and the small steam engine that drove it lost power; as the *Great Eastern* tried to hold its position in a stiff wind, the cable chafed against part of the bow. When the hauling in resumed, the damaged cable suddenly parted under the strain and flew overboard. "It is all over – It is gone," Canning exclaimed, and even Field's spirits were at least briefly shaken.

The expedition had not contemplated attempting to recover a lost cable from the depths of the Atlantic, but Canning quickly resolved to try to do just that. He ordered a grappling hook lowered into the sea, and for the

---

[31] W. H. Russell, "The Laying of the Atlantic Cable. Diary of Events," *Daily News* (August 19, 1865).

[32] "The Atlantic Telegraph," *New York Times* (August 10, 1865), 1.

[33] Russell, *Atlantic Telegraph*, 72.

[34] The following account is drawn from Russell, "Diary of Events."

Figure 6.1 The *Great Eastern* grappling for the broken end of the Atlantic cable in August 1865.
(From Louis Figuier, *Merveilles de la Science*, Vol. 2: 281, 1868.)

next nine days the *Great Eastern* moved slowly back and forth across the line of the cable, hoping to hook it and raise it from 2000 fathoms deep (Figure 6.1). Three times it managed to snag the cable and raise it toward the surface, but each time the tackle broke from the strain. Canning finally had to give up on August 9, but only because he had exhausted his supply of rope. Given their repeated failures, one might have expected Canning and his crew to be downcast as the *Great Eastern* steamed toward home, but they in fact expressed strong confidence that, with proper equipment, they could recover and complete the broken cable.

London knew none of this at the time, of course, only that contact with the *Great Eastern* had been lost. Shares of both TC&M and the Atlantic company fell amid fears that the cable had been irretrievably lost, or even that the ship had sunk.[35] The Astronomer Royal, George Biddell Airy, speculated that the interruption may have been caused by earth currents from a violent magnetic storm, but as the silence wore on, and as the electricians at Valentia reported the results of their most searching tests, it became clear that the cable had parted and been lost. The *Times* declared it "almost a national disaster," while the *Derby Mercury* observed loftily that the failure "ought to teach us that with all our self-sufficient pride of knowledge and mechanical genius, there are many things yet beyond the reach of our scientific power." But that lesson evidently did not sink in: as the *Mercury* noted, the backers of the cable project had already declared their determination to try yet again, "if the money can be raised for the next experiment."[36]

When the *Great Eastern* finally returned home, Anderson declared that he was now more sure than ever that the project would ultimately succeed and reported that Field, "buoyant and hopeful beyond us all," wanted to gather more grappling gear and head right back out to fish up the lost cable.[37] It soon became clear, however, that it would take months to repair and refit the *Great Eastern* and improve its hauling-in gear. Moreover, even before this latest setback, leaders of the Atlantic company had announced plans to lay a second cable to handle the expected crush of transatlantic traffic. They therefore set about trying to raise enough money both to make and lay a new cable the next summer and to grapple up and complete the one that had broken.[38]

At this point relations between the Atlantic Telegraph Company and TC&M grew testy. Directors of the Atlantic company questioned whether TC&M had lived up to its contractual obligation to "take all proper precautions" in laying the cable and were on the verge of suing for negligence. TC&M sought to head this off by offering both to make and lay a second cable and to recover and repair the broken one for £500,000, plus an extra £100,000 in shares in case of success.[39] The Atlantic company agreed to the deal and set about selling new shares, which would have a first claim on all profits up to an annual return of 12 percent. The investing public still held back, however, and in December the project received a new blow: the attorney general said that the Atlantic

[35] "The Atlantic Cable," *Times* (August 3, 1865), 7; "Money Market," *Daily News* (August 4, 1865).

[36] "The Atlantic Cable," *Times* (August 8, 1865), 7; "Baffled Science," *Derby Mercury* (August 9, 1865).

[37] "The Atlantic Cable," *Times* (August 21, 1865), 10.

[38] "Atlantic Telegraph Company," *Daily News* (August 10, 1865), reporting on the shareholders' meeting held on August 9, before the return of the *Great Eastern*.

[39] Saward, *Trans-Atlantic Submarine Telegraph*, 64–65; "Atlantic Telegraph Company," *Times* (August 31, 1865), 10.

company could not sell new preferred shares without a new act of Parliament, which would be impossible to secure in time for the summer laying season. At this impasse Pender and Gooch stepped forward with a fresh expedient: they would organize a new firm, the Anglo-American Telegraph Company, to act as agents of the Atlantic company in arranging to make and lay a new cable and recover and complete the old one, in return for the first £125,000 of the expected annual profits. Holders of the original Atlantic shares worried that, after standing by the project through its darkest days, they would now be cut out from most of its profits, as indeed proved to be the case. Seeing no other way forward, however, they voted in March 1866 to approve the deal. Even then, however, few outside investors were willing to buy shares in the new company, which in the end was financed mainly by cable industry insiders, led by Pender.[40]

Once its finances were set, the 1866 expedition proceeded with little drama, especially when compared to the 1858 and 1865 attempts. The design of the cable was slightly modified to make it less liable to be damaged by a broken armoring wire; Canning and his engineers installed improved hauling-in machinery and gathered ample supplies of grappling gear; and Willoughby Smith devised a new testing system that enabled technicians to monitor the electrical condition of the cable continuously, rather than just at set intervals, while also allowing signals to be freely exchanged between ship and shore.[41] When the *Great Eastern* set out from Ireland on July 13, the operators at Valentia were able to keep up a regular line of chatter with those on board, while also sending the ship daily news reports from London and the Continent, which were printed up and distributed in a shipboard newspaper.[42] The great ship was again accompanied by HMS *Terrible*, along with two chartered vessels, the *Albany* and the *Medway*, which carried additional cable as well as grappling gear to use in recovering and completing the 1865 cable.

The laying of the cable commenced soberly, without the speechmaking and huzzahs that had marked the earlier attempts, and proceeded

---

[40] On the financing of the Anglo-American company, see Cookson, *The Cable*, 145–48; see also Cyrus W. Field, *Report of Mr. Cyrus W. Field to the President and Directors of the New York, Newfoundland, and London Telegraph Company, London, March 8th, 1866* (London: William Brown and Co., 1866), detailing the arrangements with the Anglo-American company and giving optimistic estimates of the eventual returns to the Atlantic and New York, Newfoundland, and London companies.

[41] Smith, *Rise and Extension*, 142–43; Field, *Story of the Atlantic Telegraph*, 302–4.

[42] "Atlantic Telegraph Expedition," *Times* (July 17, 1866), 8. The news reports sent via the cable appear in Deane, "Diary," *Daily News* (August 18, 1866), and are reproduced in Smith, *Rise and Extension*, 366–70; they include accounts of the Austro-Prussian War then being fought on the Continent and of disturbances in Hyde Park by protesters demanding electoral reform.

with what Henry Field later called "the monotony of success."[43] After two uneventful weeks at sea, the *Great Eastern* reached Newfoundland on July 27. Smith and the TC&M electricians handed the ends of the cable over to J. C. Laws at Heart's Content and Latimer Clark in Valentia, who tested it on behalf of the Atlantic and Anglo-American companies and pronounced it perfect. The cable could carry six words per minute in initial tests, Clark said, and with improved apparatus and the use of codes, he expected "fully thrice that speed will be obtained." On July 29 the cable was thrown open for business, which soon poured in, and Clark closed his report by congratulating the companies on "the wonderful prospects of pecuniary success" which now opened before them.[44]

Public response to the completion of the cable was muted, both from wariness that it might yet fail as its predecessor had in 1858, and because the cable connecting Newfoundland to Nova Scotia had broken down and was awaiting repair.[45] As Clark observed, had the cable to the American mainland been in working order, it "would have added greatly to the *éclat* of the opening."[46] Once that link was restored on August 13, messages no longer had to wait to be carried by steamer across Cabot Strait but could pass electrically from New York to London, or indeed from almost anywhere in North America to anywhere in Europe.[47] The Anglo-American company initially set its tariffs high – £1 per word, with a twenty-word minimum – to keep the cable from being blocked up with traffic, but agitation to lower them began almost immediately, as did calls for more cables.[48]

Of course, TC&M was already under contract to provide a second cable by recovering and completing the one that had broken the year before, and almost as soon as the *Great Eastern* could refill its bunkers with coal and take on more cable from the *Medway*, it steamed back to the mid-Atlantic. There the *Terrible* and the *Albany* were already at work, and for the next three weeks the ships grappled for the lost cable. On August 17 they brought it all the way to the surface, but the crew's cheers had scarcely died away before the cable parted from the strain and slipped away. After thirty frustrating attempts, Canning's men finally managed to haul the cable aboard the *Great Eastern* in the early hours of September 2. Smith brought its end into the testing room, where Field, Anderson, Gooch,

---

[43] "Atlantic Telegraph Expedition," *Daily News* (July 11, 1866); Field, *Story of the Atlantic Telegraph*, 320.

[44] Clark to Saward, July 30, 1866, in Saward, "To the Editor," *Times* (August 6, 1866), 12.

[45] On wariness that the cable might yet fail, see *Boston Daily Advertiser* (July 30, 1866); on the broken cable across Cabot Strait, see "The Atlantic Cable," *Times* (July 31, 1866), 3.

[46] Clark to Saward, July 30, 1866, in Saward, "To the Editor," *Times* (August 6, 1866), 12.

[47] On the repair of the Cabot Strait cable, see "America," *Daily News* (August 13, 1866).

[48] "The Atlantic Cables," *Daily News* (August 3, 1866).

Figure 6.2 On September 2, 1866, the crew of the *Great Eastern* grappled up the broken end of the 1865 cable and brought it to the ship's testing room. In this detail from a watercolor by Robert Charles Dudley, Cyrus Field and Captain James Anderson can be seen looking on as Willoughby Smith anxiously awaits a reply to his first signal to the Valentia station.
(Detail from Robert Charles Dudley, "Awaiting the Reply," 1866; courtesy of the Metropolitan Museum of Art, New York.)

Thomson, and others watched anxiously as he connected it to the instruments and signaled through to Valentia (Figure 6.2). For months the staff there had been patiently monitoring a mirror galvanometer connected to the broken cable. They now watched as its spot of light began to move and spell out words, and then sent the *Great Eastern* a simple reply: "Understand, Query." As cheers spread through the ship, Canning sent a brief message to Valentia: "I have much pleasure in speaking to you through the 1865 cable."[49]

[49] "Recovery of the Atlantic Cable of 1865," *Daily News* (September 3, 1866); "The Atlantic Telegraph Expedition, 1866," *Times* (September 11, 1866), 10; "The Atlantic Cable," *Leeds Mercury* (September 11, 1866); Smith, *Rise and Extension*, 187–88.

The *Great Eastern* made its splice, steamed westward to Newfoundland, and on September 8 landed its second cable, which was also was soon opened to traffic. Grappling the lost cable from the bottom of the Atlantic was widely hailed as a "marvellous feat," in many ways even more impressive than laying the new cable had been.[50] Moreover, by establishing that cables could be recovered and repaired even in the deepest waters, it reduced the financial risk of oceanic submarine telegraphy and made it far more attractive as a commercial enterprise.

The success of the Atlantic cable project was testimony, *Lloyd's Weekly* declared, to the "English pluck" and "pocket courage" of its backers, as well as to the expertise of its engineers and electricians – several of whom, including Canning and Thomson, were soon rewarded with knighthoods.[51] Persevering through repeated reverses, they had not only united the Old World with the New but had brought to light intriguing new electrical phenomena and stimulated the development of important new measurement techniques. The Atlantic cables opened a new era in submarine telegraphy, one whose far-reaching consequences for commerce, empire, and electrical science would play out over the coming decades.

The Cable Boom

The *Times* hailed the laying of the Atlantic cable as "the most wonderful achievement of this victorious century," while other newspapers in Britain and America praised it as "a great scientific and mechanical triumph," "an organ of the civilization of our age," and "one of the grandest victories of peace."[52] Few could top *Freedom's Champion* of Atchison, Kansas: under the headline "A Great Triumph!," it called the laying of the cable "the greatest scientific triumph of our century" – "a glorious triumph of Civilization over Nature, of Science over the Elements, of Perseverance over Difficulty, of Mind over Matter, of Man over the Ocean and its dangers"; it marked "the dawn of a new era of improvement and progress," the editors declared, and would be "incal-

---

[50] "The Atlantic Cable of 1865," *Freeman's Journal* (September 4, 1866).

[51] "The Atlantic Cable," *Lloyd's Weekly* (August 5, 1866); on the knighthoods conferred on Anderson, Canning, and Thomson, see Thompson, *Kelvin, 1*: 499–500. Pender was passed over at the time because of accusations he had bought votes in Totnes, from which he had been elected to Parliament in July 1865. His election was overturned in May 1866, but he was later cleared of the most serious bribery charges and was eventually knighted in 1888; Ash, *Cable King*, 109–13, 322–23.

[52] *Times* (July 30, 1866), 8; *Manchester Times* (July 28, 1866); *Daily News* (July 30, 1866); "Laying of the Atlantic Cable," *Vermont Chronicle* (August 4, 1866).

culably beneficial to the whole human race."[53] When the news reached far-off California (by overland telegraph, of course) on July 30, the *Daily Evening Bulletin* of San Francisco hailed the completion of the cable as "The Triumph of Science and Faith" – and then turned immediately to analyzing its probable effects on the silver trade.[54]

Grand encomiums to the Atlantic cable as a harbinger of peace and progress were all very fine, but the real interest lay in its value to commerce. That value took two main forms: the transformation of trade and finance through the rapid dissemination of prices and other market information, and the profitability of the cable industry itself, on which the further extension of the network would depend. Some of the commercial effects of the cable were felt almost immediately. Just two days after first reporting its completion, the *New York Times* noted that "The successful working of the Atlantic Cable is already equalizing the quotations for American Stocks in London and New-York" and had reduced the price differential on foreign exchange markets "below the point at which it would pay to ship ordinary gold coin."[55] Later analyses of prices for US Treasury bonds in New York and London showed that the Atlantic cable eliminated almost all of the previous difference in prices in the two great financial centers and turned them into what was in effect a single integrated market.[56] To this day, more than a century and a half later, foreign exchange traders still call the pound-to-dollar rate "the cable." By simplifying the placing of long-distance orders and reducing uncertainty about fluctuating prices and exchange rates, the growing network of submarine cables smoothed and accelerated global trade in the last decades of the nineteenth century and facilitated the large international flows of capital that marked the period.[57]

Of course, that network would be unlikely to continue to grow unless investors believed there was money to be made in laying and operating submarine cables, and it soon became clear that if they got into the business at the right time, there was plenty. When the shareholders of

[53] "A Great Triumph! Atlantic Cable Successfully Laid!," *Freedom's Champion* (August 2, 1866).

[54] "The Triumph of Science and Faith" and "The Atlantic Cable and the Silver Market," *Daily Evening Bulletin* (July 30, 1866).

[55] "Monetary Affairs," *New York Times* (August 1, 1866), 3.

[56] Kenneth D. Garbade and William L. Silber, "Technology, Communication and the Performance of Financial Markets: 1840–1975," *Journal of Finance* (1978) *33*: 819–32; Claudia Steinwender, "Real Effects of Information Frictions: When the States and the Kingdom Became United," *American Economic Review* (2018) *108*: 657–96.

[57] Kevin H. O'Rourke and Jeffrey G. Williamson, *Globalization and History: The Evolution of a Nineteenth-Century Atlantic Economy* (Cambridge, MA: MIT Press, 1999); Jürgen Osterhammel, *The Transformation of the World: A Global History of the Nineteenth Century* (Princeton: Princeton University Press, 2014), 719–21.

the Anglo-American Telegraph Company held their first general meeting in February 1867, it was, chairman Charles Stewart said, with "satisfaction unalloyed." Both cables were working even better than promised, he reported, far exceeding the eight words per minute specified in the original contract; indeed, the clerks who received messages by tracking a spot of light on their mirror galvanometer found that "the cables transmit signs more rapidly than the human eye and brain can see and follow them," so that they had to ask the sending clerks to slow down.[58] The cables proved lucrative from the first; with the initial high tariff, and with newspapers willing to pay dearly to file their first stories "by ocean telegraph," the receipts for a single day in August had come to over £2000, including two messages that reportedly cost their senders £800 each, "one the speech of the King of Prussia, and the other an account of some sporting match."[59] Willingness to pay £1 per word soon faded, however, and in November, after the second cable had been completed and the links to the North American mainland repaired, Anglo-American cut its tariff in half, resulting in increased traffic and stabilizing the average revenue at about £800 per day, almost all for business and press messages.[60] This was more than enough for Anglo-American to pay its investors their promised 25 percent annual dividend, though once the company set aside money for a reserve fund to pay for repairs to its cables – the 1866 cable was repeatedly broken, reportedly by icebergs grounding on the seafloor near Newfoundland – the balance left to be passed along to the original shareholders of the Atlantic Telegraph Company proved, as Saward later observed, "very slender." This led to ongoing tensions between the two companies, which were not resolved until they were combined in 1870 on terms

---

[58] *Anglo-American Telegraph Company, Limited, First Ordinary General Meeting of Shareholders, Monday, February the 4th, 1867* (London: Metchin and Son, 1867), 1, 4.

[59] See the remarks of Stuart-Wortley at the September 27, 1866, meeting of the Atlantic Telegraph Company, in *Times* (September 28, 1866), 4. The King of Prussia's speech, transmitted "By Ocean Telegraph," appeared in the *New York Times* (August 9, 1866), 5; on its handling, see W. F. G. Shanks, "How We Get Our News," *Harper's Magazine* (March 1867) *34*: 511–22, on 517.

[60] See the remarks of Charles Stewart in *Anglo-American Telegraph Company*, 11; see also "Atlantic Telegraph Company, Report of the Directors to the Tenth Ordinary Annual Meeting of the Shareholders," March 12, 1867, WC–NYPL. Because some commercial customers were willing to pay dearly for urgent messages, Anglo-American found it got its best return by charging high tariffs, even if this left its cables idle for much of the day. An early shareholder of the Atlantic company said he had learned that "the present business does not occupy in transmission more than four hours per day, and that upon one cable only," though this evidently referred to the period before Anglo-American reduced its initial very high tariff; see "An Old Atlantic," "The Atlantic Telegraph Company" (letter), *Times* (December 24, 1866), 5.

very favorable to the Anglo-American shareholders, prominently including John Pender.[61]

There had long been talk of laying cables to connect Britain not just to North America but to all corners of the world, and one might have expected the success of the Atlantic cables to set off an immediate wave of new projects. In fact there was a prolonged lull; as Willoughby Smith later observed, "for two years all was gloom and anxiety."[62] This lull was not rooted in anything particular to the cable industry, however, but in the unsettled state of the financial markets. The collapse in early May 1866 of the banking house of Overend, Gurney and Company set off a panic – "Black Friday" – that virtually froze London capital markets for the next two years. Many companies went under and it became almost impossible to launch any new initiatives requiring as much capital as a submarine cable. Indeed, Stewart told the shareholders of Anglo-American in early 1867 that in assembling the capital for their company the previous April, they had gotten in just under the wire: after Black Friday, "they would have gotten nothing," and given the prevailing lack of public confidence, it would, he believed, have been many years before anyone tried again to lay a cable across the Atlantic.[63]

By 1868, capital markets had begun to recover and a boom in cable laying soon followed. This was given a substantial boost by the passage of the Telegraph Act of 1868, which called for the British Post Office to acquire and nationalize the inland telegraph system. Over the next few years the government would pay out over £10 million to shareholders of the Electric, Magnetic, and other British telegraph companies, and many of these telegraph-savvy investors immediately put their newly freed-up capital into overseas cable projects, which remained in private hands.[64] A flurry of cable laying ensued, and by the mid-1870s British companies had not just added additional links in the Mediterranean and across the North Atlantic but had extended new lines to India, China, Australia, and Brazil.

The first important cable of this new wave attracted relatively little public notice when it was laid from Malta to Alexandria in September 1868, but it set a pattern that would later be repeated many

---

[61] Stewart, in *Anglo-American Telegraph Company*, 2. On the fraught relations between the Atlantic and Anglo-American companies, see Saward, *Trans-Atlantic Submarine Telegraph*, 73–80. On the repeated breaking of the 1866 cable, see Bright, *Submarine Telegraphs*, 105.

[62] Smith, *Rise and Extension*, 209.

[63] Stewart, in *Anglo-American Telegraph Company*, 6.

[64] On the Telegraph Act of 1868 and the government payments to shareholders of the private companies, see Jeffrey Kieve, *The Electric Telegraph: A Social and Economic History* (Newton Abbot: David & Charles, 1973), 154–75.

times. Earlier that year Pender and a group of his associates – members of what historian Simone Müller has called "the Class of 1866," alumni of the successful Atlantic project – floated the Anglo-Mediterranean Telegraph Company with a capital of £260,000.[65] They then arranged for the cable to be manufactured and laid by TC&M, in which, of course, Pender also held a large stake. With Canning handling engineering duties and Smith serving as chief electrician, TC&M laid the cable without incident and the Anglo-Mediterranean company opened it to traffic in early October 1868. The company also took over operation of the string of cables the British government had laid along the North African coast from Malta to Alexandria in 1861, intending to use this route to back up its direct cable in case of any temporary outages. The shallow-water coastal cables were so often damaged by ships' anchors, however, and the inter-mediate stations proved so expensive to maintain, that in November 1870 Anglo-Mediterranean replaced them with a second direct cable.[66]

By then British investors were in the grip of what J. Wagstaff Blundell called "a mania for submarine telegraph companies."[67] A London accountant and a close observer of the telegraph industry, Blundell had published a pamphlet in February 1869 on *Telegraph Companies Considered as Investments; with Remarks on the Superior Advantages of Submarine Cables*, followed in 1871 and 1872 by successive editions of his *Manual of Submarine Telegraph Companies*.[68] His subject, he said, was "the use of the science of Telegraphy considered financially," and he sought to detail the financial condition and prospects of cable companies for the benefit of potential investors in the industry.[69] The number of those investors, and of companies for them to consider, was then growing rapidly. No fewer than twelve submarine telegraph companies were suc-cessfully floated between the summer of 1868 and the spring of 1870, four of them in January 1870 alone.[70] Two were to prove especially significant: the French Atlantic Telegraph Company (*Société du Cable Transatlantic Français*), launched in August 1868, and the British-Indian Submarine Telegraph Company, formed in January of the following year. Both

---

[65] Simone M. Müller, *Wiring the World: The Social and Cultural Creation of Global Telegraph Networks* (New York: Columbia University Press, 2016), 9; J. Wagstaff Blundell, *The Manual of Submarine Telegraph Companies* (London: Rixon and Arnold, 1871), 20–23.

[66] "The Anglo-Mediterranean Telegraph," *Times* (September 17, 1868), 5; Bright, *Submarine Telegraphs*, 106.

[67] Blundell, *Manual of Submarine Telegraph Companies* (1871), 6.

[68] J. Wagstaff Blundell, *Telegraph Companies Considered as Investments; with Remarks on the Superior Advantages of Submarine Cables* (London: Effingham Wilson, 1869), 3; Blundell, *Manual of Submarine Telegraph Companies* (1871), and 2nd ed., "published by the author," January 1872.

[69] Blundell, *Telegraph Companies Considered as Investments*, 6.

[70] Blundell, *Manual of Submarine Telegraph Companies* (1871).

would figure prominently in the cable boom, and both would become major components of Pender's growing cable empire.

Although the French Atlantic company was chartered in Paris and its main cable ran between French termini – from Brest in Brittany to St. Pierre, a small French island just south of Newfoundland, from which an additional cable ran to Duxbury, Massachusetts – it was in other ways a thoroughly British enterprise. Most of its capital was subscribed in London and, as a Boston newspaper observed soon after the project was launched, this "great national work," though putatively French, was "to be fashioned in an English manufactory, and intrusted to English seaman, who, from the decks of an English vessel, will commit it to the depths of the Atlantic."[71] Not only would the cable be manufactured by TC&M on the banks of the Thames and laid by the *Great Eastern*, but its engineers and electricians would all be British, most of them veterans of the Atlantic cable expeditions of 1865 and 1866: Canning, Smith, Clark, Thomson, and Varley, along with Fleeming Jenkin and Charles Hockin, who had recently helped James Clerk Maxwell establish the value of the ohm.[72]

Blundell pointed to the publicity surrounding the French Atlantic company and the large dividends paid out by Anglo-American as the bait that drew investors to submarine telegraphy.[73] Indeed, in 1869 the *Times* said "There can be no doubt that the most popular outlet now for commercial enterprise is to be found in the construction of submarine lines of telegraph"; their rising share prices and hefty dividends "show that these schemes have proved more remunerative than ever was contemplated by their promoters, except perhaps a few more enthusiastic or more far-seeing individuals."[74] The French project also stoked hopes that, by breaking Anglo-American's monopoly on transatlantic telegraphy, the new cable might lead to cheaper rates – as for a time it did. In June 1869 Anglo-American lowered its tariff to £2 per ten words and, when the French cable opened for business that August, proceeded to match its rate of £1 10 s. But the French company proved less interested in breaking up the monopoly than in joining it, and when its shareholders met in London that November, their chairman announced that "they were working on friendly terms with the Anglo-American Company, and arranging with them a mutual tariff book."[75] That is, the companies were colluding to rig

[71] "England," *Boston Daily Advertiser* (September 11, 1868).

[72] Smith, *Rise and Extension*, 229; Thompson, *Kelvin*, 1: 552.

[73] Blundell, *Manual of Submarine Telegraph Companies* (1871), 6.

[74] "West India and Panama Telegraph," *Times* (August 26, 1869), 4.

[75] "French Atlantic Cable Company," *Daily News* (November 18, 1869). On the rates charged, see Blundell, *Manual of Submarine Telegraph Companies* (1872), 18, and the advertisement for "Société du Cable Transatlantic Français" in *Pall Mall Gazette* (September 19, 1869).

their prices, and in January 1870 they formed a "pool" in which they agreed to split the transatlantic traffic and share the revenue proportionally, while covering for each other in case of temporary breakdowns. When both of the Anglo-American cables failed in December 1870, diverting all traffic to the French cable, the companies raised their rate to £3 per ten words before settling at £2 once the cables were repaired. As Blundell noted, the companies took in about £1250 per day in 1870 when their rates were at their lowest, and considerably more after they raised them, "as the public, having been induced to use the cables [at the low rate] continue to use them at the higher rate."[76] It seemed that once people got a taste of rapid communications, they were hooked. The money continued to roll in, and in 1873 Anglo-American fully absorbed the French company.

In a leading article on cable telegraphy, the *Times* declared "The execution of Imperial undertakings by means of private enterprise" to be "among the noblest features of this age."[77] Whether noble or not, it was certainly central to Britain's cable industry. When Pender launched the British-Indian Submarine Telegraph Company in January 1869, he had his eye firmly on private profits, but he clearly meant the project to serve imperial purposes as well. Its prospectus called for "constructing a Submarine Telegraph Line between Suez, Aden, and Bombay, so as to complete a direct and reliable line of telegraphic communication between Europe and India, and with a view to future extensions to China and Australia."[78] When combined with the newly laid cable from Malta to Alexandria and the string of cables Pender's Falmouth, Gibraltar and Malta Telegraph Company was planning to lay the following year, the Indian cables would form the spinal cord of an undersea network designed to tie the British Empire more closely together.[79]

---

[76] Blundell, *Manual of Submarine Telegraph Companies* (1872), 18.

[77] *Times* (December 27, 1866), 6.

[78] "The British-Indian Submarine Telegraph Company" (prospectus), *Pall Mall Gazette* (February 2, 1869). The existing Ottoman overland line was slow and unreliable; messages often took a week to pass between London and India and arrived in garbled form. Werner Siemens's Indo-European landline through Russia and Persia, opened in 1870, was better, delivering readable messages in about a day and, by 1873, in a few hours; see Daniel R. Headrick, *The Invisible Weapon: Telecommunications and International Politics, 1851–1945* (Oxford: Oxford University Press, 1991), 21–22. In 1877 Pender drew the Indo-European company into a "joint purse" agreement to rig prices and share revenues; see Wilfried Feldenkirchen, *Werner von Siemens: Inventor and International Entrepreneur* (Columbus: Ohio State University Press, 1994), 97–98.

[79] After its founding, the company shifted its Cornish terminus from Falmouth to the isolated village of Porthcurno, where cables were less likely to be snagged by anchors. Porthcurno became the chief British terminus for the Eastern Telegraph Company and the site of its training college; it is now home to the PK Porthcurno Museum of Global Communications and the archives of Cable & Wireless, the successor to Pender's Eastern Telegraph group.

Pender proposed to carry out this new project very much on the pattern of the Anglo-American and French Atlantic cables: the new cables would be made and largely financed by his manufacturing company, TC&M, which would take much of its payment in shares of the British-Indian company; they would be laid by the *Great Eastern*; and their engineers and electricians would again be veterans of the earlier Atlantic projects.[80] The success of those projects had made cable laying seem almost routine, and as the *Great Eastern* set off for Bombay in November 1869, the *Times* noted that public interest in such expeditions was beginning to flag. The *Times* reminded its readers, however, that this "apparent ease is a result of the combination of the most profound scientific knowledge with the highest degree of mechanical skill," and cast the success of cable telegraphy as a matter of national and imperial pride:

A chain which rests beneath the sea, beyond the reach of hostile interference, which in this position of safety will unite us to British India, and, before very long, to the Australian colonies also, may well be a source of pride to Englishmen. The pride will not be diminished by the reflection that the enterprise and its success are wholly our own. In such achievements no other nation has even started in the race, and the last evidence of England's Empire over the sea has been her use of its greatest depths as the resting-place of the links by which she is binding together the uttermost parts of the world in sympathetic and instantaneous intercommunication.[81]

Having long ruled the waves, Britannia would now extend that rule beneath the waves as well.

Britain's domination of cable telegraphy was a source not just of pride but of power, both political and economic. When the Indian cable reached Suez in March 1870, Pender wired officials in both England and India to announce "the completion of this great work, so important to the imperial and commercial interests of both countries."[82] After his cables connecting Cornwall to Gibraltar and Malta were completed that June, thus providing an undersea route all the way from England to India (apart from a short landline in Egypt), Pender celebrated with a grand reception at his Mayfair mansion, attended not only by Clark, Varley, Jenkin, and other pioneers of cable telegraphy but by an array of dignitaries reaching up to the Prince of Wales, "whose interest in all scientific undertakings," the *Times* reported, "is well known." Pender had a corner of the room fitted up as a telegraph office, from which guests exchanged

---

[80] On the financing of the British-Indian company, see the prospectus in *Pall Mall Gazette* (February 2, 1869); on its engineers and electricians, see J. C. Parkinson, *The Ocean Telegraph to India: A Narrative and a Diary* (London: William Blackwood and Sons, 1870), 110–11.

[81] "The British Indian Submarine Telegraph," *Times* (November 9, 1869), 5.

[82] "The British-Indian Telegraph Expedition," *Times* (March 24, 1870), 9.

messages with officials in Egypt, India, and the United States and marveled at the rapidity of the responses. The "telegram of the evening," according to the *Times*, was one the prince sent to the viceroy of India congratulating him on the completion of the cables. "I feel assured," the prince said, "this grand achievement will prove of immense benefit to the welfare of the Empire. Its success is thus a matter of Imperial interest." The viceroy replied in similar terms and also sent affectionate greetings to his wife, whose own message had taken only nine minutes to reach him at Simla.[83]

The prince also sent congratulations to Ismail Pasha, the Khedive of Egypt, and thanked him for allowing the landline to cross his territory, thus "facilitating the communication of England with her Eastern Empire." The khedive replied that these "new ways of communication, that are due to the genius of the English people," would do much to "unite us" and to promote beneficial commercial relations between their two countries.[84] He perhaps little foresaw that those commercial relations, and the debts they would lead him to incur, would contribute nine years later to his own ouster at the behest of the British.

Delivering on his pledge to extend his network beyond India to other British colonial outposts, Pender organized a series of new companies that opened cables to Penang and Singapore late in 1870, Hong Kong in June 1871, and Darwin, Australia, in November. In China, his network bumped up against one being built by C. F. Tietgen's Danish-based Great Northern Telegraph Company, which operated a landline that stretched from St. Petersburg to the Pacific and in 1871 laid a series of cables from Vladivostok to Shanghai, Hong Kong, and Nagasaki.[85] Like the backers of the French Atlantic company, Tietgen was less interested in competing with Pender than in sharing his profits. Indeed, even before it had laid its East Asian cables, the Great Northern company made a deal with Pender's China Submarine Telegraph Company: in return for the British firm agreeing not to lay its own cable from Hong Kong to Shanghai, Great Northern would hand over half of its revenue on the route.[86] Significantly, all of Great Northern's cables were manufactured

---

[83] "The British Indian Submarine Telegraph," *Times* (June 24, 1870), 12.

[84] "The British Indian Submarine Telegraph," *Times* (June 24, 1870), 12; the khedive's remarks were translated from French.

[85] On the Great Northern system, see "Telegraph Circuits," *Mechanics' Magazine* (April 14, 1871) 25: 248–49, and Kurt Jacobsen, *The Story of GN: 150 Years of Technology, Big Business and Global Politics* (Copenhagen: Historika/Gads Forlag, 2019).

[86] Jorma Ahvenhainen, *The Far Eastern Telegraphs: The History of Telegraphic Communications between the East, Europe and America before the First World War* (Helsinki: Suomalainen Tiedeakatemia, 1981), 49–52. On the telegraph in China, see Erik Baark, *Lightning Wires: The Telegraph and China's Technological Modernization, 1860–1890* (Westport, CT: Greenwood Press, 1997).

in Britain, and their early ones were laid and tested mainly by British engineers and electricians.[87] Even when cables were owned by companies from other countries, the expertise needed to make and run them was mainly drawn from Britain.

After the burst of cable laying between 1868 and 1871, the industry paused for a time to regroup. Pender consolidated his many firms into two main companies: the Eastern Telegraph Company, which covered the route from Britain to India, and the Eastern Extension, Australasia and China Telegraph Company, which as its name implied handled cables that extended further east. Both companies continued to lay many more cables and later absorbed other firms, so that together the Eastern group came to control nearly half of the world's cable mileage; in the words of Daniel Headrick and Pascal Griset, "Outside the North Atlantic, the history of world telegraphy is largely the history of Eastern."[88] In light of Pender's role in the Anglo-American company, which long dominated the North Atlantic route, and his control of TC&M, by far the biggest cable manufacturer, the history of submarine telegraphy in the last third of the nineteenth century was in fact largely the history of the "cable king" and his many enterprises (Figure 6.3).

After their brief pause, companies organized by Pender and other British entrepreneurs began to expand their networks into Britain's "informal empire" in South America in the mid-1870s, and in the 1880s, amid the "scramble for Africa," followed with cables along the east and west coasts of that continent, backed in part by subsidies from colonial governments.[89] The growing global cable network was quickly integrated into the world economy and information system, with far-reaching effects on financial and commodity markets, the dissemination of news, and the conduct of international affairs. Not everyone was pleased with the big cable companies' monopolistic practices or their prohibitively high rates – Sydney Buxton, the future postmaster general, once compared the Eastern companies to "a submarine octopus" out to

[87] The first Great Northern cables were armored by Newall and later ones by Siemens Brothers of London; their cores, insulated with India rubber, were made by William Hooper of Mitcham. Great Northern later used its own ships to lay its cables; see Bright, *Submarine Telegraphs*, 112–16, and Jacobsen, *Story of GN*, 72, 78, and 93.

[88] Daniel R. Headrick and Pascal Griset, "Submarine Telegraph Cables: Business and Politics, 1838–1939," *Business History Review* (2001) 75: 543–78; on the formation of the Eastern and Eastern Extension companies, see "Anglo-Mediterranean Telegraph Company," *Times* (May 1, 1872), 10, and Hugh Barty-King, *Girdle Round the Earth: The Story of Cable and Wireless* (London: Heinemann, 1979), 39 and 53.

[89] On Britain's "informal empire," the classic reference is John Gallagher and Ronald Robinson, "The Empire of Free Trade," *Economic History Review* (1953) 6: 1–15; on the cables laid to and around South America and Africa, see Bright, *Submarine Telegraphs*, 116–18, 124–28, and 130–35.

Figure 6.3 *Vanity Fair* published this caricature of "cable king" John Pender in October 1871, just as his network of cables was first reaching Australia.
(From *Vanity Fair*, October 28, 1871, Vol. 3.)

"spread their tentacles all over the world," while J. Henniker Heaton, a great campaigner for cheap telegrams, denounced "the cable monopolists who have seized the God-given gift of electricity and devoted it to the sole use of millionaires to the exclusion of the millions."[90] But even Buxton and Heaton acknowledged that the cable network had bound the British Empire more closely together; they just thought cheaper rates

---

[90] Buxton made his remark in Parliament on May 22, 1900; see *Hansard's Parliamentary Debates, Fourth Series* (London: Wyman, 1900) *83*: 991. Heaton made his in "Penny-a-Word Telegrams for Europe," *Financial Review of Reviews* (June 1908) 5: 40–43, a sequel to his blistering article "The World's Cables and the Cable Rings," *Financial Review of Reviews* (May 1908) 5: 5–26.

would help make those bonds even closer and stronger. British firms' control of the great majority of the world's submarine cables was widely recognized to be one of the chief bulwarks of Britain's continued imperial and economic power in the last decades of the nineteenth century and the first decades of the twentieth.[91] As Cecil Rhodes, who had a keen sense of what served the interests of Britain's empire, once observed, "Pender was 'imperialising the map' while I was just feeling my way."[92]

The global cable network continued to grow rapidly throughout this period, jumping from about 2000 nautical miles in 1865 to nearly 50,000 just ten years later, and rising to over 107,000 in 1885 and 162,000 in 1895. About two-thirds of that total had been made and laid by TC&M, and almost all of the rest by a handful of other London-based firms.[93] Even as it grew so enormously, the cable system retained much of the character it had acquired during the boom years of the late 1860s and early 1870s. The business of submarine telegraphy continued to be carried on in much the same way and by many of the same companies – above all by TC&M and the Eastern group – and apart from some refinements in sending and receiving apparatus, the underlying technology, after maturing rapidly in the 1860s, would change relatively little over the next half century or more.[94] But while its technology remained comparatively stable, the great and growing scale of the cable industry came to exert strong effects on work in electrical science in Britain, the original and continued home of submarine telegraphy.

Meeting the "Demand for Electrical Knowledge"

In the opening pages of his *Treatise on Electricity and Magnetism*, Maxwell noted how the rise of the telegraph industry had "reacted on pure science" by "giving a commercial value to accurate electrical measurements." This

---

[91] Headrick and Griset, "Submarine Telegraph Cables," 559.

[92] Sir Edward Wilshaw, the chairman of Cable & Wireless Ltd., quoted Rhodes in a May 1924 speech to the Country Conference of the Chartered Institute of Secretaries, in *The Cable and Wireless Communications of the World: Some Lectures and Papers on the Subject, 1924–1939* (London: Cable & Wireless Ltd., 1939), 1–14, on 3.

[93] Headrick, *Invisible Weapon*, table 3.1, 29; see also figure 2, 30; Daniel R. Headrick, *The Tentacles of Progress: Technology Transfer in the Age of Imperialism, 1850–1940* (Oxford: Oxford University Press, 1988), 102. Headrick drew these data from Maxime de Margerie, *Le réseau anglais de câbles sous-marins* (Paris: A. Pedone, 1909).

[94] Bernard Finn, "Submarine Telegraphy: A Study in Technological Stagnation," in Bernard Finn and Daqing Yang, eds., *Communications Under the Seas: The Evolving Cable Network and Its Implications* (Cambridge, MA: MIT Press, 2009), 9–24. Richard Noakes, "Industrial Research at the Eastern Telegraph Company, 1872–1929," *British Journal for the History of Science* (2013) 47: 119–46, emphasizes that Eastern continued to fund research on sending and receiving apparatus.

had in turn produced a new "demand for electrical knowledge," he said, or more to the point, for those possessing such knowledge.[95] The ability to maintain and operate electrical instruments, to make accurate measurements with them, and to use the resulting data to analyze the workings of electrical circuits – skills that had formerly been of purely scientific use and interest – had acquired a practical value for which there was now a growing market. Between the 1860s and the 1880s that market was centered in the cable industry. As Clark had observed, "no very exact measurements are required to made of overhead lines," and the land-based telegraph systems of the United States and Continental Europe gave only a weak stimulus to the development of electrical measurement techniques.[96] British cable engineers, on the other hand, had learned through hard experience that precise and reliable electrical measurements were crucial to the successful manufacture and operation of submarine cables. In particular, they had found they needed to be able to measure the resistance of the conductors and insulation of their cables with great precision and over a remarkably wide range, both to ensure the quality of their cables while they were being made and to locate faults after they had been laid. The need to make and compare such measurements had stimulated the development of electrical units and standards, as we have seen, as well as new instruments and techniques, and would continue to do so. It also stimulated the *teaching* of methods of electrical measurement, and it is no coincidence that the first physics teaching laboratories in Britain appeared amid the cable boom of the late 1860s and early 1870s or that they focused largely on instructing students in the principles and techniques of precision electrical measurement.[97]

George Saward once remarked that though the 1858 Atlantic cable was itself a failure, it had helped create "a school of telegraphic electricians" whose work had been "a source of fame and fortune to themselves, and a great and valuable boon to the world of science."[98] Most of the key members of that "school" – Clark, Varley, Smith, and a handful of others – were essentially self-taught; though they drew sparingly on what Wheatstone and others had written on the principles of electrical measurement, they mainly learned by doing. This inevitably led to some

---

[95] Maxwell, *Treatise*, 1: vii.

[96] Latimer Clark and Robert Sabine, *Electrical Tables and Formulae for the Use of Telegraph Inspectors and Operators* (London: E. and F. N. Spon, 1871), 7.

[97] Gooday, "Precision Measurement"; Graeme J. N. Gooday, "Precision Measurement and the Genesis of Physics Teaching Laboratories in Victorian Britain," PhD dissertation, University of Kent at Canterbury, 1989. See also Romualdas Sviedrys, "The Rise of Physics Laboratories in Britain," *Historical Studies in the Physical Sciences* (1976) 7: 405–36.

[98] Saward, *Trans-Atlantic Submarine Telegraph*, 40.

diversity of approaches in the early days, as the case of Wildman Whitehouse illustrates, but by the early 1860s, in the wake of the failure of the Atlantic and Red Sea cables and the publication of the *Joint Committee Report*, measurement practices in cable telegraphy had largely converged on a shared consensus. Clark's long contribution to the *Joint Committee Report* was especially influential in disseminating that consensus, soon finding its way into textbooks and forming the basis for Clark's own widely used *Elementary Treatise on Electrical Measurement*, published in 1868.[99]

The other great British contributor to the theory and practice of electrical measurement in the late 1850s was of course William Thomson. Unlike Clark, Varley, and the other telegraph engineers, Thomson came to the subject as one of the leading mathematical and experimental physicists of the day and he drew on those skills, along with his remarkable inventive abilities, not just to devise new methods of electrical measurement but to design the requisite instruments and to interpret the resulting data. The mirror galvanometer that Thomson introduced in 1858 was the most famous of those instruments and the one most crucial to the success of cable telegraphy. He followed it in 1869 with his siphon recorder, in which the moving spot of light was replaced by a wavy line of ink on a long paper tape, and also developed and patented a whole series of electrometers and other instruments for precision electrical measurement, all made by the Glasgow firm headed by James White, in which Thomson later became a partner.[100]

Thomson also took the lead in preparing the next generation of cable engineers and electricians, both as a mentor to Jenkin and others who had already entered the field and through his work with his students at the University of Glasgow. The small laboratory Thomson set up in a disused room in the old college buildings in 1855 has often been called the first physics teaching laboratory in Britain, and in many ways it was, but at its heart it was essentially a research laboratory; it was where Thomson carried out his own experimental investigations, beginning in the early

---

[99] Latimer Clark, *An Elementary Treatise on Electrical Measurement for the Use of Telegraph Inspectors and Operators* (London: E. and F. N. Spon, 1868); in Henry M. Noad, *The Student's Text-Book of Electricity* (London: Lockwood, 1867), iii, Noad said the *Joint Committee Report* "supplied me with much information, that part detailing the investigations of Mr. Latimer Clark having been especially valuable."

[100] Thomson patented his siphon recorder in 1867 but did not perfect it for practical use until 1869. It subsequently proved very lucrative, bringing in thousands of pounds in royalties every year; see Crosbie Smith and M. Norton Wise, *Energy and Empire: A Biographical Study of Lord Kelvin* (Cambridge: Cambridge University Press, 1989), 708–12. On Thomson's collaboration with White, see T. N. Clarke, A. D. Morrison-Low, and A. D. C. Simpson, *Brass and Glass: Scientific Instrument Making Workshops in Scotland* (Edinburgh: National Museums of Scotland, 1989), 252–66.

1850s with measurements of various thermoelectric effects.[101] Finding he needed help to complete this work, Thomson invited a few of his more advanced and enthusiastic students to join him in it, which they did in a very informal way. As W. E. Ayrton, himself later a distinguished physicist and electrical engineer, said in looking back on his days as one of those students in the 1860s, "There was no special apparatus for students' use in the laboratory, no contrivances such as would to-day be found in any polytechnic, no laboratory course, no special hours for the students to attend, no assistants to supervise or explain, no marks given for laboratory work, no workshop and even no fee to be paid."[102] Instead of receiving lessons geared for undergraduates, Thomson's students assisted in his original investigations, as when in 1857–1858 they had helped measure the resistance of different samples of "commercial copper" destined for the Atlantic cable and found how widely they differed.[103]

A few of Thomson's early students moved directly into work in cable telegraphy, notably James Burn Russell, who served as Thomson's electrical assistant during the laying and the short life of the 1858 cable.[104] More of Thomson's students did so during the boom of the late 1860s and early 1870s; W. F. King and J. D. H. Dickson, for example, carried out an important determination of the ratio of electrical units in Thomson's laboratory in 1867–1868 and then helped install his first siphon recorders on the French Atlantic cable in 1869.[105] In 1872, after his laboratory had moved to more spacious quarters at the new university campus at Gilmorehill, Thomson told his sister-in-law that "There is quite an epidemic among the laboratory students to become *telegraph engineers*."[106] Among those bitten by the bug was his nephew David

[101] Crosbie Smith, "'Nowhere But in a Great Town': William Thomson's Spiral of Classroom Credibility," in Crosbie Smith and Jon Agar, eds., *Making Space for Science: Territorial Themes in the Shaping of Knowledge* (London: Macmillan, 1998), 118–46, on 130–33.

[102] W. E. Ayrton, "Kelvin in the Sixties," *Times* (January 8, 1908), 3.

[103] William Thomson, "On the Electrical Conductivity of Commercial Copper of Various Kinds," *Proc. RS* (June 1857) *8*: 550–55, repr. in Thomson, *MPP 2*: 112–17.

[104] Edna Robertson, *Glasgow's Doctor: James Burn Russell, MOH, 1837–1904* (East Linton: Tuckwell Press, 1998), 29–41.

[105] Thompson, *Kelvin*, *2*: 524–25, 575. R. Kalley Miller, who took over teaching Thomson's classes in the spring of 1870 during the final illness of Thomson's first wife, was referring to King and Dickson when he wrote that "at the laying of the French Atlantic Cable two of the best practical and scientific electricians were young men selected from the Glasgow class." Miller added that "the success of the Atlantic Cable is in great measure the result of years of patient work in the Glasgow Laboratory." See Miller, "The Proposed Chair of Natural Philosophy," *Cambridge University Reporter* (November 23, 1870) *1*: 118–19.

[106] William Thomson to Jessie Crum, March 29, 1872, quoted in Thompson, *Kelvin 2*: 622. John Pender had helped raise money for the new buildings; see Smith, "Nowhere," 139.

Thomson King, who had reportedly been "prepared specially by Sir William Thomson for telegraphic engineering work" and later assisted Thomson and Jenkin in testing cables for the Western and Brazilian Telegraph Company. Sadly, King perished, along with several other electricians and most of the crew, when the cable ship *La Plata* sank in a storm en route to Brazil in November 1874.[107]

By the early 1870s, Thomson's physics laboratory had been joined by several others at colleges and universities across Britain, most of them devoted to teaching and all focused on techniques of precision measurement. As Graeme Gooday has noted, the genesis of these laboratories between 1866 and 1874 resulted from the confluence of "an industrially-generated 'demand' for practical scientific education" with a "research-generated 'supply'' of academic expertise in precision measurement."[108] This wave of new laboratories coincided with the crest of the cable boom, and the telegraph industry was the chief source of the new demand for competence in precision electrical measurement to which their founders were responding. This can be seen in the laboratories established by George Carey Foster at University College London (UCL) in 1866 and by P. G. Tait at the University of Edinburgh in 1868, both of whom explicitly modeled their laboratories on Thomson's at Glasgow, though with the addition of organized courses of elementary instruction. It is also reflected, though less directly, in the laboratories opened in 1870 by R. B. Clifton at Oxford and in 1872 by Frederick Guthrie at the Royal School of Mines in London. These focused less on performing original research or preparing their students to do so than on training prospective physics teachers, with Clifton's students destined to teach at elite "public" schools and most of Guthrie's at more plebeian institutions. Few of Clifton or Guthrie's students would go directly into the cable industry, but much of the rising demand in this period for school instruction in physics, and particularly in electrical measurement, can be traced to the public sense of its value to telegraphy.[109]

Apart from Thomson's laboratory in Glasgow, the British academic physics laboratory that most directly served the needs of the telegraph industry was the one William Grylls Adams established at King's College

[107] "The Foundering of the La Plata," *Glasgow Herald* (December 4, 1874); Thompson, *Kelvin*, 2: 654–55.

[108] Gooday, "Precision Measurement" (diss.), abstract.

[109] See the chapters in Gooday, "Precision Measurement" (diss.) on the laboratories established by Foster, Tait, Clifton, and Guthrie. As Gooday notes, 6.34–37, Clifton opened a makeshift laboratory in a borrowed room at Oxford in 1866 before the purpose-built Clarendon Laboratory opened in 1870. On the recruitment of former public school boys into the Eastern Telegraph Company training college at Porthcurno, see Barty-King, *Girdle Round the Earth*, 120.

London (KCL) in 1868. Maxwell had taught at KCL from 1860 to 1865 and it had been the site of the important measurements he, Jenkin, Balfour Stewart, and Charles Hockin had performed in 1863–1864 for the British Association to determine the value of the ohm. Maxwell seems, however, to have made no effort to engage his students in this or any of his other experimental work in London, nor did he take any steps toward establishing a teaching laboratory there. There is some evidence that Maxwell came to find his teaching duties at KCL uncongenial; in any case, he resigned his professorship early in 1865 and "retired" at the age of 33 to take up the life of a country gentleman on his estate in Scotland. He would later play an important part in the development of physics laboratories in Britain, but not in the 1860s and not at KCL.

A Cambridge graduate, Adams had served as a lecturer at KCL since 1863 and on Maxwell's resignation took over the professorial duties as well. He soon sought to add a teaching laboratory. The chair of natural philosophy was part of KCL's Department of Applied Sciences and served mainly young men (the minimum age of entry was only sixteen) who were preparing for careers in engineering or industry.[110] In the late 1860s telegraphy would have appealed to them – and to their parents – as a growing field and college officials, eager to boost enrollments, had every wish to accommodate the resulting demand. The KCL *Calendar* for 1867–1868 noted that the "ever-increasing demand for high scientific engineering, both at home and in the British Dominions abroad," was drawing the attention of increasing numbers of prospective students, and Adams urged the college to offer them proper facilities and instruction.[111] In February 1868 he told the college council that "Recently several amongst the Candidates preparing themselves for service in the Telegraph Department in India applied here, but left us for want of special *practical* instruction in Electrical Science and have since entered themselves at other institutions."[112] KCL needed to keep up with the likes of UCL and the Universities of Glasgow and Edinburgh, Adams said, and to do so it must establish a physics laboratory where students could be given hands-on and up-to-date instruction in techniques of precision measurement.

The college council responded favorably to Adams's appeal, granting him several rooms, an assistant, and substantial funds for apparatus. He

---

[110] John S. Reid, "Maxwell at King's College, London," in Raymond Flood, Mark McCartney, and Andrew Whitaker, eds., *James Clerk Maxwell: Perspectives on His Life and Work* (Oxford: Oxford University Press, 2014), 43–66, on 16–17.

[111] Gooday, "Precision Measurement" (diss.), 5.21.

[112] W. G. Adams and W. A. Miller to King's College London Council, February 14, 1868, in Gooday, "Precision Measurement" (diss.), 5.23.

duly opened the new laboratory in the fall of 1868 and soon found himself flooded with even more students than he had expected.[113] He offered them a curriculum that strongly emphasized measurement techniques, culminating in lessons on "Measurement of the Strength of Currents," "Measurement of Electro-motive force and Resistances," and "The Electric Telegraph."[114] Students who pursued electricity as their special subject were also taught how to make "delicate measurements with Thomson's Galvanometers and Electrometers" and how to "perform all the tests and measurements required in connection with Telegraph lines and cables."[115] By the mid-1870s, Adams's laboratory was turning out a steady stream of young men well equipped to join the telegraph industry.

The last and later the most celebrated of the first wave of British physics laboratories was Cambridge's Cavendish Laboratory.[116] Opened under Maxwell's direction in 1874, it would eventually become perhaps the most famous and productive scientific laboratory in the world, but it took some time to find its feet. One might expect Cambridge to excel in all areas of physics; it was, after all, "Newton's university," and had long given pride of place to the study of natural philosophy. By the mid-nineteenth century, however, the university focused almost entirely on the mathematical side of the subject, particularly in the highly competitive Mathematical Tripos examination; Cambridge made no official provision for experimental work in physics and had no teaching laboratory.[117] In the 1850s and 1860s a growing sense that the ancient universities were out of step with the changing needs of the nation, including in the experimental sciences and technology, prompted a series of influential calls for reforms. Though strongly opposed by the old guard, these eventually served to remake both Oxford and Cambridge in far-reaching ways. At Cambridge, a concern that

---

[113] Adams had expected no more than ten students when his laboratory first opened but instead had fifteen, rising to twenty-three; Gooday, "Precision Measurement" (diss.), 5.35 and 5.40.

[114] Gooday, "Precision Measurement" (diss.), 5.35–36.

[115] Gooday, "Precision Measurement" (diss.), 5.37.

[116] On the history of the Cavendish Laboratory, see J. J. Thomson et al., *A History of the Cavendish Laboratory, 1871–1910* (London: Longmans, Green, and Co., 1910); J. G. Crowther, *The Cavendish Laboratory, 1874–1974* (New York: Science History Publications, 1974); Dong-Won Kim, *Leadership and Creativity: A History of the Cavendish Laboratory, 1871–1919* (Dordrecht: Kluwer Academic Publishers, 2002); Isobel Falconer, "Cambridge and Building the Cavendish Laboratory," in Flood, McCartney, and Whitaker, eds., *Maxwell*, 67–98, and Malcolm Longair, *Maxwell's Enduring Legacy: A Scientific History of the Cavendish Laboratory* (Cambridge: Cambridge University Press, 2016).

[117] On the intensity of the competition for high places in the Mathematical Tripos, see Andrew Warwick, *Masters of Theory: Cambridge and the Rise of Mathematical Physics* (Chicago: University of Chicago Press, 2003), 182–212.

students were offered too little instruction in heat, electricity, and magnetism, subjects that were of increasing scientific as well as practical importance, led in 1869 to an official call for the university to found a new professorship of experimental physics and, if it could find the funds, build a new laboratory.[118] The high cost of the latter, estimated to come to £6300, threatened to stall the effort until William Cavendish, the seventh Duke of Devonshire, stepped forward. A leading industrialist as well as an enormously wealthy landowner, the duke had compiled a distinguished record as an undergraduate at Cambridge, placing second on the Mathematical Tripos in 1829.[119] Moreover, in 1861 he had succeeded Prince Albert as chancellor of the university and in 1870 was chosen to head a Royal Commission on Scientific Instruction. Not long after hearing testimony from Thomson and Clifton about their new laboratories at Glasgow and Oxford, and evidently concerned that his own university was falling behind, he wrote to officials at Cambridge in October 1870 with an offer to cover the cost of building and equipping the proposed new physics laboratory.[120] Any lingering opposition to the plan quickly melted away and the search for a new professor began.

The first thought of almost everyone involved was to try to lure Thomson down from Glasgow: not only was he an old Cambridge man but he had more experience running a physics laboratory than anyone else in Britain, and after the triumph of the Atlantic cables, his public fame and scientific reputation were at their height. A combination of personal and business reasons, however, led him to decline. The recent death of his wife had given him "an invincible repugnance," Thomson said, "to the idea of beginning a new life at all"; he preferred to stay in his familiar paths at Glasgow. Moreover, he said, "the convenience of Glasgow for getting mechanical work done" gave it a great advantage over Cambridge, where he would have found it impossible to carry on his growing partnership with White's instrument-making firm, then busily turning out siphon recorders.[121] Thomson then tried to recruit Hermann Helmholtz in his

---

[118] The official report, dated February 27, 1869, is reprinted in Crowther, *Cavendish Laboratory*, 23–27.

[119] F. M. L. Thompson, "William Cavendish, Seventh Duke of Devonshire," *ODNB*. The seventh duke had a distinguished scientific pedigree, being related on his mother's side to Robert Boyle and on his father's to Henry Cavendish.

[120] Kim, *Leadership*, 3–4. Arthur Schuster later reported that when the lowest bid from the builders exceeded the initial estimate by more than £2000, the duke agreed to cover the additional expense, as well as the cost of equipping the laboratory. Telling Schuster that the duke "had done enough," Maxwell subsequently economized on purchasing instruments, so that when Rayleigh took over the laboratory early in 1880, he found it poorly equipped. See Arthur Schuster, "John William Strutt, Baron Rayleigh, 1842–1919," *Proc. RS* (1921) 98: i–l, on xix.

[121] Thompson, *Kelvin*, 1: 563.

stead, but found he was already bound for a professorship in Berlin.[122] Maxwell was next on the list, but hesitated when first approached, saying quite rightly that he had "no experience" in "teaching experimental work." Further appeals eventually persuaded him to stand, however, and in March 1871 he was duly elected Cambridge's first professor of experimental physics.[123]

Maxwell would hold what became known as the Cavendish professorship for only a little more than eight years, until his death in November 1879, and by many measures his tenure was not especially successful. He spent most of the first three years designing the new laboratory and supervising its construction while also planning how best to organize the work to be done within it. The problem, as he told J. W. Strutt (later Lord Rayleigh), who would succeed him as Cavendish professor, was how "to make Exp. Physics bite into our University system, which is so continuous and complete without it." The overarching prestige of the Mathematical Tripos posed special challenges, and Maxwell warned Strutt that "if we succeed too well, and corrupt the minds of youth, till they observe vibrations and deflections and become Senior Op.s instead of Wranglers, we may bring the whole University and all the parents about our ears."[124]

The obstacles facing the experimental sciences at Cambridge are well reflected in a letter G. T. Bettany, a young biologist, sent to *Nature* toward the end of 1874, a few months after the Cavendish opened its doors. "The great hindrance to the success of the Cavendish Laboratory at present," he said,

is the system fostered by the Mathematical Tripos. The men who would most naturally be the practical workers in the laboratory are compelled to refrain from practical work if they would gain the best possible place in the Tripos list. Very few have courage so far to peril their place or to resign their hopes as to spend any valuable portion of their time on practical work . . . . For a man to do practical work in physics at Cambridge implies considerable exercise of courage and self-sacrifice.[125]

Facing such constraints, Maxwell proceeded cautiously. Unlike Thomson at Glasgow or Adams at KCL, he did not try to draw undergraduates into experimental work but hoped instead to fill the laboratory with young

[122] Thomson to Hermann Helmholtz, January 28, 1871, in Thompson, *Kelvin, 1*: 564–66; see also David Cahan, *Helmholtz: A Life in Science* (Chicago: University of Chicago Press, 2018), 398–401.

[123] Maxwell to E. W. Blore (draft), February 15, 1871, in Maxwell, *SLP 2*: 611–13.

[124] Maxwell to J. W. Strutt, March 15, 1871, in Maxwell, *SLP 2*: 614–16, on 615. Students who placed highest on the Mathematical Tripos were called "wranglers" (with the top finisher being the "senior wrangler"); the ranks just below were called "senior optimes" and "junior optimes."

[125] G. T. Bettany, "Practical Science at Cambridge," *Nature* (December 17, 1874), *11*: 132–33, on 133.

holders of college fellowships and recent graduates looking to complete a piece of work that might help them gain such a position. Few showed up, however; by one count, over the five years Maxwell was its director, the Cavendish attracted only about sixteen researchers in all, including William Garnett, whom Maxwell hired as his demonstrator in 1874.[126] With one partial exception, recounted later, Maxwell did not seek to organize group projects for those working in the laboratory, as Thomson had done at Glasgow, nor did he try to develop a "school" devoted to exploring and testing his own theories – no one at the Cavendish attempted, for example, to produce or detect the electromagnetic waves predicted by his field theory.[127] Maxwell instead gave researchers in his laboratory free rein; as he once told Arthur Schuster, "I never try to dissuade a man from trying an experiment. If he does not find what he wants, he may find out something else."[128] Some found this *laissez-faire* approach liberating, but it left all but the most self-directed researchers at sea. Maxwell himself spent much of his time working through the unpublished electrical papers of Henry Cavendish, most of which dated from the 1770s. This was in part an act of fealty to the patron of the laboratory and in part a way for Maxwell to explore methodological issues that turned up as he repeated many of Cavendish's experiments – the main experimental work Maxwell undertook in his years at the laboratory.[129]

Eclectic as the work at the Cavendish was under Maxwell, however, one theme ran through almost all of it, as it did at other physics laboratories of the day: precision measurement. Maxwell had emphasized the central importance of accurate measurement in his inaugural lecture in October 1871, and the apparatus he gathered for use in the laboratory, including Thomson galvanometers and quadrant electrometers, a Kew magnetometer, and several sets of resistance coils – calibrated, of course, in British Association units – was mainly intended for measurement-based research and not, it is worth noting, for laboratory teaching purposes.[130]

---

[126] Kim, *Leadership*, 20. The somewhat longer list in Falconer, "Cambridge," 89–98, includes undergraduates who were simply preparing for the practical portion of the Natural Sciences Tripos in 1878.

[127] W. M. Hicks later said that while at the Cavendish in 1874 he "was fired with the desire of measuring experimentally the velocity of propagation of electromagnetic waves," but the experiment he described did not really involve electromagnetic *waves*, but simply the spreading of electric force or potential; see Arthur Schuster, "The Clerk-Maxwell Period," in Thomson et al., *Cavendish Laboratory*, 14–39, on 19.

[128] Schuster, "The Clerk-Maxwell Period," 39.

[129] James Clerk Maxwell, ed., *The Electrical Researches of the Honourable Henry Cavendish, F.R.S.* (Cambridge: Cambridge University Press, 1879).

[130] James Clerk Maxwell, "Introductory Lecture on Experimental Physics" [October 25, 1871], in Maxwell, *SP 2*: 241–55, on 244. Maxwell listed the apparatus held by the laboratory in reports published in the *Cambridge University Reporter* each spring from

Within days of his election to the new professorship, Maxwell began laying the groundwork for what would be his most strategic acquisition of precision measuring equipment and the one that would come closest to providing a shared project for the laboratory. Writing to Thomson in March 1871 about plans for the new laboratory, Maxwell said "We should get from the B.A. some of their apparatus for the Standard committee. In particular the spinning coil and the great electrodynamometer."[131] Cambridge had no special claim on the apparatus used by the British Association Committee on Electrical Standards, which might more logically have gone to Glasgow, where Thomson had been the moving spirit behind the formation of the committee, or KCL, where the experimental determinations of the ohm had actually been performed. But Maxwell and Thomson seem to have had no trouble convincing the general council of the British Association that the new Cambridge laboratory would be the proper home for the standards apparatus, and in August 1871 the council approved a resolution ordering that "the Electrical Apparatus belonging to the British Association, now in the possession of the Committee of Electrical Standards, be placed in the Physical Laboratory of Cambridge, in charge of the Professor of Experimental Physics."[132] With the arrival of the apparatus, the Cavendish would in effect become Britain's national electrical standards laboratory.

As he prepared for the opening of the laboratory in the spring of 1874, Maxwell wrote to ask Jenkin if he knew where the standards committee apparatus then was and to remind him that it was to be transferred to the Cavendish. By the end of the year Maxwell had managed to collect nearly all of the surviving apparatus and was making plans to use it to compare the values of the various original resistance coils.[133] Word of his project evidently got out, and early in 1875 he received a letter from J. A. Fleming, a former student of Foster's at UCL (and later himself a distinguished electrical engineer) who in 1877 would come to Cambridge to study with Maxwell.[134] Fleming's letter has not survived,

---

1875 to 1879; see Maxwell, *SLP 3*: 208–15, 337–43, 456–61, 625–26, and 795–96. Some of the apparatus was intended for use in lecture demonstrations, but none seems to have been designed for elementary laboratory instruction.

[131] Maxwell to Thomson, March 21, 1871, in Maxwell, *SLP 2*: 624–28, on 627.

[132] "Recommendations Adopted by the General Committee at the Edinburgh Meeting in August 1871," *BA Report* (1871), lxix.

[133] Maxwell to Fleeming Jenkin, March 17, 1874; July 22, 1874; and November 18, 1874, in Maxwell, *SLP 3*: 51–52; 89–90; and 138–39; and Charles Hockin to Maxwell, May 14, 1874, in Maxwell, *SLP 3*: 52n. See also the lists of apparatus in Maxwell, *SLP 2*: 868–75 (June 1873), *SLP 3*: 64–65 (April 29, 1874), and *SLP 3*: 86–88 (July 14, 1874).

[134] Sir Ambrose Fleming, *Memories of a Scientific Life* (London: Marshall, Morgan and Scott, 1934). In 1885 Fleming was appointed professor of electrical technology at UCL, the first such position in Britain; after the death of John Pender in 1896, it became the

but he evidently asked Maxwell whether the Cavendish planned to produce standard ohms like those Jenkin had offered for sale on behalf of the
British Association committee in the 1860s. Maxwell's reply is revealing:

The original B. A. standards of resistance are at the Cavendish Laboratory but we
have not hitherto had time to go over them and compare them with their recorded
values. When the place is in working order I hope to be able to verify any standard
already constructed and sent to me, but I do not expect or think it desirable that
a manufactory of "ohms" should be established in the building.[135]

That is, Maxwell proposed to make the Cavendish a court of appeal for
resistance standards, checking any coil submitted against the original
ohms. He made it clear, however, that the Cavendish was a laboratory,
not a workshop; it would produce knowledge, not items of commerce. As
Simon Schaffer has emphasized, this was "a crucial institutional distinction" with strong class overtones; a workshop manned by students and
fellows would be out of place in the Victorian university.[136]

Maxwell's ambition to turn the Cavendish into an electrical standards
laboratory and make it a crucial point of reference for electrical science
and technology around the world would eventually be realized, but not by
him. In 1873 he told the vice-chancellor of the university that "the most
important electrical research" at present was "the determination of the
magnitude of certain electrical quantities, and their relations to each
other," and in his one real attempt to launch a team project at the
laboratory, he tried to organize an effort to redetermine one of those
magnitudes: the value of the ohm.[137] This was a pressing concern in the
1870s, as measurements by James Joule in Manchester and by researchers
in Germany and America had raised suspicions that the original British
Association standards may have missed their intended value of ten million

Pender Chair of Electrical Engineering. Fleming made important contributions to
electrical power engineering, worked closely with Guglielmo Marconi on wireless telegraphy, and invented the thermionic diode.

[135] Maxwell to J. A. Fleming, January 14, 1875, in Maxwell, *SLP 3*: 188.

[136] Simon Schaffer, "Late Victorian Metrology and Its Instrumentation: A Manufactory of
Ohms," in Robert Bud and Susan E. Cozzens, eds., *Invisible Connections: Instruments,
Institutions, and Science* (Bellingham, WA: SPIE Optical Engineering Press, 1992),
23–56, on 25. About a year after he was elected professor of mechanism in 1875,
James Stuart established an engineering workshop at Cambridge, but the initiative was
poorly received and in the 1880s he was essentially run out of the university; see
T. J. N. Hilken, *Engineering at Cambridge University, 1783–1965* (Cambridge:
Cambridge University Press, 1967), 65–92.

[137] Maxwell to H. W. Cookson (draft), July 5, 1873, in Maxwell, *SLP 2*: 876. Properly
speaking, the ohm was defined as 10^7 meters per second and the problem was either to
determine how closely the coils issued in 1865 as "British Association units" matched
that defined value or to construct new coils whose resistance would be as close to 10^7
meters per second as possible.

meters per second by 1 or 2 percent. But the complications of recruiting researchers into the Cavendish in its early days, together with Maxwell's reluctance to direct those in the laboratory into taking on any projects they had not chosen for themselves, kept this effort from getting far. Its main product was a paper published in 1876 in which George Chrystal and S. A. Saunder compared amongst themselves the resistances of the standard coils, now held by the Cavendish, that the British Association committee had prepared ten years before.[138]

On Maxwell's death in November 1879, Rayleigh was almost immediately chosen to succeed him as Cavendish professor. He soon set about cleaning house at the laboratory. Garnett gave way to Richard Glazebrook and W. N. Shaw, who introduced a highly structured course of elementary laboratory instruction that for the first time made the Cavendish a site of practical teaching as well research. Rayleigh also launched a fund to support the purchase of badly needed new instruments and equipment. Most importantly, he set out to give the laboratory a sharp and shared focus. Arthur Schuster, whose time at the Cavendish bridged the Maxwell and Rayleigh eras, later said that Rayleigh considered it highly important "to identify the laboratory with some research planned on an extensive scale so that a common interest might unite a number of men sharing in the work."[139] Building on Maxwell's abortive effort and on the resources already available at the Cavendish, Rayleigh, too, chose the redetermination of the ohm and set about recruiting a team to tackle it. Like Maxwell, however, he soon found that ambitious young Cambridge graduates did not regard playing a supporting role in a joint effort as the best path to a college fellowship, and few volunteered to help.

Rayleigh nonetheless pressed on, enlisting his sister-in-law Eleanor Sidgwick as his chief collaborator, reportedly joined on at least one occasion by her brother Arthur Balfour, the future prime minister.[140] Along

---

[138] Simon Schaffer, "Accurate Measurement Is an English Science," in M. Norton Wise, ed., *The Values of Precision* (Princeton: Princeton University Press, 1995), 135–72; George Chrystal and S. A. Saunder, "Results of a Comparison of the British-Association Units of Electrical Resistance," *BA Report* (1876), 13–19.

[139] Schuster, "Rayleigh," xx.

[140] Helen Fowler, "Eleanor Mildred Sidgwick [*née* Balfour]," *ODNB*; J. N. Howard, "Eleanor Mildred Sidgwick and the Rayleighs," *Applied Optics* (1964) 3: 1120–22; Robert John Strutt, Fourth Baron Rayleigh, *Life of John William Strutt, Third Baron Rayleigh*, 2nd ed. (Madison: University of Wisconsin Press, 1968), 107–23. On Arthur Balfour's participation, see G. P. Thomson, *J. J. Thomson and the Cavendish Laboratory in His Day* (London: Thomas Nelson and Sons, 1964), 25. Maxwell resisted admitting women to the Cavendish; when he finally agreed to allow them in during the summer, while he was in Scotland, Garnett reported that "I had a class who were determined to go through a complete course of electrical measurements during the few weeks for which the Laboratory was open to them"; see Schuster, "The Clerk-Maxwell Period," 35–36.

with Schuster and a few others, Rayleigh and Sidgwick carried out a series of careful redeterminations of the ohm, first with the original British Association spinning coil, then with a larger version of it, and finally using a spinning disc method that had been devised by the Danish physicist Ludvig Lorenz.[141] Rayleigh also analyzed Maxwell and Jenkin's original measurements and sought to trace where they had gone astray – mainly, he found, by miscalculating the self-induction of their spinning coil. By 1882 Rayleigh's team had established that the original British Association unit was 1.3 percent smaller than its intended value of ten million meters per second. This corrected value was later confirmed and refined by other researchers, and "true or Rayleigh ohms," as Thomson dubbed them in 1883, were to provide much of the basis for a series of international accords in the 1880s and 1890s on agreed standards of electrical resistance.[142] Under Rayleigh, the Cavendish finally achieved the status as a global arbiter of electrical measurement that Maxwell had first envisioned for it.

Rayleigh did not stay long at the Cavendish. He had taken the professorship in part in response to financial setbacks during the agricultural depression of the late 1870s, and after his brother Edward established a successful dairy operation on the family estate in Essex, Rayleigh resigned his Cambridge chair in 1884 and returned to Terling, where he set up a productive home laboratory.[143] His successor at the Cavendish, J. J. Thomson, gradually shifted the focus of research at the laboratory away from electrical standards and toward problems in atomic physics. While many academic physics laboratories continued to pursue precision electrical measurement, work on standards increasingly became the province of electrical engineering departments and specialized government

By contrast, "soon after his appointment Lord Rayleigh gave authority to admit women students on the same terms as men"; Schuster, "Rayleigh," xxiv.

[141] The papers Rayleigh wrote with Schuster and Sidgwick on determining the value of the ohm, as well as what became the ampere and the volt, were published in the *Proceedings of the Royal Society* and the *Philosophical Transactions* between 1881 and 1885 and are collected in vol. 2 of John William Strutt [Baron Rayleigh], *Scientific Papers*, 6 vols. (Cambridge: Cambridge University Press, 1899–1920). On Lorenz's method, see Helge Kragh, *Ludvig Lorenz: A Nineteenth-Century Theoretical Physicist* (Copenhagen: Royal Danish Academy of Sciences and Letters, 2018), 171–81.

[142] Thomson to Rayleigh, May 25, 1883, quoted in Simon Schaffer, "Rayleigh and the Establishment of Electrical Standards," *European Journal of Physics* (1994) *15*: 277–85, on 283. On the later international conferences and agreements on electrical standards, see Larry Lagerstrom, "Constructing Uniformity: The Standardization of International Electromagnetic Measures, 1860–1912," PhD dissertation, University of California–Berkeley, 1992.

[143] Simon Schaffer, "Physics Laboratories and the Victorian Country House," in Smith and Agar, eds., *Making Space*, 149–80, on 166.

institutes.[144] By the 1880s and 1890s the demand for electrical knowledge, rooted first in the needs of telegraphy and later also in those of the nascent power and light industry, was being thoroughly met by an extensive infrastructure of instruction, instrumentation, and institutions. Physics, and especially physics education, had been transformed by its encounter with telegraphy and other electrical technologies. Even as research in physics developed along new lines in the twentieth century, the discipline continued to be shaped by its formative experiences of the 1860s and 1870s.

Oliver Heaviside and the Making of "Maxwell's Equations"

As he was supervising the completion of the Cavendish laboratory, Maxwell was also putting the finishing touches on his *Treatise on Electricity and Magnetism*, published in March 1873. A sprawling two-volume work, it was rich and suggestive but notoriously hard to follow; some parts were almost impenetrable. Moreover, as Andrew Warwick has emphasized, different groups read it very differently.[145] Cambridge coaches preparing prospective wranglers for the Mathematical Tripos treated Maxwell's book as a compendium of techniques for tackling tricky problems in spherical harmonics and the like, while researchers at the Cavendish regarded it, in George Chrystal's words, as primarily "a Treatise on Electrical *Measurement*."[146] The editors of the *Telegraphic Journal* took yet another view; calling it a "capital work," they quoted a long passage on how best to wind a galvanometer coil and declared that many of Maxwell's pages "would delight an artizan."[147] In the 1870s few readers focused, as most now do, on the comprehensive theory of the electromagnetic field that Maxwell presented (amid much else) in Part IV of his *Treatise*.

In the long run Maxwell's most important reader may have been an obscure telegrapher at a cable office in Newcastle. Self-taught in both

---

[144] Kim, *Leadership and Creativity*, 93–118; on standards work at government institutions, see David Cahan, *An Institute for an Empire: The Physicalisch-Technische Reichsanstalt, 1871–1918* (Cambridge: Cambridge University Press, 1989); Edward C. Pyatt, *The National Physical Laboratory: A History* (Bristol: Adam Hilger, 1983); and Rexmond Cochrane, *Measures for Progress: A History of the National Bureau of Standards* (Washington, DC: National Bureau of Standards, 1966).

[145] Warwick, *Masters of Theory*, 289.

[146] George Chrystal, "Clerk Maxwell's 'Electricity and Magnetism,'" *Nature* (January 12, 1882) 25: 237–40, on 238.

[147] "Notices of Books: *A Treatise on Electricity and Magnetism*. By James Clerk Maxwell," *Telegraphic Journal* (May 15, 1873) 1: 145–46.

mathematics and physics, Oliver Heaviside was in his early 20s when he first came across the *Treatise* not long after it was published. He was immediately struck, he later said, by its "prodigious possibilities," though at first he had trouble penetrating much beyond its preface and a few "bits here and there."[148] As he dug more deeply into what he called "that mine of wealth 'Maxwell,'" however, Heaviside found in it veins and nuggets, particularly concerning electromagnetic propagation and energy flow, that he would use to transform the way the theory was understood and applied. While these results were in a sense "all 'in Maxwell,'" they had by no means been recognized by Maxwell himself; as Heaviside observed, there was a big difference between "the patent and the latent, in Maxwell."[149]

Heaviside's exploration of the intricacies of the *Treatise* eventually led him to "redress" Maxwell's theory by devising new mathematical machinery specially suited to its clear expression.[150] Dissatisfied both with the cumbersome Cartesian equations that filled many pages of the *Treatise* and with the compact but quirky quaternionic expressions in which Maxwell had summarized his main results, Heaviside stripped away what he regarded as the unnecessary excrescences of the quaternionic system to produce the simple system of vector analysis that is now so ubiquitous not just in electromagnetic theory but throughout physics.[151] Most strikingly, Heaviside recast the long list of field equations Maxwell had given in his *Treatise* into the compact set of four now universally known as "Maxwell's equations." This iconic set of vector equations now appears not only in textbooks but on T-shirts; cast in bronze, the equations stand at the base of the statue of Maxwell that was erected in 2008 near his birthplace in Edinburgh (Figure 6.4). Credit for the underlying theory of the electromagnetic field certainly belongs to Maxwell – "a heaven-born genius," in

---

[148] Heaviside to Joseph Bethenod, February 24, 1918, quoted (after retranslation) in Paul J. Nahin, *Oliver Heaviside: The Life, Work, and Times of an Electrical Genius of the Victorian Age* (Baltimore: Johns Hopkins University Press, 2002), 24. See also Basil Mahon, *The Forgotten Genius of Oliver Heaviside, A Maverick of Electrical Science* (Amherst, NY: Prometheus Books, 2017), 58.

[149] Oliver Heaviside, "On Electromagnetic Waves, Especially in Relation to the Vorticity of the Impressed Forces; and the Forced Vibrations of Electromagnetic Systems," *Phil. Mag.* (February 1888) 25: 130–56, on 153, repr. in Heaviside, *EP 2*: 375–96, on 393–94. This was the first installment of a six part series.

[150] Heaviside used the phrase "Maxwell redressed" in an undated note on the back of a page of his manuscript of "Theory of Voltaic Action," Box 14, Heaviside Collection–IET.

[151] Michael J. Crowe, *A History of Vector Analysis: The Evolution of the Idea of a Vectorial System*, 2nd ed. (New York: Dover, 1985), 162–77. J. Willard Gibbs independently developed a vectorial system very similar to Heaviside's at about the same time and for similar reasons.

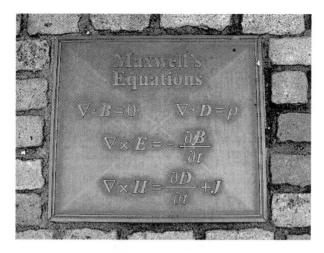

Figure 6.4 This plaque, showing Oliver Heaviside's vector form of "Maxwell's equations," is set into the pavement near the statue of James Clerk Maxwell on George Street in Edinburgh.

Heaviside's words – but the equations themselves have been taken from Heaviside.[152]

Heaviside's work was deeply rooted in cable telegraphy, particularly in the problem of tracing how pulses of current travel along submarine cables. This had been a central issue for submarine telegraphy and in many ways for British electrical science as a whole since Latimer Clark first discovered retardation in the early 1850s. By shifting attention away from the conducting wire and toward the action of the surrounding dielectric, the discovery of retardation had contributed to British physicists' and engineers' turn toward Faraday's field approach to electromagnetic phenomena. It had also had directly motivated Thomson's formulation in 1854 of his influential theory of cable transmission, as well as important experimental work. Heaviside's own experience with retardation and other propagation phenomena while working as a cable telegrapher in the late 1860s and early 1870s would shape and direct his thinking throughout his career.

Heaviside was an unusual man, with many quirks; his best friend once described him as "a first-rate oddity," though, he felt compelled to add, "never a mental invalid"[153] (Figure 6.5). Born in May 1850 in a gritty

---

[152] Heaviside to Hertz, July 13, 1889, in J. G. O'Hara and W. Pricha, eds., *Hertz and the Maxwellians* (London: Peter Peregrinus, 1987), 66–68, on 67.

[153] G. F. C. Searle, "Oliver Heaviside: A Personal Sketch," in *Heaviside Centenary Volume* (London: IEE Press, 1950), 93–96, on 96.

Figure 6.5 Oliver Heaviside at his home in Devonshire in 1893.
(Courtesy Institution of Engineering and Technology.)

part of Camden Town in north London, he had a difficult childhood, including a bout with scarlet fever that badly damaged his hearing. "I was very deaf from an early age of manhood," he later told Heinrich Hertz, "and that has influenced my whole life."[154] Perhaps as a result, he became a loner, given to working things out for himself. His father, Thomas, was a skilled wood engraver who had fallen on hard times; to help make ends meet, his mother, Rachel, opened a small school in their rented rooms. Heaviside later gave a vivid picture of his childhood:

I was born and lived 13 years in a very mean street in London, with the beer shop and bakers and grocers and coffee shop right opposite, and the ragged school and the sweeps just round the corner. Though born and bred up in it, I never took to it, and was very miserable there, all the more so because I was so exceedingly deaf that I couldn't go and make friends with the other boys and play about and enjoy myself . . .. The sight of the boozing in the pub made me a teetotaller for life. And it

[154] Heaviside to Hertz, September 13, 1889, in O'Hara and Pricha, *Hertz*, 76–78, on 76.

was equally bad indoors. [Father was] a naturally passionate man, soured by disappointment, always whacking us, so it seemed. Mother similarly soured, by the worry of keeping a school.[155]

When Oliver was thirteen, a small inheritance enabled the Heavisides to move to a better part of Camden Town and their lives improved. Oliver did well at the local grammar school, but there was no money to pay for any higher education, and he left school when he was sixteen. After a further year or so of private study, he headed north to Newcastle to join his older brother Arthur West Heaviside in the telegraph business.

That Heaviside went into telegraphy was the result of a fortunate family connection: his mother's sister had married Charles Wheatstone, one of the inventors of the telegraph.[156] As Wheatstone's "poor relations," the Heavisides looked to Sir Charles for help in finding careers for their sons, and he evidently obliged, hiring Arthur to run the Newcastle office of his Universal Private Telegraph Company.[157] After serving for a time as Arthur's informal assistant, Oliver landed a job as a telegrapher for the new Dansk-Norsk-Engelsk Telegraf-Selskab, working for a time at its station in Fredericia, Denmark, where he handled traffic coming through the Anglo-Danish cable that had been laid across the North Sea in September 1868, a link in a chain of cables and landlines that would stretch across Russia to China and Japan.

The Anglo-Danish cable was one of the few not owned by a British company; the Dansk-Norsk-Engelsk Telegraf-Selskab, absorbed in 1870 into Tietgen's Great Northern Telegraph Company, was Danish. Like almost all other cables, however, it was manufactured by a British firm and laid and tested by British engineers.[158] Much of its operating staff, too, was British, including Heaviside, and the peculiarities of cable signaling remained largely a British concern. Heaviside had ample opportunity to

[155] Heaviside to G. F. FitzGerald, June 3, 1897, in Bruce J. Hunt, *The Maxwellians* (Ithaca: Cornell University Press, 1991), 49–51.

[156] Emma and Charles Wheatstone's first child, Charles Pablo Wheatstone, was born less than three months after what was described as their "very quiet wedding" in February 1847; see Brian Bowers, *Sir Charles Wheatstone* (London: Her Majesty's Stationery Office, 1975), 154.

[157] Wheatstone reportedly "took a great interest in his nephews"; Bowers, *Wheatstone*, 155. On the Universal Private Telegraph Company, see Steven Roberts's account on distant-writing.co.uk; on Arthur West Heaviside, see Elizabeth Bruton, "From Theory to Engineering Practice: Shared Telecommunications Knowledge between Oliver Heaviside and his Brother GPO Engineer Arthur West Heaviside," *Phil. Trans. A* (2018) *376*: 1–20.

[158] As noted in "The Anglo-Danish Telegraph," *Newcastle Courant* (September 11, 1868), the cable was armored and laid by R. S. Newall, then the mayor of Gateshead, near Newcastle; it marked his brief return to the cable business after having left it in the wake of the failure of his Red Sea cables.

observe those peculiarities both while at Fredericia and after his transfer in 1870 to Newcastle. He was struck, for example, by the fact that when a fault appeared in the cable, the clarity of the signals actually improved, though their amplitude decreased.[159] Perhaps, he suggested, the prevailing insistence on perfect insulation was misguided and signaling rates could be increased by deliberately introducing small "faults" into cables.

Cable operators were always looking for ways to boost their signaling rates, of course, especially on the Anglo-Danish cable, where Wheatstone automatic transmitters had been used from the first. Fed with messages coded into punched paper tapes, Wheatstone automatics could transmit hundreds of words per minute on overhead landlines. On the Anglo-Danish cable, however, the pulses of current blurred together and became unreadable at speeds above about seventy words per minute.[160] Frustrated by this apparent inability to use his uncle's device to its full advantage, Heaviside formed an ambition to design a perfect cable that would be able to carry readable signals at an arbitrarily high rate. He soon realized that to accomplish this, he would need to make a deep dive into telegraphic theory.

Heaviside's first scientific paper, signed "O.," appeared in the *English Mechanic* in July 1872. In it he described a simple arrangement – essentially a variant of a Wheatstone bridge – for comparing electromotive forces.[161] He followed it in February 1873 with one in the *Philosophical Magazine* on the most sensitive arrangement of a Wheatstone bridge, a well-known problem that had defeated those less willing than he to tackle the heavy algebra involved. Not long after the paper appeared, Thomson happened to visit Newcastle and, as Heaviside later said, "*mentioned* it, so I gave him a copy" – no doubt a gratifying experience for the

[159] Heaviside, "On the Theory of Faults in Cables," *Phil. Mag.* (July 1879) *8*: 60–74 and (August 1879) *8*: 163–77, repr. in Heaviside, *EP 1*: 71–95, on 77; Heaviside, "The Beneficial Effect of Leakage in Submarine Cables," *Elec.* (September 15, 1893) *31*: 518–19 and Heaviside, "Short History of Leakage Effects on a Cable Circuit," *Elec.* (October 6, 1893) *31*: 603–4, repr. in Heaviside, *EMT 1*: 417–20 and 420–24. Heaviside said that just before the cable failed altogether, the signals received on a siphon recorder, though severely attenuated, were "exceedingly clear," in fact "the best ever got at the speed"; *EMT 1*: 424.

[160] On the Wheatstone automatic, see Bowers, *Wheatstone*, 176–80; on its use on the Anglo-Danish cable, see N. J. Holmes, "Submarine Telegraphs," *Nature* (May 4, 1871) *4*: 8–10. Holmes worked for Wheatstone's Universal Private Telegraph Company. On the speeds achieved on the Anglo-Danish cable, see Oliver Heaviside, "On the Speed of Signalling through Heterogeneous Telegraph Circuits," *Phil. Mag.* (March 1877) *3*: 211–21, repr. in Heaviside, *EP 1*: 61–70, on 62n.

[161] "O." [Oliver Heaviside], "Comparing Electromotive Forces," *English Mechanic* (July 5, 1872) *15*: 411, repr. in Heaviside, *EP 1*: 1. On his authorship, see Oliver Heaviside, "Voltaic Constants," *Telegraphic Journal* (May 15, 1873) *1*: 146, repr. in Heaviside, *EP 1*: 2–3.

young telegrapher. He also sent a copy to Maxwell, who noted it approvingly in the second edition of his *Treatise*. All in all it was, as Heaviside later said, "a good beginning."[162]

As Heaviside's focus shifted from telegraph instruments to deeper questions of telegraphic propagation, and as his confidence in his abilities increased, he began to chafe at the demands of his job. In May 1874, a combination of health problems and a desire to devote all of his time to his theoretical investigations led him to resign from the Great Northern company and return to London to live with his parents.[163] He was just twenty-four when he "retired" after six years working as a telegrapher. He would never again hold a regular job.

Over the next few years, Heaviside essentially remade himself into a mathematical physicist. Drawing on the small number of books available to him and on copies of journals passed along by his brother Arthur, by then a leading engineer in the newly nationalized Post Office telegraph system, he mastered calculus, differential equations, and Fourier analysis and began to delve into Thomson's electrical writings, particularly his theory of cable transmission. In a series of five papers published in the *Philosophical Magazine* and the *Journal of the Society of Telegraph Engineers and Electricians* between 1874 and 1881, Heaviside applied Thomson's theory to the working of cables through condensers, to the effects of faults on signal transmission, and to the peculiarities of circuits that included both cables and landlines.[164] His analyses of arrival curves were far more elaborate than anything Thomson himself had published and showed Heaviside's willingness and ability to follow the theory deep into its mathematical consequences.

The most important of Heaviside's early papers, "On the Extra Current," appeared in the *Philosophical Magazine* in August 1876.[165] In formulating his transmission theory in 1854, Thomson had focused on the resistance and capacitance of the cable and quite legitimately left self-induction aside as inconsequential for the slowly varying currents he was then considering. The resulting theory was purely diffusional, depicting current and voltage spreading along an electrical conductor in much the same way that heat diffuses along an iron bar. Telegraphers knew, however, that when a cable was coiled on itself or the current in it changed very

---

[162] Heaviside notebook 3A: 4, Heaviside Collection, IET; Maxwell, *Treatise*, 2nd ed., *1*: § 350.

[163] Nahin, *Heaviside*, 22–23; Rollo Appleyard, "A Link with Oliver Heaviside," *Electrical Communication* (October 1931) *10*: 53–59.

[164] These five papers were reprinted in Heaviside, *EP 1*: 47–95 and 116–41.

[165] Heaviside, "On the Extra Current," *Phil. Mag.* (August 1876) *2*: 135–45, repr. in Heaviside, *EP 1*: 53–61.

rapidly, self-induction could come into play. In "On the Extra Current," Heaviside took such cases into account by adding a new term to Thomson's original equation for how the voltage v varied along a line. The result was a form of what came to be called "the telegrapher's equation":

$$\partial^2 v/\partial x^2 = ck\, \partial v/\partial t + sc\, \partial^2 v/\partial t^2,$$

where c is capacitance, k resistance, and s inductance, all per unit length.[166] This equation cast telegraphic transmission into a sharp new light, very different from Thomson's diffusional theory. Self-induction acts as a sort of flywheel to oppose any change in the strength of a current; thus, as Heaviside noted, "in virtue of the property of the electric current which Professor Maxwell terms its 'electromagnetic momentum,' whenever any sudden change of current or charge takes place in a circuit possessing an appreciable amount of self-induction, the new state of equilibrium is arrived at through a series of oscillations in the strength of the current."[167] That is, instead of simply diffusing along a cable, a pulse of current skitters back and forth from one end to the other, its oscillations gradually dying out as it passes to and fro. This was Heaviside's first step toward conceiving of telegraph signals as electromagnetic waves. It was also his first reference in print to "Professor Maxwell," whose ideas he was just beginning to explore in depth.

Heaviside's early papers on telegraphic theory focused on the line equations, tracing how current and voltage vary along a conductor. Though mathematically demanding, they said little about the physical state of the cable and nothing about the surrounding field. In 1882 he wrote a paper on how currents propagate along wires that would remain unpublished until he included it in his collected *Electrical Papers* ten years later. There he said it provided "a sort of missing link" between his early papers on telegraphic transmission based on the line equations and his later writings, "in which the subject is discussed on the basis of Maxwell's theory of the ether as a dielectric."[168] Led by Maxwell, Heaviside was moving out of the conducting wire into the surrounding field, where he

---

[166] In the full form of the "telegrapher's equation," leakage through the insulation (r) is also taken into account, yielding $\partial^2 v/\partial x^2 = sc\, \partial^2 v/\partial t + (ck + sr)\, \partial v/\partial t + rkv$. Heaviside had discussed the effects of leakage as early as 1874, but he did not explicitly introduce it into his equations until 1887, when he treated it in connection with his "distortionless circuit"; see Heaviside, "Preliminary to Investigations Concerning Long-distance Telephony and Connected Matters," *Elec.* (June 3, 1887) *19*: 79–81, repr. in Heaviside, *EP 2*: 119–24, on 123.

[167] Heaviside, "On the Extra Current," *Phil. Mag.* (August 1876) *2*: 135–45, repr. in Heaviside, *EP 1*: 53–61, on 59.

[168] Heaviside, "Contributions to the Theory of the Propagation of Currents in Wires," *EP 1*: 141–79, on 141n.

would find a much fuller and more physically satisfying theory of telegraphic propagation.

Writing to Heinrich Hertz in July 1889 in the wake of the young
German physicist's experimental discovery of electromagnetic waves,
Heaviside said "I have been so long familiar myself with waves in dielectrics that your experimental results I take without surprise, almost as
a matter of course. But it is very different with many people" – and easily
could have been with Heaviside himself. "Take a suppositious case," he
said:

Supposing, after my first reading of Maxwell, when I imagined I had pretty well
taken it all in, I had gone to sleep for many years; and on waking, had had your
experiments to read about, and the doctrine in explanation thereof. It would have
been a perfect revelation to me! So I think it must be even to many readers of
Maxwell, who have read, but have done no more; in particular, have not sat down
and developed that chapter of his on Propagation into its consequences; no easy
matter unaided, for the equations of propagation, as they stand, are practically
unworkable.[169]

Heaviside's task and achievement was not just to work through Maxwell's
chapter on propagation for himself but to make it workable for others.

Just as Heaviside was setting out on this task in 1882, he was shifting to
a new venue for publishing his work. Although he continued to send
articles to the *Philosophical Magazine*, including two long and important
series later in the 1880s, most of his voluminous writings in this period
appeared in the *Electrician*, a weekly trade journal owned by Pender and
Anderson of the Eastern Telegraph Company. Its pages were mainly
filled with business notices and advertisements for electrical equipment,
but the editor, C. H. W. Biggs, also sought to include articles of scientific
interest, some quite advanced, and after Heaviside sent in a short piece on
"Dimensions of a Magnetic Pole" in June 1882, Biggs invited him to
become a regular contributor.[170] That November Heaviside began
a series on "The Relations between Magnetic Force and Electric
Current" in which, amid much else, he first presented the rudiments of
his system of vector analysis.[171] Apart from a three-year gap between
December 1887 and January 1891, the *Electrician* would carry articles
from his pen every few weeks for the next twenty years.

---

[169] Heaviside to Hertz, July 13, 1889, in O'Hara and Pricha, *Hertz*, 66–67.
[170] Heaviside, "Dimensions of a Magnetic Pole," *Elec.* (June 3, 1882) 9: 63–64, repr. in
Heaviside, *EP I*: 179–81; Biggs to Heaviside, September 1 and December 5, 1882,
Heaviside Collection, IET. Pender and Anderson's *Electrician*, launched in 1878, was
not related to the earlier publication of the same name, which had folded in 1864.
[171] Heaviside, "The Relations between Magnetic Force and Electric Current," *Elec.*
(November 18–December 16, 1882), repr. in Heaviside, *EP I*: 195–231.

In one of the first of these articles, Heaviside said that while most readers no doubt found the electrical writings of such "eminent mathematical scientists" as Ampère and Maxwell quite forbidding, he believed the underlying ideas "could be to a great extent stripped of their usual symbolical dress, and in their naked simplicity made to appeal to the sympathies of the many."[172] If this was his intention, however, he had a hard time sticking to it, and soon veered into deep mathematics, complete with elliptic integrals and spherical harmonics. What ordinary readers of the *Electrician* made of Heaviside's writings is not clear; no doubt most just skipped over them. But even as he appealed to Heaviside to tone down the mathematics, Biggs saw value in his writings and continued to publish his articles, while paying him about £40 a year for them.[173]

Throughout the 1880s Heaviside continued to live with his parents in Camden Town, working in almost total isolation. He kept in touch with his brother Arthur and collaborated with him on some telegraphic and telephonic investigations, but apart from a few letters stirred up by his articles in the *Electrician*, he seems to have had no other social or scientific contacts. Yet in his own way Heaviside was happy. "There was a time in my life," he later said, "when I was something like old Teufelsdröckh in his garret, and was in some measure satisfied or contented with a mere subsistence. But that was when I was making discoveries. It matters not what others may think of their significance. They were meat and drink and company to me."[174]

Heaviside's most important discovery came in the summer of 1884. The year before he had begun a series in the *Electrician* on "The Energy of the Electric Current" in which he drew on Maxwell's *Treatise*, Thomson's papers on electrostatics and magnetism, and that "stiff but thoroughgoing" guide to dynamics, Thomson and Tait's *Treatise on Natural Philosophy*, to explore how best to represent the energy associated with electric currents.[175] After considering several alternatives, including one based on the vector potential, he concluded, as Maxwell had, that $\mu\mathbf{H}^2$ (where $\mathbf{H}$ is the magnetic force and μ the permeability) was the only

[172] Heaviside, "Magnetic Force," *EP 1*: 195.

[173] On Biggs, see Nahin, *Heaviside*, 103–6; on Heaviside's income in this period, see Hunt, *Maxwellians*, 71.

[174] Heaviside to Joseph Larmor, July 18, 1908, in Hunt, *Maxwellians*, 61. "Teufelsdröckh" was the fictional German "philosopher of clothes" in Thomas Carlyle's *Sartor Resartus* (1838).

[175] Heaviside, "The Energy of the Electric Current," *Elec.* (January 20–March 23, 1883), repr. in Heaviside, *EP 1*: 231–55. Heaviside's remark about William Thomson and P. G. Tait, *Treatise on Natural Philosophy* (Oxford: Clarendon Press, 1867), appears in Heaviside, *EMT 3*: 178.

expression that located the energy properly in the surrounding field. Heaviside then took up a question Maxwell had not addressed: how does that energy move around when the current changes? He noted that when a current begins to flow in a wire, energy must somehow pass into the field around it, to be stored there in the strain and motion of the medium – perhaps, he said, echoing Maxwell's molecular vortex theory, in the rotation of tiny flywheels in the ether. If the electromotive force driving the current is then shut off, this stored energy will flow from the field into the wire, briefly producing the "extra current" associated with self-induction. It was not yet clear, however, exactly how this occurred, or how the energy got from one place to another.

Heaviside tackled that question in a series on "The Induction of Currents in Cores" that began running in the *Electrician* in May 1884.[176] Framing the problem in telegraphic terms, he asked how, once energy has passed along a wire, it enters the iron core of an electromagnet in a receiving instrument, there to power the tapping of a lever or the deflection of a needle. His breakthrough came in June while he was writing up a section on "Transmission of Energy into a Conducting Core."[177] Starting with Maxwell's expressions for the magnetic energy and for the rate of dissipation within a conductor and going through a laborious series of integrations and transformations, he hit on a remarkably simple result: the flow of energy at any point in the field is simply the vector product of the electric and magnetic forces at that point: $\mathbf{S} = \mathbf{E} \times \mathbf{H}$. This had some striking, even startling consequences. In particular, it implied that the energy delivered by an electric current does not flow along within the conducting wire, as everyone had always assumed, but instead passes through the surrounding dielectric and enters the wire from its sides, there to be dissipated as heat.

The idea that electrical energy flows through the insulation rather than the wire was (and is) counterintuitive for most people, especially those holding the common view that electricity is a fluid that flows through conductors like water in a pipe, carrying its energy along with it. But Heaviside had never "swallowed the electric fluids," as he later said, and had long looked on electric charges and currents as manifestations of strains and motions in the surrounding field; he thus found it quite natural that energy should pass through that field

[176] Heaviside, "The Induction of Currents in Cores," *Elec.* (May 3, 1884–January 3, 1885), repr. in Heaviside, *EP 1*: 353–416.

[177] Heaviside, "Transmission of Energy into a Conducting Core," *Elec.* (June 21, 1884) *13*: 133–34, repr. in Heaviside, *EP 1*: 377–78; see also Hunt, *Maxwellians*, 121.

as well.[178] He was also led to a new view of how a current rises in a wire: rather than being pushed along from one end, he said, the current is driven by actions that start on the boundary of the wire and work their way inward. A rapidly varying current is thus unable to penetrate very far into the body of the wire before the action reverses, so that most of the current is confined to the outer "skin" of the wire – a vitally important phenomenon for alternating currents and high-frequency signal transmission.

Heaviside was convinced that a result as fundamental as his energy flow theorem ought to follow directly from the basic field equations, without requiring the roundabout derivation he had originally been forced to use. He therefore worked backwards from his energy flow formula, focusing closely on the electric and magnetic force vectors **E** and **H** as the quantities that best reflected the real state of the field. He proceeded to "murder" the vector and scalar potentials that had figured so prominently in Maxwell's *Treatise*, saying they obscured what was really going on in the field, and restated Maxwell's theory simply in terms of **E** and **H**. In particular, he combined several of Maxwell's original equations to derive a new one, $-\mathrm{curl}\,\mathbf{E} = \mu\,\partial\mathbf{H}/\partial t$, to pair with Maxwell's $\mathrm{curl}\,\mathbf{H} = k\,\mathbf{E} + c\,\partial\mathbf{E}/\partial t$ (where k is the conductivity and c the permitivity, yielding terms for the conduction and displacement currents, respectively).[179] These "duplex" equations, as Heaviside called them, provided the foundation for all of his subsequent work on Maxwell's theory; combined with $\mathrm{div}\,c\mathbf{E} = \rho$ and $\mathrm{div}\,\mu\mathbf{H} = 0$ (representing electric charge and the absence of any magnetic charge), they form the now standard set of "Maxwell's equations."

Once Heaviside had these equations in hand, he was able, he later said, to do "more work in a week than in all the previous years"; in fact, he said, "I sketched out all my later work."[180] Many of his most important results first appeared in a long series on "Electromagnetic Induction and Its Propagation" that began in the *Electrician* in January 1885 and would run for the next three years.[181] He started by considering the energy carried by telegraph lines: "From London to Manchester, Edinburgh, Glasgow, and hundreds of other places" – including, he might have added, cable stations around the world – "day and night, are sent with great velocity, in rapid succession, backwards and forwards, electric

[178] Heaviside to FitzGerald, April 13, 1894, in Hunt, *Maxwellians*, 231; see also Heaviside, "On the Transmission of Energy through Wires by the Electric Current," *Elec.* (January 10, 1885) *14*: 178–80, repr. in Heaviside, *EP 1*: 434–41.
[179] On how to get from Maxwell's original equations to Heaviside's form of "Maxwell's equations," see Hunt, *Maxwellians*, 245–47.
[180] Heaviside to Hertz, July 13, 1889, O'Hara and Pricha, *Hertz*, 67.
[181] Heaviside, "Electromagnetic Induction and Its Propagation," *Elec.* (January 3, 1885–December 30, 1887), repr. in Heaviside, *EP 1*: 429–560 and *EP 2*: 39–155.

currents, to effect mechanical motions at a distance, and thus serve the material interests of man."[182] But what was really flowing, he asked, and how did it get from one place to another? After poking holes in the usual picture of electrical fluids and of energy flowing within wires, he gave a remarkably clear account of his remodeled form of Maxwell's theory, showing how it could be used to elucidate the varied phenomena of electromagnetic propagation and to improve the quality of telegraphic signaling.

Heaviside did not have access to many scientific journals and often had to piece together his knowledge of recent developments from brief mentions in the *Electrician*, the *Philosophical Magazine*, and whatever other publications his brother Arthur passed along to him. This was apparently how he learned that J. H. Poynting of Mason College in Birmingham had hit on the energy flow theorem some months before him, which is why physicists now speak of the "Poynting flux" rather than the "Heaviside flux." A Cambridge graduate who had worked with Maxwell at the Cavendish (though on measuring the gravitational constant rather than on anything electrical), Poynting had been struck by Rayleigh's treatment of the energy of sound waves in his *Theory of Sound*. In 1883 he took up the analogous question of how energy moves through the electromagnetic field and, by a roundabout route similar to the one Heaviside would soon follow, proceeded to derive the flow formula.[183] Poynting sent his paper "On the Transfer of Energy in the Electromagnetic Field" to the Royal Society of London in December 1883; an abstract appeared in its *Proceedings* the next month and the full paper in the *Philosophical Transactions* in the latter half of 1884.[184]

Having had no access to publications as august as the *Philosophical Transactions*, Heaviside probably first learned of Poynting's work from a brief mention Oliver Lodge made of it in the *Philosophical Magazine* in June 1885.[185] Heaviside was no doubt disappointed to learn that

---

[182] Heaviside, "Transmission of Energy," *EP 1*: 434–41, on 434.

[183] Richard Noakes was the first to point out that Poynting drew on Rayleigh's *Theory of Sound* in deriving his energy flow theorem; see Warwick, *Masters of Theory*, 347, n. 122.

[184] J. H. Poynting, "On the Transfer of Energy in the Electromagnetic Field," *Phil. Trans.* (1884) *175*: 343–61; an abstract appeared in *Proc. RS* (January 1884) *36*: 186–87.

[185] Oliver J. Lodge, "On the Paths of Electric Energy in Voltaic Circuits," *Phil. Mag.* (June 1885) *19*: 487–94; on 489, Lodge said Poynting's paper was "of the very greatest interest and power" and stated its main results. Heaviside almost certainly saw this paper, as it was a sequel to a long paper on voltaic circuits that Lodge had delivered at the 1884 meeting of the British Association, a copy of which he sent to Heaviside after belatedly coming across Heaviside's discussion of the same question in *Elec.* (February 2, 1884) *8*: 270; see Lodge, "On the Seat of Electromotive Force in the Voltaic Cell," *Phil. Mag.* (May 1885) *19*: 340–65, on 357n. Heaviside's reply to Lodge (January 15, 1885, Lodge Collection, UCL) is one of the few letters he wrote to another scientist before 1888.

someone else had found the energy flow theorem before him; writing later to Lodge about his theory of the skin effect, he said he had not initially emphasized it, as "I thought much more of my other discovery, the transfer of energy, which Prof. Poynting is to be congratulated upon, as I later found."[186] But while Poynting used his flow theorem to clarify some important points about the workings of the electromagnetic field, he seems to have regarded it more as a nice mathematical theorem than as a clue to refashioning the basic equations of Maxwell's theory. That was left to Heaviside, whose struggles with cable problems had made him especially attuned to anything involving electromagnetic propagation and the relationship between wires and fields.

In the many installments of "Electromagnetic Induction and Its Propagation," Heaviside used his new equations to explore how electromagnetic waves propagate both in free space and along wires, with wires functioning as waveguides rather than as pipelines. He also showed theoretically how distortion on a cable could be reduced or even eliminated by properly balancing its resistance, capacitance, inductance, and leakage, thus fulfilling the ambition that had led him to take up telegraphic theory in the first place.[187] His claim that one could reduce distortion by "loading" a line with extra inductance put him in conflict with W. H. Preece, chief of the Post Office telegraph department (and effectively Arthur Heaviside's boss), who had publicly declared self-induction to be the enemy of clear signaling. In 1887 Preece blocked publication of a paper the Heaviside brothers had written on the inductive loading of telephone lines, and he may also have been behind Biggs's dismissal from the *Electrician* a short time later; in any case, Heaviside's long-running series in the journal was abruptly terminated, bringing his work for a time to a dead

Heaviside followed with "Remarks on the Volta-Force," *Journal of the Society of Telegraph-Engineers and Electricians* (April 1885) *15*: 269–96, repr. in Heaviside, *EP 1*: 416–28, in which he mentioned his energy flow theorem, saying later he was "Not aware of Poynting at time"; see Heaviside notebook 3A: 88, Heaviside Collection, IET. Heaviside first mentioned Poynting in print in "On the Self-Induction of Wires," *Phil. Mag.* (August 1886) *22*: 118–38, repr. in Heaviside, *EP2*: 168–85, on 172; he stated his energy flow theorem and said "I found I had been anticipated by Prof. Poynting."

[186] Heaviside to Lodge, June 27, 1888, Lodge Papers, UCL MS Add. 89.

[187] On distortionless propagation, see Nahin, *Heaviside*, 147–51, and Hunt, *Maxwellians*, 132–36. Inductive loading later became important for long-distance telephony and, from the 1920s, for cable telegraphy. Ido Yavetz, *From Obscurity to Enigma: The Work of Oliver Heaviside, 1872–1889* (Boston: Birkhäuser, 1995) emphasizes (212–13) that the distortionless condition can be extracted from the line equations without reference to field effects, though it is notable that Heaviside in fact arrived at this condition, and through it his proposals for inductive loading, only after he had developed his field ideas on energy flow, the skin effect, and the way currents rise in wires.

stop.[188] Heaviside would hold a grudge against Preece for the rest of his life. Early in 1888, however, he persuaded the *Philosophical Magazine* to begin carrying a new series of articles on "Electromagnetic Waves," and the timing proved fortunate.[189] Lodge was then drawing wide attention with a series of dramatic experiments on lightning protection, in some of which he used Leyden jar discharges to send electromagnetic waves surging back and forth along wires. Later that year Hertz published his even more impressive experiments on electromagnetic waves in free space. These were widely hailed as the definitive confirmation of Maxwell's theory, and in the hands of Guglielmo Marconi, as well as Lodge and others, within a decade they would provide the basis for practical systems of wireless telegraphy. As physicists began clamoring to learn all they could about electromagnetic waves, they were delighted to find that Heaviside had already worked out the theory of such waves in great detail – in connection with telegraph problems.

After his long years of isolation and obscurity, recognition came to Heaviside with remarkable swiftness. In 1888 he began what would become a long and active correspondence with Lodge, Hertz, G. F. FitzGerald, and other leading "Maxwellians"; early in 1889, Lodge and FitzGerald visited him in Camden Town, and Sir William Thomson heaped him with praise in an address to the newly renamed Institution of Electrical Engineers; and in 1891 the Royal Society of London elected him a Fellow.[190] That same year, the *Electrician* welcomed him back to its pages, launching a series on "Electromagnetic Theory" that would run for more than a decade and be reprinted in three large volumes, while in 1892 his earlier writings were collected and published by Macmillan in two volumes of *Electrical Papers*.

By the mid-1890s, Heaviside's form of Maxwell's equations was finding its way into standard textbooks, notably August Föppl's *Einführung in die Maxwell'sche Theorie der Elektricität*, from which Albert Einstein and

---

[188] On Heaviside's long feud with Preece, see Nahin, *Heaviside*, 59–78 and 139–85, and Hunt, *Maxwellians*, 137–44 and 171–73.

[189] The first installment of Heaviside's series on "Electromagnetic Waves" included a note by William Thomson on "the velocity of electricity in a wire," *Phil. Mag.* (February 1888) *25*: 155n; this no doubt drew wider attention to Heaviside's paper, but as Heaviside later said, it showed Thomson to be "all at sea on the subject"; see Heaviside notebook 3A: 150, Heaviside Collection, IET. Heaviside added five more installments of "Electromagnetic Waves" later that year, repr. in Heaviside, *EP 2*: 375–467. In January 1889 he had fifty copies of the set printed and bound at his own expense and distributed them to a carefully selected list of British and foreign physicists; see Hunt, *Maxwellians*, 145, 179–80.

[190] On Heaviside's growing ties to other Maxwellians and the increasing recognition of his work, see Nahin, *Heaviside*, 161–75; Mahon, *Forgotten Genius*, 139–53; and Hunt, *Maxwellians*, 175–208.

many others first learned Maxwell's theory.[191] But as Heaviside's work was being incorporated more deeply into mainstream physics, its roots in cable telegraphy were fading from view. In a review of Heaviside's *Electrical Papers*, FitzGerald wrote that "Maxwell's treatise is cumbered with the *débris* of his brilliant lines of assault, of his entrenched camps, of his battles. Oliver Heaviside has cleared those away, has opened up a direct route, has made a broad road, and has explored a considerable tract of country."[192] In clearing the field, however, Heaviside had made it easier for others to overlook not only Maxwell's earlier battles but also his own path into the study of electromagnetism. As time went on, physicists increasingly came to look upon Heaviside's version of Maxwell's theory, and particularly his canonical set of "Maxwell's equations," as simply the natural expression of the workings of the electromagnetic field and to forget or ignore the long struggle with problems of cable propagation that had led Heaviside to it.

Ridiculed in the early 1860s as an abject failure, by the end of the next decade oceanic submarine telegraphy was recognized as a transformative success that had changed the way commerce was conducted, international relations were managed, and news was disseminated. In Britain, it had also exerted deep and pervasive effects on the scale and focus of work in electrical science. By creating a demand for precision electrical measurement, it had fostered the creation of physics laboratories for both teaching and research, and by confronting physicists and engineers with phenomena of electrical propagation, it had shifted their attention to the role of the surrounding dielectric and the actions of the electromagnetic field.

The technology that drove all of these changes was, however, sunk beneath the waves and largely out of sight. After the dramas surrounding the first Atlantic cables had been surmounted, submarine telegraphy quickly became just another part of the unseen infrastructure of the modern world, taken for granted and scarcely thought of except by a small community of specialists. Apart from occasional complaints

[191] August Föppl, *Einführung in die Maxwell'sche Theorie der Elektricität* (Leipzig: Teubner, 1894). Föppl also disseminated Heaviside's vector methods; see Crowe, *History of Vector Analysis*, 226–27. Heaviside praised Föppl's book as "the clearest and most advanced" he had seen; see his review of Charles Emerson Curry, *Electricity and Magnetism* (London: Macmillan, 1897), in *Elec.* (September 10, 1897) *39*: 643–44, repr. in Heaviside, *EMT 3*: 503–7. On Einstein's use of Föppl's book, see Arthur I. Miller, *Albert Einstein's Special Theory of Relativity: Emergence (1905) and Early Interpretation (1905–1911)* (Reading, MA: Addison-Wesley, 1981), 151; see also Hunt, *Maxwellians*, 207–8.

[192] G. F. FitzGerald, review of "Heaviside's *Electrical Papers*," *Elec.* (August 11, 1893) *31*: 389–90, repr. in Joseph Larmor, ed., *Scientific Writings of the Late G. F. FitzGerald* (Dublin: Hodges & Figgis, 1902), 292–300, on 294.

about excessive rates and concerns about possible strategic vulnerabilities, after the 1870s the wider public paid relatively little attention to the web of cables that was spreading around them. Nonetheless, this vast network continued to have far-reaching effects, not just on commerce and international affairs but on the ways physicists conceived of and represented the fundamental workings of the world.

Epilogue
Full Circle

In 1902, a consortium of British imperial powers laid a string of cables across the Pacific, connecting Canada to Fiji, Australia, and New Zealand. The new cables completed the "All Red Line," circling the globe while touching only on British-controlled territories, and set the capstone to the worldwide British cable network[1] (Figure 7.1). That network would remain of vital strategic and economic importance for decades to come, but as the twentieth century dawned, both physics and electrical technology found themselves moving in new directions. Cable telegraphy had nourished the rise of field theory, but that theory had led in its turn to the discovery of electromagnetic waves and then to the development and promotion by Oliver Lodge, Guglielmo Marconi and others of practical systems of wireless telegraphy. By the 1910s and 1920s, that new technology, exploited by companies in several countries around the world, finally began to break the British monopoly on global telecommunications.[2] At the same time, attention in physics laboratories had largely shifted from the niceties of precision electrical measurement to the exploration of the subatomic realm. University programs in electrical technology, many of which had begun in physics departments, had been spun off as independent departments of electrical engineering, and physicists increasingly came to

---

[1] R. Bruce Scott, *Gentlemen on Imperial Service: A Story of the Trans-Pacific Telecommunications Cable* (Victoria, BC: Sono Nis Press, 1994); George Johnson, ed., *The All Red Line: The Annals and Aims of the Pacific Cable Project* (Ottawa: James Hope & Sons, 1903). On the controversies surrounding the Pacific cable project and the reasons why it was eventually undertaken by a consortium of colonial governments rather than Eastern Telegraph or another private firm, see Robert W. D. Boyce, "Imperial Dreams and the National Realities: Britain, Canada and the Struggle for a Pacific Telegraph Cable, 1879–1902," *English Historical Review* (2000) *115*: 39–70, and Stewart Ash, *The Cable King: The Life of John Pender* (London: Stewart Ash, 2018), 312–18 and 360–76.

[2] Daniel R. Headrick, *The Invisible Weapon: Telecommunications and International Politics, 1851–1945* (Oxford: Oxford University Press, 1991), 116–37, 173–217; Hugh G. J. Aitken, *The Continuous Wave: Technology and American Radio, 1900–1932* (Princeton: Princeton University Press, 1985), 250–301.

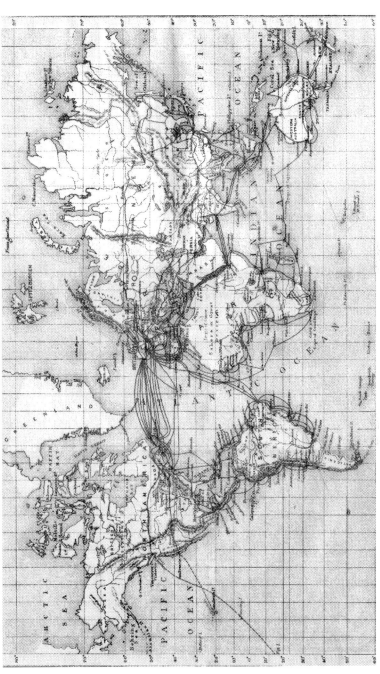

Figure 7.1 A map of the world's main cable routes, *circa* 1901. The Pacific cable, then about to be laid from Canada to Australia and New Zealand, is shown as a dashed line, as is a planned cable from San Francisco to Hawaii. Almost all of the cables on this map were made and laid by British firms. (From William Clauson–Thue, *The ABC Universal Commercial Electric Telegraphic Code*, 5th ed., 1901; courtesy Bill Burns.)

look upon their electrical apparatus as tools to use in other researches rather than as objects of investigation in themselves.[3]

Even as physics and telegraphy drifted apart, however, the close ties they had forged in the second half of the nineteenth century continued to have lasting consequences. Many of the successive shifts in telecommunications technologies that continue to reshape our world today still follow lines that were first laid down in the nineteenth century. The replacement of copper wires by fiber optic cables since the 1980s, for example, reflects Maxwell's insight of a century and a half ago that light itself consists of very short electromagnetic waves, so that there is less difference than might first appear between the bundles of glass fibers that now carry internet data across the oceans and the copper wires covered with gutta-percha that carried dots and dashes in the nineteenth century. Indeed, as Poynting and Heaviside might point out, in both cases the energy of the signals is carried by the insulating dielectric, whether glass or gutta-percha, rather than by any electrical conductor.

At a deeper level, physicists' understanding of how the world works continues to be shaped in subtle and persistent ways by the context in which their reigning ideas were first formed – in the case of field theory, by the cable industry that knit together the Victorian British Empire. Scientific knowledge is inescapably contingent: the picture scientists form of the world depends on the evidence they encounter and the weight they give to it. This evidence is typically mediated through various technologies – whether scientific instruments or technologies devised for more practical ends – that consequently shape the theories scientists (and others) form to account for what they see.[4] Field theory is not the only logically defensible way to formulate the known phenomena of electromagnetism; indeed, theories based on direct action at a distance between electrical particles held sway through most of the nineteenth century, especially in Germany.[5] Yet in the second half of the century, the field approach scored a decisive victory, first and most fully in Britain. To understand how and why this occurred – to understand how and why field

---

[3] Robert Rosenberg, "Academic Physics and the Origins of Electrical Engineering in America," PhD dissertation, Johns Hopkins University, 1990.

[4] M. Norton Wise, "Mediating Machines," *Science in Context* (1988) *2*: 81–117.

[5] Of the many alternatives to ordinary field theory, perhaps the best known is the one based on advanced and retarded potentials laid out in J. A. Wheeler and R. P. Feynman, "Classical Electrodynamics in Terms of Direct Interparticle Interaction," *Reviews of Modern Physics* (July 1949) *21*: 425–33. On German action-at-a-distance theories of electrodynamics in the nineteenth century, see André Koch Torres Assis, *Weber's Electrodynamics* (Dordrecht: Kluwer Academic, 1994), and Christa Jungnickel and Russell McCormmach, *Intellectual Mastery of Nature: Theoretical Physics from Ohm to Einstein*, 2 vols. (Chicago: University of Chicago Press, 1986), *1*: 137–46 and *2*: 74–77.

theory emerged and won out at that time and in that place – we must look beyond the purely internal history of physics and consider as well the demands and opportunities presented by the Victorian cable industry, particularly the propagation phenomena that so bedeviled and fascinated British physicists and engineers. It is when we look to this wider context that we grasp the sense in which electricity was truly an "imperial science."

Bibliography

Primary Sources

Archival Collections

Cambridge University Library
  Airy Correspondence, Royal Greenwich Observatory papers, RGO 6
  Kelvin Collection, CUL Add. MS 7342
  Maxwell Papers, CUL Add. MS 7655
Institution of Engineering and Technology Archives, London
  Latimer Clark notes, UK0108 SC/MSS 22
  J. J. Fahie correspondence, UK0108 SC/MSS 009
  Heaviside Collection, UK0108 SC/MSS 005
  W. H. Preece papers, UK0108 NAEST 017
National Museums Liverpool (Merseyside Maritime Museum)
  Atlantic Telegraph Company Minute Book, BICC Archive, B/BICC/I/2/7
New York Public Library
  Wheeler Collection of Electricity and Magnetism, Rare Book Collection, Astor, Lenox, and Tilden Foundations
PK Porthcurno Museum of Global Communications
  Alcatel Archive
  Cable and Wireless Archive
Trinity College Dublin
  William Rowan Hamilton Corrrespondence, TCD MS 7767
University College London, Special Collections
  Lodge Papers, MS Add. 89
University of Glasgow
  Kelvin Papers, GB 247 Kelvin
  Thomson Family Papers, GB 247 MS Gen. 1752

Newspapers

Boston Daily Advertiser
Brighton Gazette

Bristol Mercury
The Constitution, or Cork Advertiser
Daily Evening Bulletin (San Francisco)
Daily News (London)
Derby Mercury
Freedom's Champion (Atchison, Kansas)
Freeman's Journal (Dublin)
Glasgow Herald
Glasgow Sentinel
The Globe (London)
Leeds Mercury
Lloyd's Weekly (London)
Morning Chronicle (London)
Newcastle Courant
New York Times
Pall Mall Gazette (London)
Times (London)
Vermont Chronicle

British Parliamentary Papers

Electric telegraph companies. Copies of correspondence between the electric telegraph companies under contract with the government respecting the failure to lay down or keep in working order the electric wires, British Parliamentary Papers, 1860, LXII.211 (London, 1861).

Papers explanatory of the intended transfer of the Falmouth and Gibraltar electric telegraph cable to a line from Rangoon to Singapore, British Parliamentary Papers, 1860, LXII.257 (London, 1861).

Further correspondence respecting the establishment of telegraphic communications in the Mediterranean and with India, British Parliamentary Papers, 1860, LXII.269 (London, 1861).

Further correspondence respecting the establishment of telegraphic communications in the Mediterranean and with India, British Parliamentary Papers, 1860, LXII.461 (London, 1861).

Report of the Joint Committee to Inquire into the Construction of Submarine Telegraph Cables, British Parliamentary Papers, 1860, LXII.591 (London, 1861).

Correspondence respecting the Dardanelles and Alexandria telegraph, British Parliamentary Papers, 1863, LXXIII.601 (London, 1863).

Hansard's Parliamentary Debates, Fourth Series (London: Wyman, 1900) *83*.

Books and Articles

[Airy, G. B.], "Telegraphic Longitude of Brussels," *Athenæum* (January 14, 1854), 54–55; signed "A. B. G."

"Alpha," "Red Sea Telegraph" (letter), *Times* (August 24, 1860), 8.

"An Old Atlantic," "The Atlantic Telegraph Company" (letter), *Times* (December 24, 1866), 5.

Anderson, Sir James, *Statistics of Telegraphy* (London: Waterlow and Sons, 1872), 64.

[Anon.], "Transmarine Telegraph," *The Spectator* (August 31, 1850) *23*: 831.

[Anon.], "Experiments on the Transmission of Voltaic Electricity through Copper Wires Covered with Gutta Percha," *Chemical Record and Journal of Pharmacy* (March 27, 1852) *2*: 234–35.

[Anon.], Clark Patent 2956, December 20, 1853, *Repertory of Patent Inventions*, (1854) *23*: 474.

[Anon.], "Mediterranean Telegraph," *Illustrated London News* (October 6, 1855) *27*: 423.

[Anon.], "The Atlantic Telegraph," *Engineer* (January 30, 1857) *3*: 82–83.

[Anon.], *Letter from a Shareholder to Mr. Whitehouse, and His Reply* (London: Bradbury and Evans, 1858).

[Anon.], "The Atlantic Telegraph," *Engineer* (October 8, 1858) *6*: 268.

[Anon.], "Mr. Whitehouse and His Injuries," *Mechanics' Magazine* (October 9, 1858) *69*: 342–43.

[Anon.], "Notes Upon Passing Events," *Journal of Gaslighting* (July 16, 1861) *10*: 495–97.

[Anon.], "Standards of Electrical Measurement," *Elec.* (November 9, 1861) *1*: 3.

[Anon.], "Long-Sea Telegraphs," *All the Year Round* (March 15, 1862) *7*: 9–12 and (March 22, 1862) *7*: 39–45.

[Anon.], "The Indo-European Telegraph," *Illustrated London News* (July 8, 1865) *47*: 21–22.

[Anon.], *The Atlantic Telegraph: Report of the Proceedings at a Banquet, given to Mr. Cyrus W. Field, by the Chamber of Commerce of New-York, at the Metropolitan Hotel, November 15th, 1866* (New York: privately printed, 1866).

[Anon.], *Anglo-American Telegraph Company, Limited, First Ordinary General Meeting of Shareholders, Monday, February the 4th, 1867* (London: Metchin and Son, 1867).

[Anon.], "Effect of Cold on Insulation," *Journal of the Telegraph* (April 1, 1868) *1*: (9) 4.

[Anon.], "What Is an 'Ohm'?," *Journal of the Telegraph* (April 15, 1868) *1*: (10) 4.

[Anon.], "Telegraph Circuits," *Mechanics' Magazine* (April 14, 1871) *25*: 248–49.

[Anon.], "Notices of Books: *A Treatise on Electricity and Magnetism*. By James Clerk Maxwell," *Telegraphic Journal* (May 15, 1873) *1*: 145–46.

[Anon.], "Charles Hockin" (obituary), *Elec.* (May 6, 1882) *8*: 409–10.

[Anon.], "Fleeming Jenkin" (obituary), *Proc. ICE* (1885) *82*: 365–77.

[Anon.], "Charles Victor De Sauty" (obituary), *Elec.* (April 14, 1893) *30*: 685.

[Anon.], "Mr. Latimer Clark" (obituary), *Science* (November 18, 1898) *8*: 704–5.

[Anon.], "W. F. King" (obituary), *Proceedings of the Royal Society of Edinburgh* (1929) *49*: 386–87.

Ayrton, W. E., "Kelvin in the Sixties," *Times* (January 8, 1908), 3.

Bettany, G. T., "Practical Science at Cambridge," *Nature* (December 17, 1874), *11*: 132–33.

Blundell, J. Wagstaff, *Telegraph Companies Considered as Investments; with remarks on the Superior Advantages of Submarine Cables* (London: Effingham Wilson, 1869).

The Manual of Submarine Telegraph Companies (London: Rixon and Arnold, 1871; 2nd ed., "published by the author," January 1872).

Brett, John W., *On the Origin and Progress of the Oceanic Electric Telegraph* (London: W. S. Johnson, 1858).

"The Atlantic Telegraph" (letter), *Morning Post* (September 23, 1858), 2.

"Atlantic Telegraph" (letter), *Engineer* (October 8, 1858), 6: 267.

Brewster, David, "The Atlantic Telegraph," *North British Review* (November 1858) 29: 519–55.

Briggs, Charles F. and Augustus Maverick, *The Story of the Telegraph, and a History of the Great Atlantic Cable* (New York: Rudd & Carleton, 1858).

Bright, Sir Charles T., "Inaugural Address of the New President," *Journal of the Society of Telegraph-Engineers and Electricians* (January 1887), 16: 7–40.

Brine, Frederic, *Map of Valentia, Shewing the Positions of the Various Ships and Lines of Cable Connected with the Atlantic Telegraph* (London: Edward Stanford, 1859).

Christie, Samuel Hunter, "Experimental Determination of the Laws of Magneto-Electric Induction in Different Masses of the Same Metal, and of Its Intensity in Different Metals" *Phil. Trans.* (1833) 123: 95–142.

Chrystal, George, "Clerk Maxwell's 'Electricity and Magnetism,'" *Nature* (January 12, 1882) 25: 237–40.

Chrystal, George and S. A. Saunder, "Results of a Comparison of the British-Association Units of Electrical Resistance," *BA Report* (1876), 13–19.

Clark, Latimer, *Experimental Investigation of the Laws which Govern the Propagation of the Electric Current in Long Submarine Telegraph Cables* (London: George E. Eyre and William Spottiswoode, 1861).

(letter), *Elec.* (January 17, 1862) 1: 129.

An Elementary Treatise on Electrical Measurement for the Use of Telegraph Inspectors and Operators (London: E. and F. N. Spon, 1868).

Clark, Latimer, and Sir Charles Bright, "On the Principles Which Should be Observed in the Formation of Standards of Measurement of Electrical Quantities and Resistance," in Clark, *Experimental Investigation of the Laws which Govern the Propagation of the Electric Current in Long Submarine Telegraph Cables* (London: George E. Eyre and William Spottiswoode, 1861) 49–50; repr. as "Measurement of Electrical Quantities and Resistance" in *Elec.* (November 9, 1861) 1: 3–4.

Clark, Latimer, and Robert Sabine, *Electrical Tables and Formulae for the use of Telegraph Inspectors and Operators* (London: E. and F. N. Spon, 1871).

Clauson-Thue, William, *The ABC Universal Commercial Electric Telegraphic Code*, 5th ed. (London: Eden Fisher and Co., 1901).

Deane, John C., "Diary of the Atlantic Telegraph Expedition, 1866," *Daily News* (August 18, 1866).

Everett, J. D., "First Report of the Committee for the Selection and Nomenclature of Dynamical and Electrical Units," *BA Report* (1873), 222–25.

Faraday, Michael, "A Speculation Touching Electric Conduction and the Nature of Matter," *Phil. Mag.* (February 1844) 24: 136–44.

"On the Use of Gutta Percha in Electrical Insulation," *Phil. Mag.* (March 1848) 32: 165–67.

"On Electric Induction—Associated cases of Current and Static Effects," *Proceedings of the Royal Institution* (1854)*1*: 345–55, and in *Phil. Mag.* (March 1854) *7*: 197–208.

"On Subterraneous Electro-telegraph Wires," *Phil. Mag.* (June 1854) *7*: 396–98.

"Experimental Researches in Electricity, Eleventh Series," *Phil. Trans.* (1838) *128*: 1–40.

Experimental Researches in Electricity, 3 vols. (London: Taylor/Quaritch, 1839–55).

Field, Cyrus W., *Prospects of the Atlantic Telegraph. A Paper Read Before the American Geographical and Statistical Society, at Clinton Hall, New York, May 1, 1862* (New York: privately printed, 1862).

Report of Mr. Cyrus W. Field to the President and Directors of the New York, Newfoundland, and London Telegraph Company, London, March 8th, 1866 (London: William Brown and Co., 1866)

FitzGerald, G. F., "Heaviside's *Electrical Papers*," *Elec.* (August 11, 1893) *31*: 389–90, repr. in Joseph Larmor, ed., *Scientific Writings of the Late G. F. FitzGerald* (Dublin: Hodges & Figgis, 1902), 292–300.

Fizeau, H., "Sur une expérience relative à la vitesse de propagation de la lumière," *Comptes Rendus* (1849) *29*: 90–92.

Föppl, August, *Einfuhrung in die Maxwell'sche Theorie der Elektricität* (Leipzig: Teubner, 1894).

Forde, H. C., "The Malta and Alexandria Submarine Telegraph Cable," *Proc. ICE* (May 1862) *21*: 493–514; discussion, 531–40.

Galbraith, Joseph A., and Samuel Haughton, *Manual of Astronomy* (London: Longmans, 1855).

[Galton, Douglas], "Ocean Telegraphy," *Edinburgh Review* (January 1861) *113*: 113–43.

[Gisborne, Francis], "Telegraphic Communication with India," *Cambridge Essays* (London: J. W. Parker, 1857), 106–24.

Glaisher, James, "On the Meteorology of England, during the Quarter Ended June 30th, 1857," *Journal of the Statistical Society of London* (December 1857) *20*: 450–51.

Goldsmid, Frederic John, *Telegraph and Travel: A Narrative of the Formation and Development of Telegraphic Communication between England and India* (London: Macmillan, 1874).

Gooch, Daniel, *Diaries of Sir Daniel Gooch* (London: Kegan Paul, Trench, Trübner, & Co., 1892).

Guthrie, Frederick, "Teaching Physics," *Journal of the Society of Arts* (May 1886) *34*: 659–63.

Hearder, J. N., "Description of a Magnetometer and Appendages," *Annual Report of the Cornwall Polytechnic Society*, (1844) *12*: 98–100.

"On the Atlantic Cable," *Phil. Mag.* (January 1859) *17*: 27–42.

"The Atlantic Cable" (letter), *Engineer* (April 1, 1859) *7*: 224.

Heaton, J. Henniker, "The World's Cables and the Cable Rings," *Financial Review of Reviews* (May 1908) *5*: 5–26.

"Penny-a-Word Telegrams for Europe," *Financial Review of Reviews* (June 1908) *5*: 40–43.

[Heaviside, Oliver], "Comparing Electromotive Forces," *English Mechanic* (July 5, 1872) *15*: 411, signed "O.," repr. in Heaviside, *EP 1*: 1.

Heaviside, Oliver, "Voltaic Constants," *Telegraphic Journal* (May 15, 1873) *1*: 146, repr. in Heaviside, *EP 1*: 2–3.

"On the Extra Current," *Phil. Mag.* (August 1876) *2*: 135–45, repr. in Heaviside, *EP 1*: 53–61.

"On the Speed of Signalling through Heterogeneous Telegraph Circuits," *Phil. Mag.* (March 1877) *3*: 211–21, repr. in Heaviside, *EP 1*: 61–70.

"On the Theory of Faults in Cables," *Phil. Mag.* (July 1879) *8*: 60–74 and (August 1879) *8*: 163–77, repr. in Heaviside, *EP 1*: 71–95.

"Contributions to the Theory of the Propagation of Currents in Wires," written in 1882, first published in *EP 1*: 141–79.

"Dimensions of a Magnetic Pole," *Elec.* (June 3, 1882) *9*: 63–64, repr. in Heaviside, *EP 1*: 179–81.

"The Relations Between Magnetic Force and Electric Current," *Elec.* (November 18–December 16, 1882), repr. in Heaviside, *EP 1*: 195–231.

"The Energy of the Electric Current," *Elec.* (January 20–March 23, 1883), repr. in Heaviside, *EP 1*: 231–55.

"The Induction of Currents in Cores," *Elec.* (May 3, 1884–January 3, 1885), repr. in Heaviside, *EP 1*: 353–416.

"Transmission of Energy into a Conducting Core," *Elec.* (June 21, 1884) *13*: 133–34, repr. in Heaviside, *EP 1*: 377–78.

"On the Transmission of Energy through Wires by the Electric Current," *Elec.* (January 10, 1885) *14*: 178–80, repr. in Heaviside, *EP 1*: 434–41.

"Electromagnetic Induction and Its Propagation," *Elec.* (January 3, 1885–December 30, 1887), repr. in Heaviside, *EP 1*: 429–560 and *EP 2*: 39–155.

"Remarks on the Volta-Force," *Journal of the Society of Telegraph-Engineers and Electricians* (April 1885) *15*: 269–96, repr. in Heaviside, *EP 1*: 416–28.

"On the Self-Induction of Wires," *Phil. Mag.* (August 1886) *22*: 118–38, repr. in Heaviside, *EP 2*: 168–85.

"Preliminary to Investigations Concerning Long-distance Telephony and Connected Matters," *Elec.* (June 3, 1887) *19*: 79–81, repr. in Heaviside, *EP 2*: 119–24.

"On Electromagnetic Waves, Especially in Relation to the Vorticity of the Impressed Forces; and the Forced Vibrations of Electromagnetic Systems," *Phil. Mag.* (February 1888) *25*: 130–56, repr. in Heaviside, *EP 2*: 375–96.

"The Beneficial Effect of Leakage in Submarine Cables," *Elec.* (September 15, 1893) *31*: 518–19, repr. in Heaviside, *EMT 1*: 417–20.

"Short History of Leakage Effects on a Cable Circuit," *Elec.* (October 6, 1893) *31*: 603–4, repr. in Heaviside, *EMT 1*: 420–24.

"Reviews" [review of Charles Emerson Curry, *Theory of Electricity and Magnetism* (London: Macmillan, 1897)], *Elec.* (September 10, 1897) *39*: 643–44, repr. in Heaviside, *EMT 3*: 503–7.

Electrical Papers, 2 vols. (London: Macmillan, 1892).

Electromagnetic Theory, 3 vols. (London: Electrician Co., 1893–1912).

Holmes, N. J., "Submarine Telegraphs," *Nature* (May 4, 1871) *4*: 8–10.

Howe, Henry, *Adventures and Achievements of Americans* (New York: Geo. F. Tuttle, 1858).

Hurd, Percy A., "Our Telegraphic Isolation," *Contemporary Review* (June 1896) *69*: 899–908.

Jenkin, Fleeming, "On the Insulating Properties of Gutta-Percha," *Joint Committee Report*, 464–81.

"On the Insulating Properties of Gutta Percha," *Proc. RS* (March 1860) *10*: 409–15.

"Provisional Report of the Committee appointed by the British Association on Standards of Electrical Resistance," *BA Report* (1862), 125–35, repr. in Smith, *Reports*, 1–16.

"Experimental Researches on the Transmission of Electric Signals through Submarine Cables.—Part I. Laws of Transmission through various lengths of one Cable," *Phil. Trans.* (1862), *152*: 987–1017.

"Report on Electrical Instruments," *Reports by the Juries, International Exhibition, 1862* (London: Society of Arts, 1863), 44–98.

"Report of the Committee on Standards of Electrical Resistance," *BA Report* (1864), 345–49, repr. in Smith, *Reports*, 159–66.

"Electrical Standard" (letter, dated February 7, 1865), *Phil. Mag.* (March 1865) *29*: 248.

"Report of the Committee on Standards of Electrical Resistance," *BA Report* (1865), repr. in Smith, *Reports*, 190–95,

"Report on the New Unit of Electrical Resistance Proposed and Issued by the Committee on Electrical Standards Appointed in 1861 by the British Association," *Proc. RS* (April 1865) *14*: 154–64, repr. in *Phil. Mag.* (June 1865) *29*: 477–86, and in Smith, *Reports*, 277–90.

"Reply to Dr. Werner Siemens's Paper 'On the Question of the Unit of Electrical Resistance,'" *Phil. Mag.* (September 1866) *32*: 161–77.

"The Atomic Theory of Lucretius," *North British Review* (March 1868) *48*: 211–42, repr. as "Lucretius and the Atomic Theory" in Sidney Colvin and J. A. Ewing, eds., *Papers Literary, Scientific, &c. by the late Fleeming Jenkin, F. R. S.*, 2 vols. (London: Longmans, 1887), *1*: 177–214.

Electricity and Magnetism (London: Longmans, Green, 1873).

Johnson, George, ed., *The All Red Line: The Annals and Aims of the Pacific Cable Project* (Ottawa: James Hope & Sons, 1903).

King, W. F., "Description of Sir Wm. Thomson's Experiments made for the Determination of v, the Number of Electrostatic Units in the Electromagnetic Unit," *BA Report* (1869), 434–36.

Lindsay, J. B., "On some Experiments upon a Telegraph for communicating across Rivers and Seas, without the employment of a submerged Cable," *BA Report* (1854), part 2, 157.

Lodge, Oliver J., "On the Paths of Electric Energy in Voltaic Circuits," *Phil. Mag.* (June 1885) *19*: 487–94.

"On the Seat of Electromotive Force in the Voltaic Cell," *Phil. Mag.* (May 1885) *19*: 340–65.

"Modern Views of Electricity," *Nature* (January 31, 1889) *39*: 319–22.

Modern Views of Electricity (London: Macmillan, 1889; 1893; 1907).

Longridge, J. A., and C. H. Brooks, "On Submerging Telegraphic Cables," *Proc ICE* (February 1858) *17*: 221–61; discussion, 298–366.

[Mann, R. J.], *The Atlantic Telegraph: A History of Preliminary Experimental Proceedings, and a Descriptive Account of the Present State & Prospects of the Undertaking, Published by Order of the Directors of the Company* (London: Jarrold and Sons, 1857).

"De la Rive on Electrical Science," *Edinburgh Review* (July 1857) *106*: 26–62.

"The Atlantic Telegraph," *Chambers's Journal* (June 27, 1857) *7*: 401–4.

Marshman, John, "The Red Sea Telegraph" (letter), *Times* (September 3, 1857), 8.

Telegraph to India, Extracts from Correspondence Relative to the Establishment of Telegraphic Communication between Great Britain and India, China and the Colonies (London: J. E. Adlard, 1858).

Matthiessen, Augustus and C. Vogt, "On the Influence of Traces of Foreign Metals on the Electrical Conducting Power of Mercury," *Phil. Mag.* (March 1862) *23*: 171–79.

Maury, M. F., *Explanations and Sailing Directions to Accompany the Wind and Current Charts*, 2 vols. (Washington, DC: William A. Harris, 1858).

Maxwell, James Clerk, "On Faraday's Lines of Force" [read December 10, 1855 and February 11, 1856], *Transactions of the Cambridge Philosophical Society* (1864) *10*: 27–83, repr. in Maxwell, *SP 1*: 155–229.

"On Physical Lines of Force," *Phil. Mag.* (March 1861) *21*: 161–75, (April 1861) *21*: 281–91, (May 1861) *21*: 338–48, (January 1862) *23*: 12–24, and (February 1862) *23*: 85–95, repr. in Maxwell, *SP 1*: 451–513.

"A Dynamical Theory of the Electromagnetic Field," *Phil. Trans.* (1865) *155*: 459–512, repr. in Maxwell, *SP 1*: 526–97.

"On Governors," *Proc. RS* (March 1868) *16*: 270–83, repr. in Maxwell, *SP 2*: 105–20.

"On a Method of Making a Direct Comparison of Electrostatic and Electromagnetic Force; with a Note on the Electromagnetic Theory of Light," *Phil. Trans.* (1868) *158*: 643–57, repr. in Maxwell, *SP 2*: 125–43.

"Introductory Lecture on Experimental Physics" [October 25, 1871], in Maxwell, *SP 2*: 241–55.

Treatise on Electricity and Magnetism, 2 vols. (Oxford: Clarendon Press, 1873; 1881; 1892).

"Longman's Text-Books of Science" [review of Fleeming Jenkin, *Electricity and Magnetism*], *Nature* (May 15, 1873) *8*: 42–43, repr. in Maxwell, *SLP 2*: 842–44.

"Report on the Cavendish Laboratory for 1874," *Cambridge University Reporter* (April 27, 1875), 352–54, repr. in Maxwell, *SLP 3*: 208–15.

"On Dimensions," *Encyclopedia Britannica*, 9th ed. (Edinburgh: Adam and Charles Black, 1875–89) *7*: 240–41, repr. in Maxwell, *SLP 3*: 517–20.

ed., *The Electrical Researches of the Honourable Henry Cavendish, F.R.S.* (Cambridge: Cambridge University Press, 1879).

Scientific Papers of James Clerk Maxwell, ed. W. D. Niven, 2 vols. (Cambridge: Cambridge University Press, 1890).

Scientific Letters and Papers of James Clerk Maxwell, ed. P. M. Harman, 3 vols. (Cambridge: Cambridge University Press, 1990–2002).

Maxwell, James Clerk and Fleeming Jenkin, "On the Elementary Relations between Electrical Measurements," *BA Report* (1863), 130–63, repr. in *Report of the Committee appointed by the British Association on Standards of Electrical Resistance* (London: Taylor and Francis, 1864); repr. with minor revisions in *Phil. Mag.* (June 1865) *29*: 436–60 and (supp. 1865) *29*: 507–25; repr. with more extensive changes in Fleeming Jenkin, ed., *Reports of the Committee on Electrical Standards appointed by the British Association for the Advancement of Science* (London: Spon, 1873), 59–96, and in Smith, *Reports*, 86–140.

Maxwell, James Clerk, Balfour Stewart, and Fleeming Jenkin, "Description of an Experimental Measurement of Electrical Resistance, made at King's College," Appendix D to [Thomson], "Report, 1863," in *BA Report* (1863), 163–76, repr. in Smith, *Reports*, 140–58.

Maxwell, James Clerk, Fleeming Jenkin, and Charles Hockin, "Description of a further Experimental measurement of Electrical Resistance made at King's College," Appendix A to [Jenkin], "Report, 1864," in *BA Report* (1864), 350–51, repr. in Smith, *Reports*, 166–67.

Miller, R. Kalley, "The Proposed Chair of Natural Philosophy," *Cambridge University Reporter* (November 23, 1870) *1*: 118–19.

"Momus," (letter), *Elec.* (January 17, 1862) *1*:128–29.

Mullaly, John, *The Laying of the Cable, or The Ocean Telegraph* (New York: Appleton, 1858).

Noad, Henry M., *The Student's Text-Book of Electricity* (London: Lockwood, 1867).

"P. Q.," (letter), *Elec.* (January 24, 1862) *1*: 143–44.

Parkinson, J. C., *The Ocean Telegraph to India: A Narrative and a Diary* (London: William Blackwood and Sons, 1870).

Peel, George, "Nerves of Empire," in *The Empire and the Century* (London: John Murray, 1905), 249–87.

Peel, C. L., "Submarine Cables" (letter), *Times* (February 29, 1860), 9.

Poynting, J. H., "On the Transfer of Energy in the Electromagnetic Field" (abstract), *Proc. RS* (January 1884) *36*: 186–87.

"On the Transfer of Energy in the Electromagnetic Field," *Phil. Trans.* (1884) *175*: 343–61.

[Preece, W. H.], "The Atlantic Telegraph" (letters), *Engineer* (November 5, 1858) *6*: 355, (November 19, 1858), *6*: 391–92, and (December 3, 1858), *6*: 431, all signed "A Telegraph Engineer and Practical Electrician."

"Inaugural Address," *Journal of the Instititution of Electrical Engineers* (January 1893) *22*: 36–68.

Prescott, G. B., *History, Theory, and Practice of the Electric Telegraph* (Boston: Ticknor and Fields, 1863; 3rd ed., 1866)

Pullen, W. J. J., "The Red Sea Telegraph Line," *Times* (May 21, 1858), 12.

Rayleigh, Lord [John William Strutt], and Arthur Schuster, "On the Determination of the Ohm in Absolute Measure," *Proc. RS* (May 1881) *32*: 104–41.

Rosa, E. B., and N. E. Dorsey, "A Comparison of the Various Methods of Determining the Ratio of the Electromagnetic to the Electrostatic Unit of Electricity," *Bulletin of the Bureau of Standards* (1907) *3*: 605–22.

[Russell, James Burn], "Paying-Out the Atlantic Cable," *Sydney Morning Herald* (February 8, 1859), 5, excerpt in Thompson, *Kelvin, 1*: 360–64.

"Atlantic Cable: Leaves from the Journal of an Amateur Telegrapher," *West of Scotland Magazine and Review* (1859).

Russell, W. H., "The Laying of the Atlantic Cable. Diary of Events," *Daily News* (August 19, 1865).

The Atlantic Telegraph (London: Day and Son, 1865).

Rutter, John O. N., *Human Electricity: The Means of Its Development, Illustrated by Experiments* (London: John W. Parker and Son, 1854).

Saward, George, "The Atlantic Cable" (letter), *Times* (September 6, 1858), 7.

"The Atlantic Telegraph" (letter), *Times* (September 15, 1858), 7.

"The Atlantic Telegraph" (letter), *Morning Chronicle* (September 22, 1858), repr. in *Times* (September 24, 1858), 7.

The Trans-Atlantic Submarine Telegraph: A Brief Narrative of the Principal Incidents in the History of the Atlantic Telegraph Company (London: privately printed, 1878).

Shaffner, Tal. P., *The Telegraph Manual* (New York: Pudney and Russell, 1859).

Shanks, W. F. G., "How We Get Our News," *Harper's Magazine* (March 1867) *34*: 511–22.

Siemens, Charles William, "On the Electrical Tests Employed during the Construction of the Malta and Alexandria Telegraph, and on Insulating and Protecting Submarine Cables," *Proc. ICE* (May 1862), *21*: 515–30; discussion, *21*: 531–40.

Siemens, Werner, "Sur la telegraphie electrique," *Comptes Rendus* (1850) *30*: 434–37.

"Ueber die electrostatische Induction und die Verzögerung der Stroms in Flaschendrähten," *Annalen der Physik* (September 1857) *102*: 66–122, trans. as "On Electrostatic Induction and Retardation of the Current in Cores" in Werner Siemens, *Scientific and Technical Papers, 1*: 87–135.

"Vorschlag eines reproducirbaren Widerstandsmaasses," *Annalen der Physik* (June 1860) *110*: 1–20, trans. as "Proposal for a new reproducible Standard Measure of Resistance to Galvanic Currents" in *Phil. Mag.* (January 1861) *21*: 25–38 and as "Proposal for a Reproducible Unit of Electrical Resistance" in Werner Siemens, *Scientific and Technical Papers, 1*: 162–80.

"To the Committee appointed by the British Association to report on Standards of Electrical Resistance," *BA Report* (1862), 152–55, repr. in Smith, *Reports*, 39–44.

Scientific and Technical Papers of Werner von Siemens, 2 vols. (London: Murray, 1892–95).

Inventor and Entrepreneur: Recollections of Werner von Siemens (London: Lund Humphries, 1966).

Siemens, Werner and C. W. Siemens, "Outline of the Principles and Practice involved in dealing with the Electrical Conditions of Submarine Electric Telegraphs," *BA Report* (1860), 32–34.

"Outline of the Principles and Practice involved in Testing the Electrical Condition of Submarine Telegraph Cables," in Werner Siemens, *Scientific and Technical Papers, 1:* 137–59.

Smith, F. E., ed., *Reports of the Committee on Electrical Standards Appointed by the British Association for the Advancement of Science* (Cambridge: Cambridge University Press, 1913).

Smith, Willoughby, "Diary of the Atlantic Telegraph Expedition, 1866," *Daily News* (August 18, 1866).

The Rise and Extension of Submarine Telegraphy (London: J. S. Virtue, 1891).

Strutt, John William [Baron Rayleigh], *Scientific Papers,* 6 vols. (Cambridge: Cambridge University Press, 1899–1920).

Tait, P. G., *Sketch of Thermodynamics* (Edinburgh: Edmonston and Douglas, 1868; 1877).

Thomson, William, "On the Uniform Motion of Heat in Homogeneous Solid Bodies, and Its Connection with the Mathematical Theory of Electricity," *Cambridge Mathematical Journal* (February 1842) *3:* 71–84, repr. in *Phil. Mag.* (supp. 1854) *7:* 502–15 and in Thomson, *Reprint of Papers on Electrostatics and Magnetism* (London: Macmillan, 1872), 1–14.

"On the Mathematical Theory of Electricity in Equilibrium," *Cambridge and Dublin Mathematical Journal* (November 1845) *1:* 75–95, repr. with additions in *Phil. Mag.* (July 1854) *8:* 42–62, and in Thomson, *Reprint of Papers on Electrostatics and Magnetism* (London: Macmillan, 1872), 15–37.

"Applications of the Principle of Mechanical Effect to the Measurement of Electro-motive Forces, and of Galvanic Resistances, in Absolute Units," *Phil. Mag.* (December 1851) *2:* 551–62, repr. with additions in Thomson, *MPP 1:* 490–502.

"On the Theory of the Electric Telegraph," *Proc. RS* (May 1855) *7:* 382–99, repr. in *Phil. Mag.* (February 1856) *11:* 146–60 and in Thomson, *MPP 2:* 61–76.

"On the Electro-statical Capacity of a Leyden Phial and of a Telegraph Wire Insulated in the Axis of a Cylindrical Conducting Sheath," *Phil. Mag.* (supp. 1855) *9:* 531–35, repr. in Thomson, *Reprint of Papers on Electrostatics and Magnetism* (London: Macmillan, 1872), 38–41.

"On Peristaltic Induction of Electric Currents in Submarine Telegraph Wires," *BA Report* (1855), Section A, 21–22, repr. in Thomson, *MPP 2:* 77–78.

"Telegraphs to America" (letter), *Athenæum* (November 1, 1856), 1338–39, repr. in Thomson, *MPP 2:* 94–102.

"On Practical Methods for Rapid Signalling by the Electric Telegraph," *Proc. RS* (November 1856) *8:* 299–303 and (December 1856) *8:* 303–307, repr. in Thomson, *MPP 2:* 103.

"On Mr. Whitehouse's Relay and Induction Coils in action on Short Circuit," *BA Report* (1857), 21.

"On the Electro-dynamic Qualities of Metals," *Phil. Trans.* (1856) *146:* 649–751, repr. with additions in Thomson, *MPP 2:* 189–407.

"On the Electrical Conductivity of Commercial Copper of Various Kinds," *Proc. RS* (June 1857) *8:* 550–55, repr. in Thomson, *MPP 2:* 112–17.

"The Atlantic Cable. Professor Thomson in Reply to Mr. Whitehouse" (letter), *Morning Chronicle* (October 18, 1858).

"Analytical and Synthetical Attempts to Ascertain the Cause of the Differences of Electrical Conductivity Discovered in Wires of Nearly Pure Copper," *Proc. RS* (February 1860), *10*: 300–9, repr. in Thomson, *MPP 2*: 118–28.

"Telegraph, Electric," *Encyclopedia Britannica*, 8th ed. (Edinburgh: Adam and Charles Black, 1860), *21*: 94–116.

"On the Measurement of Electric Resistance," *Proc. RS* (June 1861) *11*: 313–28.

"Report of the Committee appointed by the British Association on Standards of Electrical Resistance," *BA Report* (1863), 111–24, repr. in Smith, *Reports*, 58–78.

"On the Forces Concerned in the Laying and Lifting of Deep-Sea Cables," *Proceedings of the Royal Society of Edinburgh* (December 1865) *5*: 495–509, repr. in Thomson, *MPP 2*: 153–67.

"Presidential Address," *BA Report* (1871), lxxxiv–cv, repr. in Thomson, *Popular Lectures and Addresses*, 3 vols. (London: Macmillan, 1889–1894) *2*: 132–205.

Reprint of Papers on Electrostatics and Magnetism (London: Macmillan, 1872).

"Electrical Units of Measurement," in *Practical Applications of Electricity. A Series of Lectures Delivered at the Institution of Civil Engineers* (London: Institution of Civil Engineers, 1884), 149–74, repr. in Thomson, *Popular Lectures and Addresses* (London: Macmillan, 1889–1894) *1:* 73–136.

"Note on the Contributions of Fleeming Jenkin to Electrical and Engineering Science," in Sidney Colvin and J. A. Ewing, eds., *Papers Literary, Scientific, &c. by the late Fleeming Jenkin, F. R. S.*, 2 vols. (London: Longmans, 1887), *1:* clv–clix.

"Ether, Electricity, and Ponderable Matter," *Journal of the Institution of Electrical Engineers* (January 1889) *18*: 4–35, repr. in Thomson, *MPP 3*: 484–515.

Mathematical and Physical Papers, 6 vols. (Cambridge: Cambridge University Press, 1882–1911).

Popular Lectures and Addresses, 3 vols. (London: Macmillan, 1889–1894).

Thomson, William and P. G. Tait, *Treatise on Natural Philosophy* (Oxford: Clarendon Press, 1867).

Thoreau, Henry David, *Walden; or, Life in the Woods* (Boston: Ticknor and Fields, 1854).

Varley, C. F., "On Improvements in Submarine and Subterranean Telegraph Communications," *BA Report* (1854), part 2, 17–18.

"Report on the State of the Atlantic Telegraph Cable," *Morning Chronicle* (September 22, 1858); also in *Times* (September 22, 1858), 7; *Daily News* (September 22, 1858); *Birmingham Daily Post* (September 23, 1858); *Lloyd's Weekly* (September 26, 1858); *Aberdeen Journal* (September 29, 1858); *Liverpool Mercury* (September 23, 1858); *Newcastle Courant* (September 24, 1858); and *Manchester Times* (September 25, 1858).

"On some of the Methods used for ascertaining the Locality and Nature of Defects in Telegraphic Conductors," *BA Report* (1859), 252–55.

Varley, S. A., "On the Electrical Qualifications Requisite in Long Submarine Telegraph Cables," *Proc. ICE* (March 1858) *17*: 368–85.

Webb, Frederick Charles, "On the Practical Operations Connected with Paying Out and Repairing Submarine Telegraph Cables," *Proc. ICE* (February 1858) *17*: 262–97; discussion, 298–366.

"Old Cable Stories Retold–VII," *Elec.* (June 5, 1885) *14*: 65–66 and (October 16, 1885) *15*: 428–29.

Weber, Wilhelm, "Messungen galvanischer Leitungswiderstände nach einem absoluten Maasse," *Annalen der Physik* (March 1851) *82*: 337–69, trans. as "On the Measurement of Electric Resistance," *Phil. Mag.* (September 1861) *22*: 226–40 and (October 1861) *22*: 261–69.

Wheatstone, Charles, "An Account of Several New Instruments and Processes for Determining the Constants of a Voltaic Circuit," *Phil. Trans.* (1843) *133*: 303–27.

Wheeler, J. A. and R. P. Feynman, "Classical Electrodynamics in Terms of Direct Interparticle Interaction," *Reviews of Modern Physics* (July 1949) *21*: 425–33.

Whitehouse, Wildman, *Report on a Series of Experimental Observations on two lengths of Submarine Electric Cable, containing, in the aggregate, 1,125 miles of wire, being the substance of a paper read before the British Association for the Advancement of Science, at Glasgow, Sept. 14th, 1855* (Brighton: privately printed, 1855).

"Experimental Observations on an Electric Cable," 1855 *BA Report*, Section A, 23–4.

"On the Construction and Use of an Instrument for determining the Value of Intermittent or Alternating Electric Currents for purposes of Practical Telegraphy," *BA Report* (1856), 19–21 and in *Engineer* (September 26, 1856) *2*: 523.

"The Law of Squares – Is It Applicable or Not to the Transmission of Signals in Submarine Circuits?," *BA Report* (1856), 21–3 and in *Athenæum* (August 30, 1856), 1092.

"The Atlantic Telegraph" (letter), *Athenæum* (October 11, 1856), 1247.

"Atlantic Telegraph" (letter), *Athenæum* (November 8, 1856), 1371.

"Experiments on the Retardation of Electric Signals, Observed in Submarine Conductors," *Engineer* (January 23, 1857) *3*: 62–3.

"The Atlantic Telegraph" (letter), *Times* (September 7, 1858), 7.

"The Atlantic Telegraph" (letter), *Times* (September 16, 1858), 7.

The Atlantic Telegraph: The Rise, Progress, and Development of Its Electrical Department (London: Bradbury & Evans, 1858); also in *Morning Chronicle* (September 20, 1858) and *Engineer* (September 24, 1858) *6*: 230–32.

"The Atlantic Cable. Mr. Whitehouse's Reply to the Statement of the Directors of the Atlantic Telegraph Company," *Morning Chronicle* (September 28, 1858); also in *Daily News* (September 29, 1858) and *Engineer* (October 15, 1858), *6*: 287–88.

"Professor Whitehouse and the Atlantic Telegraph," *Daily News* (September 29, 1858).

Reply to the Statement of the Directors of the Atlantic Telegraph Company (London: Bradbury & Evans, 1858).

"A Few Thoughts on the Size of Conductors for Submarine Circuits," *Engineer* (October 29, 1858), *6*: 329; also in *Mechanics' Magazine* (October 16, 1858) *69*:

366–67, (October 23, 1858) *69*: 395–96, and (October 30, 1858) *69*: 417–18. and in *Civil Engineer and Architect's Journal* (December 1858), *21*: 408–10.

"On some of the Difficulties in testing Submarine Cables," *Athenæum* (October 30, 1858), 553.

"Contributions on the Submarine Telegraph" (title only), *BA Report* (1858), 25.

"The Atlantic Telegraph" (letter), *Engineer* (November 26, 1858) 6: 410.

Recent Correspondence between Mr. Wildman Whitehouse and the Atlantic Telegraph Company (London: Bradbury & Evans, 1858).

Wilshaw, Sir Edward, "Speech to the Country Conference of the Chartered Institute of Secretaries," in *The Cable and Wireless Communications of the World: Some Lectures and Papers on the Subject, 1924–1939* (London: Cable & Wireless Ltd., 1939), 1–14.

Window, F. R., "On Submarine Electric Telegraphs," *Proc. ICE* (January 1857) *16*: 188–202; discussion, 203–25.

[Woods, Nicholas], "The Atlantic Telegraph Expedition," *Times* (July 15, 1858), 10, repr. in Bright, *Atlantic Cable*, 91–105.

Secondary Sources

Ahvenhainen, Jorma, *The Far Eastern Telegraphs: The History of Telegraphic Communications between the East, Europe and America Before the First World War* (Helsinki: Suomalainen Tiedeakatemia, 1981).

Aitken, Hugh G. J., *The Continuous Wave: Technology and American Radio, 1900–1932* (Princeton: Princeton University Press, 1985).

Appleyard, Rollo, "A Link with Oliver Heaviside," *Electrical Communication* (October 1931) *10*: 53–59.

The History of the Institution of Electrical Engineers, 1871–1931 (London: Institution of Electrical Engineers, 1939).

Arthur, John W., *Brilliant Lives: The Clerk Maxwells and the Scottish Enlightenment* (Edinburgh: John Donald, 2016).

Assis, André Koch Torres, *Weber's Electrodynamics* (Dordrecht: Kluwer Academic, 1994).

Ash, Stewart, *The Cable King: The Life of John Pender* (London: Stewart Ash, 2018).

Baark, Erik, *Lightning Wires: The Telegraph and China's Technological Modernization, 1860–1890* (Westport, CT: Greenwood Press, 1997).

Baker, E. C., *Sir William Preece, F.R.S., Victorian Engineer Extraordinary* (London: Hutchinson, 1976).

Barty-King, Hugh, *Girdle Round the Earth: The Story of Cable and Wireless* (London: Heinemann, 1979).

Beckert, Sven, *Empire of Cotton: A Global History* (New York: Vintage, 2014).

Bektas, Yakup, "The Sultan's Messenger: Cultural Constructions of Ottoman Telegraphy, 1847–1880," *Technology and Culture* (2000) *41*: 669–96.

Bence Jones, Henry, *Life and Letters of Faraday*, 2 vols. (London: Longmans, Green, and Co., 1870).

Bowers, Brian, *Sir Charles Wheatstone* (London: Her Majesty's Stationery Office, 1975).

Boyce, Robert W. D., "Imperial Dreams and the National Realities: Britain, Canada and the Struggle for a Pacific Telegraph Cable, 1879–1902," *English Historical Review* (2000) *115*: 39–70.

Bright, Charles, *Submarine Telegraphs: Their History, Construction, and Working* (London: Lockwood, 1898).

The Story of the Atlantic Cable (New York: Appleton, 1903).

The Life Story of Sir Charles Tilston Bright, revised and abridged edition (London: Archibald Constable, 1908).

"The Atlantic Cable of 1858," *Electrical World* (September 30, 1909) *54*: 788–91.

Bright, Edward Brailsford, and Charles Bright, *The Life Story of the Late Sir Charles Tilston Bright*, 2 vols. (London: Archibald Constable, 1899).

Brockway, Lucile, *Science and Colonial Expansion: The Role of the British Royal Botanic Garden* (New York: Academic Press, 1979).

Bruton, Elizabeth, "From Theory to Engineering Practice: Shared Telecommunications Knowledge between Oliver Heaviside and His Brother GPO Engineer Arthur West Heaviside," *Phil. Trans. A* (2018) *376*: 1–20.

Buchwald, Jed Z., *From Maxwell to Microphysics: Aspects of Electromagnetic Theory in the Last Quarter of the Nineteenth Century* (Chicago: University of Chicago Press, 1985).

The Creation of Scientific Effects: Heinrich Hertz and Electric Waves (Chicago: University of Chicago Press, 1994).

Burke, John G., "Bursting Boilers and the Federal Power," *Technology and Culture* (1966) *7*: 1–23.

Cahan, David, *An Institute for an Empire: The Physicalisch-Technische Reichsanstalt, 1871–1918* (Cambridge: Cambridge University Press, 1989).

Helmholtz: A Life in Science (Chicago: University of Chicago Press, 2018).

Campbell, Lewis and William Garnett, *The Life of James Clerk Maxwell* (London: Macmillan, 1882).

Cantor, Geoffrey, *Michael Faraday: Sandemanian and Scientist* (Basingstoke: Macmillan, 1991).

Cardwell, D. S. L., *From Watt to Clausius: The Rise of Thermodynamics in the Early Industrial Age* (Ithaca: Cornell University Press, 1971).

Cat, Jordi, *Maxwell, Sutton, and the Birth of Color Photography: A Binocular Study* (New York: Palgrave Macmillan, 2013).

Channell, David, "The Harmony of Theory and Practice: The Engineering Science of W. J. M. Rankine," *Technology and Culture* (1982) *23*: 39–52.

Clarke, T. N., A. D. Morrison-Low, and A. D. C. Simpson, *Brass and Glass: Scientific Instrument Making Workshops in Scotland* (Edinburgh: National Museums of Scotland, 1989).

Cochrane, Rexmond, *Measures for Progress: A History of the National Bureau of Standards* (Washington, DC: National Bureau of Standards, 1966).

Constable, Thomas, *Memoir of Lewis D. B. Gordon* (Edinburgh: privately printed, 1877).

Cookson, Gillian, *The Cable: The Wire That Changed the World* (Port Stroud: Tempus, 2003).

Cookson, Gillian and Colin A. Hempstead, *A Victorian Scientist and Engineer: Fleeming Jenkin and the Birth of Electrical Engineering* (Aldershot: Ashgate, 2000).

Crilly, A. J., "Hugh Blackburn," *ODNB*.

Cronon, William, *Nature's Metropolis: Chicago and the Great West* (New York: W. W. Norton, 1991).

Crowe, Michael J., *A History of Vector Analysis: The Evolution of the Idea of a Vectorial System*, 2nd ed. (New York: Dover, 1985).

Crowther, J. G., *The Cavendish Laboratory, 1874–1974* (New York: Science History Publications, 1974).

d'Agostino, Salvo, "Esperimento e teoria nell'opera di Maxwell: Le misure per le unita assolute elettromagnetiche e la velocita della luce" ("Experiment and Theory in Maxwell's Work: The Measurements for Absolute Electromagnetic Units and the Velocity of Light"), *Scientia* (1978) *113*: 453–80.

Darrigol, Olivier, *Electrodynamics from Ampère to Einstein* (Oxford: Oxford University Press, 2000).

Daston, Lorraine and Peter Galison, *Objectivity* (Cambridge, MA: MIT Press, 2007).

de Clark, Sybil G., "The Dimensions of the Magnetic Pole: A Controversy at the Heart of Early Dimensional Analysis," *Archive for History of Exact Sciences* (2016) *70*: 293–324.

de Cogan, Donard, "Dr. E. O. W. Whitehouse and the 1858 Trans-Atlantic Telegraph Cable," *History of Technology* (1985) *10*: 1–15.

They Talk Along the Deep: A Global History of the Valentia Island Telegraph Cables (Norwich: Dosanda Publications, 2016).

Dibner, Bern, *The Atlantic Cable* (Norwalk, CT: Burndy Library, 1959).

Du Boff, Richard B., "Business Demand and the Development of the Telegraph in the United States, 1844–1860," *Business History Review* (1980) *54*: 459–79.

"The Telegraph and the Structure of Markets in the United States, 1845–1890," *Research in Economic History* (1983) *8*: 253–77.

"The Telegraph in Nineteenth-Century America: Technology and Monopoly," *Comparative Studies in Society and History* (1984) *26*: 571–86.

Everitt, C. W. F., *James Clerk Maxwell: Physicist and Natural Philosopher* (New York: Charles Scribner's Sons, 1975).

Fairley, Rob, "Jemima Blackburn," *ODNB*.

Falconer, Isobel, "Cambridge and Building the Cavendish Laboratory," in Raymond Flood, Mark McCartney, and Andrew Whitaker, eds., *James Clerk Maxwell: Perspectives on His Life and Work* (Oxford: Oxford University Press, 2014), 67–98.

Feldenkirchen, Wilfried, *Werner Siemens: Inventor and International Entrepreneur* (Columbus: Ohio State University Press, 1994).

Field, Henry M., *The Story of the Atlantic Telegraph* (New York: Charles Scribner's Sons, 1893).

Field, Isabella Judson, *Cyrus W. Field, His Life and Work, 1819–1892* (New York: Harper & Brothers, 1896).

Figuier, Louis, *Les Merveilles de la Science ou Description Populaire des Inventions Modernes*, 6 vols. (Paris: Furne, Jouvet, et Cie., 1867–1870).

Finn, Bernard, "Submarine Telegraphy: A Study in Technological Stagnation," in Bernard Finn and Daqing Yang, eds., *Communications Under the Seas: The Evolving Cable Network and Its Implications* (Cambridge, MA: MIT Press, 2009), 9–24.

Fleming, Sir Ambrose, *Memories of a Scientific Life* (London: Marshall, Morgan and Scott, 1934).

Fowler, Helen, "Eleanor Mildred Sidgwick [*née* Balfour]," *ODNB*.

Gallagher, John and Ronald Robinson, "The Empire of Free Trade," *Economic History Review* (1953) *6*: 1–15.

Galison, Peter, *Image and Logic: A Material Culture of Microphysics* (Chicago: University of Chicago Press, 1997).

Garbade, Kenneth D., and William L. Silber, "Technology, Communication and the Performance of Financial Markets: 1840–1975," *Journal of Finance* (1978) *33*: 819–32.

Glazebrook, R. T., *James Clerk Maxwell and Modern Physics* (London: Cassell, 1901).

Godfrey, Helen, *Submarine Telegraphy and the Hunt for Gutta Percha: Challenge and Opportunity in a Global Trade* (Leiden: Brill, 2018).

Gooday, Graeme J. N., "Precision Measurement and the Genesis of Physics Teaching Laboratories in Victorian Britain," PhD dissertation, University of Kent at Canterbury, 1989.

"Precision Measurement and the Genesis of Physics Teaching Laboratories in Victorian Britain," *British Journal for the History of Science* (1990) *23*: 25–51.

The Morals of Measurement: Accuracy, Irony, and Trust in Late Victorian Electrical Practice (Cambridge: Cambridge University Press, 2004).

Gordon, John Steele, *A Thread Across the Ocean: The Heroic Story of the Transatlantic Cable* (New York: Walker and Co., 2002).

Green, Allan, "Dr. Wildman Whitehouse and His 'Iron Oscillograph'; Electrical Measurements Relating to the First Atlantic Cable," *International Journal for the History of Engineering and Technology* (2012) *82*: 68–92.

Hankins, Thomas L., *Sir William Rowan Hamilton* (Baltimore: Johns Hopkins University Press, 1980).

Harman, P. M., "Through the Looking-Glass, and What Maxwell Found There," in A. J. Kox and Daniel M. Siegel, eds., *No Truth Except in the Details: Essays in Honor of Martin J. Klein* (Dordrecht: Kluwer, 1995), 78–93.

The Natural Philosophy of James Clerk Maxwell (Cambridge: Cambridge University Press, 1998).

Harris, Christina Phelps, "The Persian Gulf Submarine Telegraph of 1864," *Geographical Journal* (1969) *135*: 169–90.

Harris, Robert Dalton, and Diane DeBlois, *An Atlantic Telegraph: The Transcendental Cable* (Cazenovia, NY: Ephemera Society of America, 1994).

Headrick, Daniel R., "Botany, Chemistry, and Tropical Development," *Journal of World History* (1996) *7*: 1–20.

"Gutta-Percha: A Case of Resource Depletion and International Rivalry," *IEEE Technology and Society Magazine* (December 1987) *6*: 12–16. ·

The Tentacles of Progress: Technology Transfer in the Age of Imperialism, 1850–1940 (Oxford: Oxford University Press, 1988).

The Invisible Weapon: Telecommunications and International Politics, 1851–1945 (Oxford: Oxford University Press, 1991).

Headrick, Daniel R. and Pascal Griset, "Submarine Telegraph Cables: Business and Politics, 1838–1939," *Business History Review* (2001) *75*: 543–78.

Hearder, Ian G., "Jonathan Nash Hearder," *ODNB*.

Hearn, Chester G., *Circuits in the Sea: The Men, the Ships, and the Atlantic Cable* (Westport, CT: Praeger, 2004).

Hempstead, C. A., "The Early Years of Oceanic Telegraphy: Technology, Science, and Politics," *Proceedings of the Institution of Electrical Engineers* (1989) *136A*: 297–305.

"An Appraisal of Fleeming Jenkin (1833–1885), Electrical Engineer," *History of Technology* (1991) *13*: 119–44.

Hendry, John, *James Clerk Maxwell and the Theory of the Electromagnetic Field* (Bristol: Adam Hilger, 1986).

Hevly, Bruce W., "The Heroic Science of Glacier Motion," *Osiris* (1996) *11*: 66–86.

Hilken, T. J. N., *Engineering at Cambridge University, 1783–1965* (Cambridge: Cambridge University Press, 1967).

Hon, Giora and Bernard R. Goldstein, *Reflections on the Practice of Physics: James Clerk Maxwell's Methodological Odyssey in Electromagnetism* (London: Routledge: 2020).

Hopley, I. B., "Maxwell's Work on Electrical Resistance: The Redetermination of the Absolute Unit of Resistance," *Annals of Science* (1957) *13*: 265–72.

"Maxwell's Determination of the Number of Electrostatic Units in One Electromagnetic Unit of Electricity," *Annals of Science* (1959) *15*: 91–108.

Houghton, Walter E., et al., eds., *Wellesley Index to Victorian Periodicals, 1824–1900*, 5 vols. (Toronto: University of Toronto Press, 1966–1989).

Howard, J. N., "Eleanor Mildred Sidgwick and the Rayleighs," *Applied Optics* (1964) *3*: 1120–22.

Howse, Derek, *Greenwich Time and the Discovery of the Longitude* (Oxford: Oxford University Press, 1980).

Hunt, Bruce J., "Michael Faraday, Cable Telegraphy, and the Rise of British Field Theory," *History of Technology* (1991) *13*: 1–19.

The Maxwellians (Ithaca: Cornell University Press, 1991).

"The Ohm is Where the Art Is: British Telegraph Engineers and the Development of Electrical Standards," *Osiris* (1994) *9*: 48–63.

"Scientists, Engineers, and Wildman Whitehouse: Measurement and Credibility in Early Cable Telegraphy," *British Journal for the History of Science* (1996) *29*:155–69.

"Insulation for an Empire: Gutta-Percha and the Development of Electrical Measurement in Victorian Britain," in Frank A. J. L. James, ed., *Semaphores to Short Waves* (London: Royal Society of Arts, 1998), 85–104.

"Lord Cable," *Europhysics News* (2004) *35*: 186–88.

Pursuing Power and Light: Technology and Physics from James Watt to Albert Einstein (Baltimore: Johns Hopkins University Press, 2010).

"Maxwell, Measurement, and the Modes of Electromagnetic Theory," *Historical Studies in the Natural Sciences* (2015) *45*: 303–39.

Jackson, Roland, "John Tyndall and the Early History of Diamagnetism," *Annals of Science* (2015) *72*: 435–89.

Jacobsen, Kurt, *The Story of GN: 150 Years of Technology, Big Business and Global Politics* (Copenhagen: Historika/Gads Forlag, 2019).

James, Frank A. J. L., ed., *Correspondence of Michael Faraday*, 6 vols. (London: IEE/IET, 1991–2011).

Jungnickel, Christa and Russell McCormmach, *Intellectual Mastery of Nature: Theoretical Physics from Ohm to Einstein*, 2 vols. (Chicago: University of Chicago Press, 1986).

Kargon, Robert, "Model and Analogy in Victorian Science: Maxwell's Critique of the French Physicists," *Journal of the History of Ideas* (1969) *30*: 423–36.

Kershaw, Michael, "'A Thorn in the Side of European Geodesy': Measuring Paris-Greenwich Longitude by Electric Telegraph," *British Journal for the History of Science* (2014) *47*: 637–60.

Kieve, Jeffrey, *The Electric Telegraph: A Social and Economic History* (Newton Abbot: David & Charles, 1973).

Kim, Dong-Won, *Leadership and Creativity: A History of the Cavendish Laboratory, 1871–1919* (Dordrecht: Kluwer Academic Publishers, 2002).

King, Agnes Gardner, *Kelvin the Man: A Biographical Sketch by his Niece* (London: Hodder and Stoughton, 1925).

Kirby, M. W., "Robert Stephenson," *ODNB*.

Kline, Ronald, "Science and Engineering Theory in the Invention and Development of the Induction Motor, 1880–1900," *Technology and Culture* (1987) *28*: 283–313.

Knudsen, Ole, "The Faraday Effect and Physical Theory, 1845–1873," *Archive for History of Exact Sciences* (1976) *15*: 235–81.

Kragh, Helge, *Ludvig Lorenz: A Nineteenth-Century Theoretical Physicist* (Copenhagen: Royal Danish Academy of Sciences and Letters, 2018).

Lagerstrom, Larry, "Putting the Electrical World in Order: The Construction of an International System of Electromagnetic Measures," PhD dissertation, University of California–Berkeley, 1992.

Latour, Bruno, *Science in Action: How to Follow Scientists and Engineers through Society* (Cambridge, MA: Harvard University Press, 1987).

Lawford, G. L. and L. R. Nicholson, *The Telcon Story, 1850–1950* (London: Telegraph Construction & Maintenance Co., 1950).

Layton, Edwin, "Mirror-Image Twins: The Communities of Science and Technology in 19th-Century America," *Technology and Culture* (1971) *12*: 562–89.

Lee, Jenny Rose, "Empire, Modernity, and Design: Visual Culture and Cable & Wireless' Corporate Identities, 1924–1955," PhD dissertation, University of Exeter, 2014.

Longair, Malcolm, "'. . . a paper. . . I hold to be great guns': A Commentary on Maxwell (1865) 'A Dynamical Theory of the Electromagnetic Field,'" *Phil. Trans. A* (2015) *373*: 20140473.

   Maxwell's Enduring Legacy: A Scientific History of the Cavendish Laboratory (Cambridge: Cambridge University Press, 2016).

Löser, Wolfgang, "Der Bau unterirdischer Telegraphenlinien in Preussen von 1844–1867," *NTM* (1969) *6*: 52–67.

Lynch, A. C.,"History of the Electrical Units and Early Standards," *Proceedings of the Institution of Electrical Engineers* (1985)*132A*: 564–73.

Mahon, Basil, *The Forgotten Genius of Oliver Heaviside, A Maverick of Electrical Science* (Amherst, NY: Prometheus Books, 2017).

Margerie, Maxime de, *Le réseau anglais de câbles sous-marins* (Paris: A. Pedone, 1909).

Martin, Thomas, ed., *Faraday's Diary*, 7 vols. (London: G. Bell and Sons, 1932–36).

Miller, Arthur I., *Albert Einstein's Special Theory of Relativity: Emergence (1905) and Early Interpretation (1905–1911)* (Reading, MA: Addison-Wesley, 1981).

Mitchell, Daniel Jon, "What's Nu? A Re-Examination of Maxwell's 'Ratio-of-Units' Argument, from the Mechanical Theory of the Electromagnetic Field to 'On the Elementary Relations between Electrical Measurements,'" *Studies in History and Philosophy of Science* (2017) *65–66*: 87–98.

Morus, Iwan R., "Telegraphy and the Technology of Display: The Electricians and Samuel Morse," *History of Technology* (1991) *13*: 20–40.

Müller, Simone M., *Wiring the World: The Social and Cultural Creation of Global Telegraph Networks* (New York: Columbia University Press, 2016).

Nahin, Paul J., *Oliver Heaviside: The Life, Work, and Times of an Electrical Genius of the Victorian Age* (Baltimore: Johns Hopkins University Press, 2002).

Nersessian, Nancy J., *Faraday to Einstein: Constructing Meaning in Scientific Theories* (Dordrecht: Kluwer, 1984).

Nickles, David Paull, *Under the Wire: How the Telegraph Changed Diplomacy* (Cambridge, MA: Harvard University Press, 2003).

Noakes, Richard, "Industrial Research at the Eastern Telegraph Company, 1872–1929," *British Journal for the History of Science* (2013) *47*: 119–46.

Physics and Psychics: The Occult and the Sciences in Modern Britain (Cambridge: Cambridge University Press, 2019).

O'Connell, Joseph, "Metrology: The Creation of Universality by the Circulation of Particulars," *Social Studies of Science* (1993) *23*: 129–73.

O'Hara, J. G. and W. Pricha, *Hertz and the Maxwellians* (London: Peter Peregrinus, 1987).

Olesko, Kathryn M., "Precision, Tolerance, and Consensus: Local Cultures in German and British Resistance Standards," in Jed Z. Buchwald, ed., *Scientific Credibility and Technical Standards* (Dordrecht: Kluwer Academic, 1996), 117–56.

O'Rourke, Kevin H. and Jeffrey G. Williamson, *Globablization and History: The Evolution of a Nineteenth-Century Atlantic Economy* (Cambridge, MA: MIT Press, 1999)

Osterhammel, Jürgen, *The Transformation of the World: A Global History of the Nineteenth Century* (Princeton: Princeton University Press, 2014).

Pancaldi, Giuliano, "The Web of Knowing, Doing, and Patenting: William Thomson's Apparatus Room and the History of Electricity," in Mario Biagioli and Jessica Riskin, eds., *Nature Engaged: Science in Practice from the Renaissance to the Present* (New York: Palgrave Macmillan, 2012), 263–85.

Peterson, Walter, "The Queen's Messenger: An Underwater Cable to Balaklava," *War Correspondent: The Journal of the Crimean War Research Society* (2008) 26: 31–39.

Pollard, A. F., "Josiah Latimer Clark," *ODNB*.

Prime, Samuel Irenaeus, *The Life of Samuel F. B. Morse* (New York: Appleton, 1875).

Pyatt, Edward C., *The National Physical Laboratory: A History* (Bristol: Adam Hilger, 1983).

Read, Donald, *The Power of News: The History of Reuters, 1849–1989* (Oxford: Oxford University Press, 1992).

Reid, John S., "Maxwell at King's College, London," in Raymond Flood, Mark McCartney, and Andrew Whitaker, eds., *James Clerk Maxwell: Perspectives on His Life and Work* (Oxford: Oxford University Press, 2014), 43–66.

Robertson, Edna, *Glasgow's Doctor: James Burn Russell, MOH, 1837–1904* (East Linton: Tuckwell, 1998).

Roche, John J., *The Mathematics of Measurement: A Critical History* (London: Athlone Press, 1998).

Rosenberg, Robert, "Academic Physics and the Origins of Electrical Engineering in America," PhD dissertation, Johns Hopkins University (1990).

Schaffer, Simon, "Late Victorian Metrology and Its Instrumentation: A Manufactory of Ohms," in Robert Bud and Susan Cozzens, eds., *Invisible Connections: Instruments, Institutions, and Science* (Bellingham, WA: SPIE Optical Engineering Press, 1992), 23–56.

"Rayleigh and the Establishment of Electrical Standards," *European Journal of Physics* (1994) 15: 277–85.

"Accurate Measurement is an English Science," in M. Norton Wise, ed., *The Values of Precision* (Princeton: Princeton University Press, 1995), 135–72.

"Physics Laboratories and the Victorian Country House," in Crosbie Smith and John Agar, eds. *Making Space for Science: Territorial Themes in the Shaping of Knowledge* (London: Macmillan, 1998), 149–80.

Scharf, J. Thomas, *History of Saint Louis City and County, From the Earliest Periods to the Present Day*, 2 vols. (Philadelphia: Louis H. Everts and Co., 1883).

Schuster, Arthur, "The Clerk-Maxwell Period," in J. J. Thomson et al., *A History of the Cavendish Laboratory, 1871–1910* (London: Longmans, Green, and Co., 1910), 14–39.

"John William Strutt, Baron Rayleigh, 1842–1919," *Proc. RS* (1921) 98: i–l.

Schwarzlose, Richard A., *The Nation's Newsbrokers. Volume 1: The Formative Years: From Pretelegraph to 1865* (Evanston, IL: Northwestern University Press, 1989).

Scott, J. D., *Siemens Brothers, 1858–1958: An Essay in the History of Industry* (London: Weidenfeld and Nicolson, 1958).

Scott, R. Bruce, *Gentlemen on Imperial Service: A Story of the Trans-Pacific Telecommunications Cable* (Victoria, BC: Sono Nis Press, 1994).

Searle, G. F. C., "Oliver Heaviside: A Personal Sketch," in *Heaviside Centenary Volume* (London: IEE Press, 1950), 93–96.

Siegel, Daniel M., "The Origin of the Displacement Current," *Historical Studies in the Physical and Biological Sciences* (1986) *17*: 99–146.

Innovation in Maxwell's Electromagnetic Theory: Molecular Vortices, Displacement Current, and Light (Cambridge: Cambridge University Press, 1991).

"Author's Response," *Metascience* (1993) *4*: 27–33.

Silverman, Kenneth, *Lightning Man: The Accursed Life of Samuel F. B. Morse* (New York: Alfred A. Knopf, 2003).

Simpson, Thomas K., *Maxwell on the Electromagnetic Field: A Guided Study* (New Brunswick, NJ: Rutgers University Press, 1997).

Figures of Thought: A Literary Appreciation of Maxwell's "Treatise on Electricity and Magnetism" (Santa Fe, NM: Green Lion Press, 2005).

Smith, Crosbie, "Engineering the Universe: William Thomson and Fleeming Jenkin on the Nature of Matter," *Annals of Science* (1980) *37*: 387–412.

"'Nowhere But in a Great Town': William Thomson's Spiral of Classroom Credibility," in Crosbie Smith and John Agar, eds., *Making Space for Science: Territorial Themes in the Shaping of Knowledge* (London: Macmillan, 1998), 118–146.

The Science of Energy: A Cultural History of Energy Physics in Victorian Britain (Chicago: University of Chicago Press, 1998).

Smith, Crosbie and M. Norton Wise, *Energy and Empire: A Biographical Study of Lord Kelvin* (Cambridge: Cambridge University Press, 1989).

Steinwender, Claudia, "Real Effects of Information Frictions: When the States and the Kingdom Became United," *American Economic Review* (2018) *108*: 657–96.

Stevenson, R. L., "Memoir of Fleeming Jenkin," in Sidney Colvin and J. A. Ewing, eds., *Papers Literary, Scientific, &c., by the late Fleeming Jenkin*, 2 vols. (London: Longmans, Green, and Co., 1887), *1*: xi–cliv.

Strutt, Robert John (Fourth Baron Rayleigh), *Life of John William Strutt, Third Baron Rayleigh*, 2nd ed. (Madison: University of Wisconsin Press, 1968).

Sviedrys, Romualdas, "The Rise of Physics Laboratories in Britain," *Historical Studies in the Physical Sciences* (1976) *7*: 405–36.

Tarrant, Donald, *Atlantic Sentinel: Newfoundland's Role in Transatlantic Cable Communications* (St. John's, NL: Flanker Press, 1999).

Thompson, F. M. L., "William Cavendish, Seventh Duke of Devonshire," *ODNB*.

Thompson, Robert Luther, *Wiring a Continent: The History of the Telegraph Industry in the United States, 1832–1866* (Princeton: Princeton University Press, 1947).

Thompson, Silvanus P., *Life of William Thomson, Baron Kelvin of Largs*, 2 vols. (London: Macmillan, 1910).

Thomson, G. P., *J. J. Thomson and the Cavendish Laboratory in His Day* (London: Thomas Nelson and Sons, 1964).

Thomson, J. J., et al., *A History of the Cavendish Laboratory, 1871–1910* (London: Longmans, Green, and Co., 1910).

Tully, John, "A Victorian Ecological Disaster: Imperialism, the Telegraph, and Gutta-Percha," *Journal of World History* (2009) *20*: 559–79.

Tunbridge, Paul, *Lord Kelvin: His Influence on Electrical Measurements and Units* (London: Peter Peregrinus, 1992).

Turner, Joseph, "Maxwell on the Logic of Dynamical Explanation," *Philosophy of Science* (1956) *23*: 36–47.

Vail, J. Cumming, *Early History of the Electro-Magnetic Telegraph, from letters and journals of Alfred Vail* (New York: Hine Brothers, 1914).

Vetch, R. H., and David Channell, "Sir Douglas Strutt Galton," *ODNB*.

Vincenti, Walter G., *What Engineers Know and How They Know It: Analytical Studies from Aeronautical History* (Baltimore: Johns Hopkins University Press, 1990).

Warwick, Andrew, *Masters of Theory: Cambridge and the Rise of Mathematical Physics* (Chicago: University of Chicago Press, 2003).

Williams, L. Pearce, *The Origins of Field Theory* (New York: Random House, 1966).

Wilson, David B., *Kelvin and Stokes: A Comparative Study in Victorian Physics* (Bristol: Adam Hilger, 1987).

Wilson, David B., ed., *Correspondence between Sir George Gabriel Stokes and Sir William Thomson, Baron Kelvin of Largs*, 2 vols. (Cambridge: Cambridge University Press, 1990).

Wise, M. Norton, "The Maxwell Literature and British Dynamical Theory," *Historical Studies in the Physical Sciences* (1982) *13*: 175–205.

"Mediating Machines," *Science in Context* (1988) *2*: 81–117.

Wise, M. Norton, and Crosbie Smith, "Measurement, Work and Industry in Lord Kelvin's Britain," *Historical Studies in the Physical and Biological Sciences* (1986) *17*: 147–73.

Woodman, Harold D., *King Cotton and His Retainers: Financing and Marketing the Cotton Crop of the South, 1800–1925* (Lexington: University of Kentucky Press, 1968).

Yavetz, Ido, *From Obscurity to Enigma: The Work of Oliver Heaviside, 1872–1889* (Boston: Birkhäuser, 1995).

Index

action-at-a-distance theories, 4, 18, 23, 25, 28, 30, 190, 199, 274
Adams, William Grylls, 245–47, 249
Aden, 113–16, 119, 236
Agamemnon, 66–69, 81–84, 87, 94, 110
Airy, George Biddell, 16–17, 19–22, 27, 30, 32–34, 226
Albert, Prince, 248
All the Year Round, 218–19
Allan, Thomas, 118, 119, 128
ampere, 3, 159, 213, 254
Ampère, A.-M., 18, 30, 264
Anderson, James, 142, 223, 226, 228, 263
Anglo-American Telegraph Company, 227–28, 232–33, 235–37, 239
Anglo-Mediterranean Telegraph Company, 176, 234
Annalen der Physik, 25, 164
Argyll, Duke of, 50
Associated Press, 6
Athenæum, 22, 52–56, 57, 77, 106, 107
Atlantic cable (1857–58)
　design, 58–62, 64, 128, 130
　failure, 37, 96, 98–99, 108, 125, 128, 140, 151, 153, 219
　laying expeditions, 37, 68–69, 81–83, 84
　manufacture, 59, 65, 76–77, 152
　origins of project, 40–46
　shore ends, 84, 87, 91, 96, 102, 218
　testing, 59, 66–67, 70, 77, 80, 99, 129, 130, 153
Atlantic cables (1865–66)
　design, 222, 227
　laying expeditions, 143, 174, 216, 223–30
　manufacture, 221–22
　shore ends, 223
　testing, 195, 223, 227
Atlantic Telegraph: A History of Preliminary Experimental Proceedings [R. J. Mann], 63–64, 73, 106

Atlantic Telegraph Company, 63, 70, 73, 85, 96, 98, 107, 173, 228
　board of directors, 37, 63, 68, 69, 81, 83, 86, 125, 218, 221
　and Thomson, 58, 76–77, 79, 151
　conflicts with Whitehouse, 80, 81, 87, 89, 91–95, 98, 99–100, 103, 104
　provisional board, 59–62, 65
　contract specifications, 60, 77, 152, 153
　electrical department, 37, 66, 80, 81, 98
　endorses thin conductor, 62–64
　finances, 44–45, 56, 57, 59, 84, 218, 221–22, 225–27, 232
　and Joint Committee, 122–23, 124–25, 128, 131, 186
　launch of, 43–46, 56–58, 60, 65
　and Red Sea cable project, 69, 110
　and Whitehouse's spending, 65–66, 79, 80
Ayrton, W. E., 244

Bain telegraph, 15, 21
Balfour, Arthur, 253
Banks, James, 46
Bartholomew, E. G., 134
Benghazi, Libya, 123
Bettany, G. T., 249
Bidder, George Parker, 124, 127, 133
Biggs, C. H. W., 263–64, 268
Biscay, Bay of, 81, 120, 123
Bishop Gutta Percha Works (New York), 41
Blundell, J. Wagstaff, 234–36
Board of Trade (UK), 122, 124, 186
Boltzmann, Ludwig, 207
Bowers, Brian, 134
Brett, Jacob, 8–12
Brett, John Watkins, 8, 30, 35, 37, 59, 89
　Atlantic cable design, 61
　Atlantic Telegraph Company, 41–44, 56–57, 83
　English Channel cable, 8, 10–12

Brett, John Watkins (cont.)
  and Field, 41, 43–45, 52, 56
  financial interests, 8, 41, 44, 57
  Joint Committee testimony, 45–46, 131
  and Whitehouse, 46, 50, 52
Bright, Charles, 138, 162
Bright, Charles Tilston, 38, 128, 138,
  176, 219
  Atlantic cable design, 61
  Atlantic Telegraph Company, 56, 63, 69,
    76, 91
  British Association Committee on
    Electrical Standards, 159–64
  electrical units and standards, 166–67,
    187–88
  Falmouth-Gibraltar cable, advice on,
    119–21
  and Field, 43, 56
  financial interests, 57
  Joint Committee Report, 97, 125
  knighthood, 84
  Magnetic Telegraph Company, 13,
    43, 149
  paying-out machinery, 68, 99
  resistance coils, 149, 151
  and Whitehouse, 56–57, 76
Bright, E. B., 30, 162
Brighton, 46, 52
British Association Committee on Electrical
  Standards, 145, 179, 192–93, 196,
  199, 200–1, 211
  experimental apparatus, 179,
    192–94, 251
  formation of, 159–63, 188
  and Maxwell, 182, 186, 188–89, 212–15
  resistance coils, 170–73, 194, 209,
    222, 250
  system of units, 165–69, 175–78, 196,
    201, 222
British Association for the Advancement of
  Science, 160, 162, 251
  meetings
    Liverpool (1854), 30
    Glasgow (1855), 47–52, 70
    Cheltenham (1856), 52, 53
    Dublin (1857), 58
    Leeds (1858), 105
    Oxford (1860), 135
    Manchester (1861), 117, 159–63,
      186–87
    Cambridge (1862), 166, 169
British Association unit of resistance,
  176–77, 188, 195–96, 254
  and Siemens's unit, 168, 170, 174–75,
    188, *See also* ohm

British Empire, 8, 10, 41, 97, 109, 117–19,
  142, 220, 236–41, 246, 272–74
  cables as nervous system, 1, 96, 217, 236
British-Indian Submarine Telegraph
  Company, 177, 234, 236
Brown, William, 83
Brunel, Isambard Kingdom, 45, 60
Brussels observatory, 17, 22, 27, 32
Buchanan, James, 100
Buxton, Sydney, 239

cable design, 112, 118, 121, 125, 139,
  142, 241
  armoring, 10, 11, 60, 61, 99, 227
  Atlantic cable (1857–58), 58–62, 64,
    128, 130
  Atlantic cables (1865–66), 222, 227
  Joint Committee inquiry, 125
  Red Sea and India cables (1859–60), 140
  thickness of conductor, 51, 53–55, 59,
    61, 62–64, 99, 128, 139
cable network
  commerce, 1, 8, 11–12, 84, 85, 97, 220,
    231, 239, 270, 271
  dissemination of news, 1, 85, 232,
    239, 270
  international affairs, 84, 118–19, 239,
    270, 271
  nervous system of British Empire, 1, 96,
    217, 236
cables
  Anglo-Danish, 259–60
  Black Sea, 35, 50–51, 54, 55, 58,
    107, 131
  Cabot Strait, 40–43, 47, 60, 228
  Candia-Alexandria, 111, 112–13
  Dover-Ostend, 13, 17, 33
  English Channel, 8–13, 16, 41, 60, 141,
    142, 147
  Falmouth-Gibraltar, 118, 119–22, 124,
    133, 186
  French Atlantic, 176, 237, 244
  Malta-Alexandria, 123, 176, 177, 195,
    219, 233–34
  Mediterranean, 35, 59, 111, 141
  Pacific, 272
  Persian Gulf, 110
    See also Atlantic cable (1857–58);
      Atlantic cables (1865–66);
      Red Sea and India cables (1859–60)
Cabot Strait, 40–43, 228
*Cambridge [and Dublin] Mathematical
  Journal*, 27, 30
Cambridge Philosophical Society, 30
Cambridge University, 27, 31, 213

Cavendish Laboratory, 179, 196, 211,
    212–13, 217, 247–55, 267
Cavendish professorship, 249, 253, 254
    Mathematical Tripos, 29, 213, 247–48,
    249, 255
Camden Town, London, 258–59, 264, 269
Campbell, Lewis, 194
Canning, Samuel, 121, 128, 224–25, 227,
    228–30, 234, 235
    repairs to cable at Valentia (1858), 91–94
Cape Breton Island, Nova Scotia, 40, 41
Cape Ray, Newfoundland, 40, 41
Carr, George, 59
Cavendish, Henry, 248, 250
Chambers's Journal, 63
Chatterton, John, 133
Chemical Record and Journal of Pharmacy, 14
China Submarine Telegraph
    Company, 238
Christie, Samuel Hunter, 75
Chrystal, George, 199, 212–14, 253, 255
Civil Engineer and Architect's Journal, 106
Civil War (US), 220
Clark, Edwin, 124, 127
Clark, Latimer, 22, 28, 34, 64, 176, 228,
    235, 237
    British Association Committee on
        Electrical Standards, 159–64,
        165, 175
    electrical measurement, 145–48,
        151, 242
    electrical units and standards, 153, 157,
        159–67, 187–88, 201–2
    *Elementary Treatise on Electrical
        Measurement* (1868), 243
    Joint Committee, 124, 127, 153, 157
    Joint Committee Report, 14, 24, 134,
        137–40, 159, 243
    naming of electrical units, 160, 169
    resistance coils, 132, 151
    retardation of signals, 14–15, 17, 20–21,
        24–25, 30, 36, 46, 138–39, 217, 257
Clifton, R. B., 245, 248
conductivity
    of commercial copper, 75–77, 151–53
    of copper, 75, 76, 132, 138, 152
    of gutta-percha, 132, 135–37, 138
Constantinople, 109, 112, 117, 219
contract specifications, 77, 121, 126, 141,
    151–53, 180, 195
    and electrical standards, 145, 158,
    176–78, 187
Cooke and Wheatstone telegraph, 5, 14
Cooke, W. F., 5
Cooper, Peter, 41, 43

Crampton, T. R., 11
Crawfurd, John, 117
Crimean War, 35

Daily Evening Bulletin (San Francisco), 231
Daily News (London), 56, 123, 220
Daniell's cell, 88, 99
    as standard of electrical tension, 160,
        165, 168, 188, 201
Dansk-Norsk-Engelsk Telegraf-Selskab,
    259
de Mauley, Lord, 16
de Sauty, C. V., 82, 84–85, 87–88, 90, 124
Derby Mercury, 226
Derby, Lord, 111, 121
Devonport (Plymouth), 69–70, 77, 80–81,
    84, 87
Devonshire, Duke of (William
    Cavendish), 248
Dickson, J. D. H., 244
dielectric, 23–25, 32, 136, 137, 257,
    262–63, 265, 270, 274
dimensional analysis, 197, 202–3
displacement current, 207, 266
Disraeli, Benjamin, 121
Dorsey, N. E., 197
Dunsink observatory (Dublin), 30

earth currents, 78, 95, 226
East India Company, 9, 109, 110
Eastern Telegraph Company, 143, 176,
    236, 239, 241, 245, 263
Edinburgh, 131, 189, 256
    University of, 212, 245, 246
Edinburgh Review, 63, 131
Egypt, 109, 113, 237–38
Einstein, Albert, 269
Electric and International Telegraph
    Company, 96, 103, 106, 122, 125,
    138, 173, 233
electric induction, 15, 22–23, 25, 36, 55,
    62, 138, 139, 142, 147
Electric Telegraph Company, 5, 13, 14, 17,
    30, 146, 148, 150, 154
    Lothbury offices, 17, 20–21
electrical measurement, 38, 74–75, 108,
    145, 178–79, 196, 212, 217, 241–43
    Joint Committee, 134, 141–42
    Whitehouse's system, 69, 79, 101,
        108, 130
electrical units and standards, 2, 3, 38,
    213, 242
    British Association system, 165–69,
        175–78, 201, 222
    naming of units, 160, 169, 177, 196

electrical units and standards (cont.)
  resistance standards, 132, 136, 141, 149,
    150, 153–55, 158
  Siemens's mercury resistance unit, 136,
    137, 156–57, 164, 170, 187,
    195, 222
    and British Association unit, 166, 168,
      174–75, 188, 201
  Thomson's absolute system, 132,
    137, 167
  Weber's absolute system, 75,
    132, 201
Electrician (1861–64), 140, 164–65
Electrician (1878–1952), 263–64, 266–67,
    268, 269
electromagnetic field, 4, 18, 36, 139, 142,
    155, 218, 256, 267–68, 270
electromagnetic theory of light, 1, 34, 36,
    183, 189, 191–92, 203–4, 205–6,
    207–10, 211
electromagnetic waves, 1, 32, 211, 218,
    250, 262, 263, 272
Elliott Brothers, 170, 194
Encyclopedia Britannica, 137
energy flow theorem, 256, 265–66,
    267–68, 274
energy, as basis for system of units, 75, 155,
    162, 163, 167, 188, 201–2
Engineer, 106, 140
English Mechanic, 260
Esselbach, Ernst, 168–69
ether, 142, 200–1, 206–7, 262, *See also*
    vortex ether theory
Euphrates Valley telegraph line,
    109–10
European Telegraph Company, 13
Everitt, Francis, 197
Exiles Club, 114

Fairbairn, William, 124, 127, 133
Falmouth, Gibraltar and Malta Telegraph
    Company, 236
farad, 144, 151, 169, 178
Faraday effect (magneto-optic rotation),
    183, 192, 208
  and Maxwell's vortex theory, 183–86,
    191–92, 206–8, 215
Faraday, Michael, 9, 17–18, 30, 31, 81,
    136, 139
  endorses thin conductor, 62
  field theory, 2, 18–20, 25, 30, 36, 142,
    199, 257
  lines of force, 18–20, 28
  "On Electric Induction" (1854), 22–23,
    26–27, 28–30

and Maxwell, 162, 190–92
retardation of signals, 17, 20–23, 25,
    36, 46
faults, methods for locating, 90, 96, 101,
    103, 105, 132, 148–50, 153,
    157–58, 187, 242
fiber optic cables, 274
field theory, 2, 4, 18–20, 25, 30, 181, 199,
    212, 272–75
Field, Cyrus, 47, 56, 84, 127, 216, 221,
    224, 228
  Atlantic cable design, 59–61, 62
  Atlantic cable project, origins, 37, 40–46
  Atlantic Telegraph Company
    board of directors, 59–62, 83
    rush to launch, 56–57, 65, 99,
      104, 128
  and Brett, 41, 43–45, 52
  and Bright, 43
  financial interests, 40, 44, 45, 57
  shielded from blame, 38, 98
  and Whitehouse, 46, 50, 57, 59
Field, Henry, 228
Field, Matthew, 40, 41
FitzGerald, G. F., 208, 213, 269–70
FitzRoy, Robert, 127
Fizeau, Hippolyte, 190, 192, 208
Fleming, J. A., 251
Föppl, August, 269
Forde, H. C., 120–22, 134
Foster, George Carey, 245, 251
Foucault, Léon, 208–10
Fourier, Joseph, 27, 31, 202
France, J. R., 86, 89–90, 92, 94
Franco-Austrian War, 118, 119, 121
Fredericia, Denmark, 259–60
Freedom's Champion (Atchison,
    Kansas), 230
French Atlantic Telegraph Company,
    234–36, 238

Galbraith, J. A., 190, 192
Galison, Peter, 198
Galton, Douglas, 133, 136, 140, 221
  formation of Joint Committee,
    122–23, 124
  Joint Committee hearings, 127, 129–31
  Joint Committee Report, 125
  "Ocean Telegraphy" (1861), 131
Garnett, William, 250, 253
Gassiot, J. P., 81
Gauss, C. F., 155, 165, 201
Gib core, 122, 123, 135
Gibbs, J. Willard, 256
Gisborne, Frederic, 40–41

Gisborne, Lionel, 120, 122, 127, 134
  Red Sea cable project, 109–13
Gladstone, William, 121, 122
Glasgow, 47–52, 57, 79, 81, 189, 215,
  248, 251
  University of, 26–27, 37, 131, 151,
    209, 246
  Thomson's laboratory, 74, 187,
    243–45, 248, 249
Glass, Elliot and Company, 43, 111,
  123–24, 219, 221–22
Glass, Richard, 60, 65–66, 128
Glazebrook, Richard, 196, 253
Goliath, 10
Gooch, Daniel, 221–22, 227, 228
Gooday, Graeme, 245
Gordon, Lewis D. B., 27, 131
Göttingen, 75
governors, theory of, 193
Great Eastern, 45, 66, 143, 222, 235, 237
  Atlantic cable expeditions (1865–66),
    174, 216, 223–26, 227–30
Great Indian Submarine Telegraph
  Company, 118
Great Northern Telegraph Company,
  238–39, 259, 261
Green, Allan, 44, 46, 52, 71, 86
Greenwich observatory, 16–17, 22, 30, 32
Griset, Pascal, 239
Grove, William Robert, 81
Guthrie, Frederick, 178, 245
Gutta Percha Company, 9, 10, 14, 59, 60,
  65, 112, 120, 150
  Atlantic cables (1865–66), 221
  commercial copper, 77, 152
  Gib core, 122
  Wharf Road works, 17, 20, 22, 36,
    129, 139
gutta-percha, 9–10, 13, 24–26, 29, 157, 274
  conductivity, 132, 135–37, 138
  heat damage, 66, 70, 110, 129
  inductive capacity, 24, 135, 192
  Joint Committee tests, 124, 133–38

Hallani, Oman, 114–16
Hamilton, Archibald, 30–31
Hamilton, William Rowan, 30
Harman, Peter, 197
Haughton, Samuel, 190, 192
Headrick, Daniel, 239
Hearder, J. N., 72, 80, 82, 87
Heart's Content, Newfoundland, 216, 228
Heaton, J. Henniker, 240
Heaviside, Arthur West, 259, 261, 264,
  267, 268

Heaviside, Oliver, 32, 274
  conflict with Preece, 268–69
  early life, 257–61
  Electrical Papers (1892), 262, 269–70
  and *Electrician* (1878–1952), 263–65
  energy flow theorem, 265–66
  and Maxwellians, 269
  Maxwell's equations, 218, 256, 266,
    269–70
  and Maxwell's *Treatise*, 256, 261,
    264, 266
  as a telegrapher, 255, 257, 259–61
Helmholtz galvanometer, 78
Helmholtz, Hermann, 75, 248
Hendry, John, 197
Henley, W. T., 81, 86
Hertz, Heinrich, 1, 211, 258,
  263, 269
Hicks, W. M., 250
Hockin, Charles, 170, 194, 204–5, 209,
  235, 246
Holmes, N. J., 260

Illustrated London News, 47
India Office, 119–20
India rubber (caoutchouc), 124, 127,
  133–34, 192, 239
Indian Rebellion, 69, 95, 109, 110,
  117, 121
Indian Telegraph Department, 119,
  173, 246
inductive capacity, 23–24, 31, 36, 135, 137,
  139, 150, 178, 191–92
inductive loading, 268
Institution of Civil Engineers, 62, 133,
  144
Institution of Electrical Engineers, 215, 269
Ismail Pasha, 238

Jacobi, Moritz, 149, 153
Jenkin, Bernard Maxwell, 198
Jenkin, Fleeming, 131–32, 176, 200,
  235, 237
  British Association Committee on
    Electrical Standards, 159, 161–63,
    165–71
  electrical measurement, 178–80
  electrical units and standards, 144–45,
    153, 157
  Electricity and Magnetism (1873), 145,
    179, 195, 199, 200
  Joint Committee Report, 134, 136–37
  Joint Committee testimony, 132,
    153, 157
  and Maxwell, 179, 182, 192–95

Jenkin, Fleeming (cont.)
  "On the Elementary Relations Between
    Electrical Measurements" (1863),
    196–99, 201–5, 209
  and Thomson, 131–32, 243
Joint Committee Report, 24, 128, 131,
    133–36, 140–43, 164, 216, 218, 243
  appendices, 133–34, 140
    Clark, 14, 137–40, 159, 243
    Jenkin, 136–37
    Siemens brothers, 134–36
  praise for, 97, 125, 140, 142
Joint Committee to Inquire into the
    Construction of Submarine
    Telegraph Cables, 97, 99, 117, 146,
    151, 218
  and electrical standards, 153,
    157–58, 186
  formation of, 122–23
  members, 124–25, 127
  testimony
    Brett, 45–46, 131
    Jenkin, 132, 153, 157
    Newall, 127, 131
    Thomson, 129–31, 157, 187
    Whitehouse, 127–29
Joule, James, 163, 195, 252
Journal of Gaslighting, 140
*Journal of the Society of Telegraph Engineers
    and Electricians*, 261
Journal of the Telegraph, 175

Karachi, 110–11, 113–16, 119
Kelvin. *See* Thomson, William
Kew observatory, 193
kinetic theory of gases, 193
King, David Thomson, 245
King, W. F., 209, 244
King's College London, 5, 124, 188,
    245–47, 249
  determination of ohm, 170, 193,
    246, 251
Kirchhoff, Gustav, 190, 195
Kline, Ronald, 200
Knightstown, Ireland, 85, 87
Kohlrausch, Rudolf, 190, 192, 203, 206,
    209–10
Kosseir, Egypt, 113
Küper and Company, 43, 47, 60, 65

laboratories, physics teaching, 74, 178, 217,
    242, 270
  Cambridge (Cavendish), 179, 212–13,
    247–55
  Edinburgh, 245

Glasgow, 74, 187, 243–45, 248, 249
  King's College London, 245–47, 249
  Oxford (Clarendon), 245, 248
  Royal School of Mines (London), 245
  University College London, 245
Lagrangian analysis, 205, 206–8
Lampson, Curtis, 83
law of squares, 32, 46, 51–54, 64, 107, 120,
    128, 132
Laws, J. C., 228
Layton, Edwin, 199–200
Lincoln, Abraham, 220
Linde, Theodor, 82
lines of force, 18–20, 28, 183, 206
Liverpool, 44, 57
Lloyd's Weekly, 230
Lodge, Oliver, 1, 213, 267, 269, 272
London International Exhibition (1862),
    155, 169
longitude, determination by telegraph, 16,
    21, 27
Lorentz, H. A., 208
Lorenz, Ludvig, 254

MacFarlane, Donald, 74, 81
Magnetic Telegraph Company, 13, 43, 56,
    149, 173, 233
Malaya, 9
Mann, R. J., 63, 73, 106
Marconi, Guglielmo, 269, 272
marine galvanometer, 79, 81, 83
Marshman, John C., 109–11, 127
Mason College, Birmingham, 267
Matthiessen, Augustus, 134, 152, 156, 158,
    162, 171, 174
Maury, Matthew Fontaine, 40, 59
Maxwell, James Clerk, 1, 4, 36, 131, 235
  British Association Committee on
    Electrical Standards, 159, 162–63,
    182, 186, 188–89, 212–15
  British Association resistance coils,
    170–71, 251–52
  displacement current, 207
  "A Dynamical Theory of the
    Electromagnetic Field" (1864), 181,
    182–83, 185, 196–97, 200–1,
    204–9, 215
  electrical measurement, 178–80, 250
  electrical units and standards, 144–45
  electromagnetic theory of light, 34, 36,
    183, 189, 191–92
  "On the Elementary Relations Between
    Electrical Measurements" (1863),
    196–99, 201–5, 209
  and Faraday, 19, 29, 162, 190–92

"On Faraday's Lines of Force" (1856), 30, 183
Glenlair estate, 189, 205
and Heaviside, 261
and Jenkin, 179, 182, 192–95
"On Physical Lines of Force" (1861–62), 181–86, 189, 191, 196–97, 201, 205–8, 214–15
ratio of units (*v*), 155, 182, 189–92, 202–6, 207, 208–12, 215
review of Jenkin, *Electricity and Magnetism* (1873), 145, 179, 195, 213
theory of the electromagnetic field, 36, 136, 155, 183, 185, 199, 212, 218, 250, 255
and Thomson, 29–30, 162, 191, 194, 205
Treatise on Electricity and Magnetism (1873), 179, 182, 185, 208, 212–15, 255–56, 270
and electrical measurement, 212–14, 241, 255
and electrical technology, 4, 213, 217
and Heaviside, 256, 261, 264, 266
vortex ether theory, 162, 181–86, 189–92, 204, 206–7, 265
Maxwellians, 213, 269
Maxwell's equations, 218, 256, 266, 269–70
Mayes, William, 132
Mechanics' Magazine, 102, 106
Mediterranean Telegraph Company, 89
metric system, 168
Miller, R. Kalley, 244
Miller, W. A., 134
mirror galvanometer, 99, 103, 130, 151, 243, 250
invention of, 77–80, 81
use on Atlantic cables, 86, 90, 91, 94, 95, 101, 102, 229, 232
Montgomerie, William, 9
Morning Chronicle, 22, 102–3, 105
Morse telegraph, 6, 14, 86
Morse, Samuel F. B., 5–6, 40, 43–44, 50, 56–57, 62, 67
Müller, Simone, 234
Muscat, 114–16

Napoleon III, 8, 117
National Bureau of Standards (US), 197
National Physical Laboratory (UK), 196
New York Times, 224, 231
New York, Newfoundland, and London Telegraph Company, 41, 227
Newall, R. S., 27, 35, 123, 133
Anglo-Danish cable, 259

armoring Atlantic cable, 65, 66
Birkenhead works, 65, 131, 136, 137
Joint Committee testimony, 127, 131
patent claims, 65, 123
Red Sea cables, 110–16
Newcastle, 255, 259–60
Newfoundland, 40–43, 47, 57, 94, 224, 230, 232
government concessions, 40, 41, 219
Heart's Content, 216, 228
Trinity Bay, 68, 84–91, 95–96
Niagara, 66–69, 81, 82–84
Niven, W. D., 197, 213
Noakes, Richard, 146, 267
Northcote, Stafford, 119, 120

Oersted, Hans Christian, 4, 5
ohm, 3, 132, 144, 151, 159, 170, 173, 175, 178, 204, 211, 213
adoption by telegraph companies, 175–78
in America, 175
and Cavendish Laboratory, 252
determination of, 179, 182, 192, 196, 201, 209, 215, 235, 246, 251
naming of, 160, 169, 177, 188, 196
and ratio of units (*v*), 209–10
redetermination of, 211, 252, 253–54, *See also* British Association unit of resistance
Ohm, G. S., 151
Ohm's law, 108, 138
Ottoman Empire, 109, 110, 113, 236
Overend, Gurney and Company, 233
Oxford University, 247
Clarendon Laboratory, 245, 248

Palmerston, Lord, 121, 221
Paris, 28, 235
Paris observatory, 16–17
Parliament, 117, 121, 227
Peel, Charles, 127
Pender, John, 195, 234, 244, 251, 263
Anglo-American Telegraph Company, 227, 233, 239
cable empire, 235, 236–41
Eastern Telegraph Company, 176, 239
election bribery accusations, 230
TC&M, 221–22, 234, 239
Philosophical Magazine, 22, 27, 34, 164, 171, 189, 194, 260–61, 263, 267
PK Porthcurno Museum of Global Communications, 176, 236
Plymouth, 69, 70, 72, 77, 80–83, 84, 87, 123
Porthcurno, Cornwall, 176, 236, 245

Post Office telegraph system, 106, 233, 261, 268
Poynting, J. H., 267–68, 274
Preece, W. H., 21, 63
    conflict with Heaviside, 268–69
    exchange with Whitehouse (1858), 106–8
Prussia, 24–25, 149, 232

Queenstown (Cobh), Ireland, 66–68, 77, 83, 84

ratio of units (*v*), 204, 244
    and Maxwell, 155, 182, 189–92, 202–6, 207, 208–12, 215
    and ohm, 209–10
    and speed of light, 155, 189–92, 205–6, 211–12, 215
    and Thomson, 33–34, 36, 55, 203
    and Weber, 55, 155, 190, 192, 203, 206, 209–10
Rayleigh, Lord (J. W. Strutt), 249, 267
    as Cavendish professor, 253–54
    redetermination of ohm, 211, 253–54
Red Sea [and India] Telegraph Company, 109–17, 118, 127, 218
Red Sea and India cables (1859–60), 109–14
    contract, 111–13
    failure, 109, 114–17, 125, 131, 140, 151, 219
    government guarantee, 109, 111–12, 116–17, 118, 218
    Joint Committee inquiry, 125, 127, 133
    strategic value, 109, 110, 119
    testing, 112, 132, 135
resistance coils, 90, 130, 132, 144–46, 149, 151, 156, 157–58, 163, 178, 187
    British Association, 166, 170–73, 174, 193, 194–95, 250, 251–52, 253
    Siemens and Halske, 154, 174
retardation of signals, 3, 14–26, 27, 30, 46, 48–50, 57, 113, 134, 142, 147, 257
    Atlantic cable design, 61–62, 64
    discovery by Clark, 14–15, 24–25, 36, 46, 138–39, 217, 257
    law of squares, 32, 46, 51–54, 64, 107, 120, 132
    thickness of conductor, 106–7, 128, 139
    Thomson's theory, 31–36, 51–54, 78
    Whitehouse's experiments, 52–55, 105, 106
Rhodes, Cecil, 241
Rosa, E. B., 197
Rowland, Owen, 134

Royal Commission on Scientific Instruction, 248
Royal Institution (London), 18, 22, 101
Royal School of Mines (London), 245
Royal Society of London, 34, 58, 74, 162, 269
    Philosophical Transactions, 75, 136, 205, 267
    Proceedings, 34, 50, 136, 267
Russell, James Burn, 70, 87, 94, 95, 244
Russell, W. H., 174, 223
Rutter, John O. N., 46

Saunder, S. A., 253
Saward, George, 63, 65, 66, 81, 220, 232, 242
    Joint Committee, 124, 127, 128, 129, 130
    letters to press (1858), 100, 101, 103
    and Whitehouse, 89, 91–92
Schaffer, Simon, 213, 252
Schuster, Arthur, 248, 250, 253
science-technology interaction, 2, 4–5, 22, 141–42, 200, 270–71, 274–75
Second Opium War, 123
self-induction, 209, 211, 254, 261–62, 265, 268
Shaw, W. N., 253
Sidgwick, Eleanor, 253
Siegel, Daniel, 197
Siemens and Halske, 112, 122, 123, 154–55, 168, 174
Siemens, Werner, 24–26, 174, 236
    Joint Committee, 134–36
    mercury resistance unit, 136, 137, 156–57, 158, 164, 170, 187, 195, 222
        and British Association unit, 166, 168, 174–75, 188, 201
    Prussian telegraph lines, 24–26, 149
    Red Sea cables, 112, 114
Siemens, William, 26, 125
    Joint Committee, 126, 127, 134–36
siphon recorder, 243, 248
skin effect, 266, 268
Smith, Crosbie, 213
Smith, Willoughby, 10, 76, 112, 137, 233, 234, 235
    cable testing, 150–51, 227
    Joint Committee, 128, 133, 157
    resistance standards, 136, 151, 222
Society of Telegraph Engineers, 215
South Atlantic Telegraph Company, 118
Spectator, 10

speed of light, 34, 203, 204, 206, 208–10
  and ratio of units (*v*), 155, 189–92,
    205–6, 211–12, 215
spinning coil resistance apparatus, 169–71,
    193–94, 200, 204, 211, 215, 251
St. John's, Newfoundland, 40, 43
St. Pierre, 176, 235
standardization of materials, 152,
    157–58
Stanley, Lord, 111
Statham, Samuel, 20, 22, 28, 59, 60
steam engine, 4, 200
Stephenson, Robert, 119–20, 122–23,
    124–25
Stewart, Balfour, 170, 193, 246
Stewart, Charles, 232, 233
Stokes, George Gabriel, 29, 31, 34–35,
    205, 208
Stuart, James, 252
Stuart-Wortley, James, 125, 127, 218,
    220–21
Suakin, Sudan, 113, 114
Submarine Telegraph Company, 12–13,
    16, 33, 41
Suez, Egypt, 113, 236, 237

Tait, P. G., 171, 186, 206, 245,
    264
Taylor, Moses, 41
Telegraph Act (1868), 233
Telegraph Construction and Maintenance
    Company (TC&M), 222–23, 228,
    234, 235, 239–41
  finances, 225–26, 237
telegraph, landlines, 3, 5–7, 109–10,
    146–47, 242, 260
  and commerce, 7
  and dissemination of news, 6–7
  underground lines, 6, 23, 32–33,
    34, 44
  in Britain, 5, 13–14, 17, 21–22, 56,
    146–49
  in Prussia, 24–26, 149
telegrapher's equation, 262
Telegraphic Journal, 255
Telegraphic Plateau, 40
thermodynamics, 4, 26, 74, 200
Thomson unit (resistance), 132, 137,
    167
Thomson, J. J., 254
Thomson, William, 26, 37, 64, 96, 123,
    136, 153, 176, 221, 229, 235
  absolute system of electrical units, 132,
    164, 167–68, 187–88, 202
  Atlantic cable design, 64, 130

Atlantic Telegraph Company board of
    directors, 37, 58, 64, 76–77, 79, 83,
    94–95, 100, 151
British Association Committee on
    Electrical Standards, 159,
    161–62, 165
Cavendish Laboratory, 248
commercial copper, 75–77, 151–53, 244
electrical measurement, 38, 178–79, 243
electrical units and standards, 38,
    144–45, 173
"Electrical Units of Measurement"
    (1883), 144, 179
"On the Electro-dynamic Qualities of
    Metals" (1856), 74–75
and electromagnetic theory of light,
    210, 211
Faraday effect (magneto-optic rotation),
    183, 206, 208
and Heaviside, 260, 269
and industry, 26, 189, 215
and Jenkin, 131–32, 243
Joint Committee testimony, 129–31,
    157, 187
knighthood, 26, 230
laboratory (Glasgow), 74, 187,
    243–45, 248
law of squares, 32, 46, 51–54, 64, 107,
    120, 128, 132
marine galvanometer, 79, 81, 83
and Maxwell, 29–30, 162, 191, 194, 205
mirror galvanometer, 89, 99, 103, 130,
    151, 243, 250
  invention of, 77–80, 81
  use on Atlantic cables, 86, 90, 91, 94,
    95, 101, 102
patents, 26, 29, 30, 79, 243
ratio of units (*v*), 33–34, 36, 55, 203
sails on *Agamemnon*, 68, 81–84, 103
siphon recorder, 243, 248
spinning coil resistance apparatus, 193
students, 70, 74, 151, 209, 243–44, 249
theory of signal propagation, 31–32, 36,
    53–56, 59, 78, 120, 132, 142, 151,
    155, 222, 257, 261–62
"Theory of the Electric Telegraph"
    (1855), 30–36
Treatise on Natural Philosophy, 264
and Whitehouse, 50–58, 66, 68, 79–80,
    85–86, 94–95, 100, 101–2, 105, 130
Thoreau, Henry David, 6
Tietgen, C. F., 238, 259
Tiffany and Co., 84
Times (London), 100, 124, 216, 223, 230,
    235–38

Treasury Office (UK), 109, 111,
      118–23, 133
Trent affair, 220
Trinity Bay, Newfoundland, 68, 84–91,
      95–96
Tripoli, Libya, 123
Tupper, Charles, 59, 60
Tyndall, John, 18, 52

units and standards, 156, 166, 169–70, 193
Universal Private Telegraph Company, 259
University College London, 245, 246, 251

Vail, Alfred, 6
Valentia Island, Ireland, 68, 82, 84, 85–96,
      100–3, 104, 223–24, 226, 227–29
Varley unit (resistance), 149, 151, 154, 158
Varley, Cromwell Fleetwood, 14, 30, 87,
      96, 153, 176, 235, 237
   electrical measurement, 145–49, 151
   electrical units and standards, 153,
      157, 175
   Joint Committee, 124–25, 127–28, 129,
      153, 157
   resistance coils, 132, 149, 151
   on Whitehouse's induction coils, 103–4
Varley, S. A., 62, 64, 131
vector analysis, 256, 263, 270
vector potential, 207, 264, 266
Verdet, Émile, 191
Victoria, Queen, 92, 100, 114
volt, 3, 159, 169, 213, 254
Volta, Alessandro, 4
vortex ether theory, 162, 181–86, 189–92,
      204, 206–7, 265, *See also* ether

Wales, Prince of, 223, 237
Walker, C. V., 64, 81
Warwick, Andrew, 213, 255
Webb, F. C., 64, 99, 116, 121, 149
Weber, Wilhelm, 195
   absolute system of electrical units, 75,
      132, 155–56, 162, 163–69, 174,
      188, 201
   action-at-a-distance theory, 18, 25, 30,
      190, 199, 274
   ratio of units (*v*), 55, 190, 192, 203, 206,
      209–10
   resistance measurements, 75, 76, 151,
      156, 169
   unit of electrical resistance, 156, 169, 201
West, Charles, 126
Western and Brazilian Telegraph
      Company, 245
Western Union Telegraph Company, 175

Wheatstone automatic telegraph, 260
Wheatstone bridge, 74, 144, 150, 151, 154,
      162, 260
Wheatstone, Charles, 5, 75, 81, 151, 242
   British Association Committee on
      Electrical Standards, 159, 162
   and Heaviside family, 259
   Joint Committee, 122–24, 127, 134
   resistance standard, 149, 153
White, James, 79, 81, 243, 248
Whitehouse, E. O. Wildman, 37, 46, 82,
      118, 124
   Atlantic cable (1857–58)
      blamed for failure, 38, 96, 98, 128
      design, 59, 61, 128
      testing, 65, 66–67, 70, 77, 80,
         129, 130
   Atlantic Telegraph Company, 37, 57,
      69, 107
   conflicts with board of directors, 80,
      81, 89, 91–95, 98, 99–100, 104
   batteries, 66, 85
   and Brett, 46, 50
   and Bright, 56–57, 76
   cable experiments, 47–55, 59, 62,
      105, 128
   commercial copper, 76–77
   exchange with Preece (1858), 106–8
   fault in Valentia harbor, fear of, 87, 90,
      100, 104
   and Field, 46, 50, 59
   financial interests, 57
   induction coils, 52, 56, 58, 66, 77, 78, 79,
      82, 85–89, 94
      effect on cable insulation, 86–89, 98,
         99, 103–4, 129–30
   Joint Committee testimony, 127–29
   letters to press (1858), 100–5
   magneto-electrometer, 70–74, 79
   patents, 52, 57–58, 65, 73, 85, 101–2
   praised by Mann, 63–64, 73
   relays (receivers), 56, 58, 66, 73, 77, 79
      use on Atlantic cable, 82, 85–91, 94,
         101–2, 104
   retardation of signals, 48–50, 52–55
   and Saward, 89, 91–92
   system of electrical measurement, 38, 69,
      79, 101, 108, 130, 243
   and Thomson, 50–58, 66, 68, 79–80,
      85–86, 94–95, 101–2, 105
Whitworth, Joseph, 221
wireless telegraphy, 269, 272
Wise, Norton, 213
Wollaston, Charlton, 10, 33, 147, 149
Woodhouse, Henry, 68

Printed in the United States
By Bookmasters